Structural Studies of Macromolecules
by
Spectroscopic Methods

Structural Studies of Macromolecules
by
Spectroscopic Methods

Edited by

K. J. Ivin

*Professor of Physical Chemistry,
Department of Chemistry,
The Queen's University of Belfast*

A Wiley–Interscience Publication

JOHN WILEY & SONS

London · New York · Sydney · Toronto

Copyright © 1976, by John Wiley & Sons, Ltd.

Library of Congress Cataloging in Publication Data:

Main entry under title:

Structural studies of macromolecules by spectroscopic
methods.

    'A Wiley–Interscience publication.'
    Includes bibliographical references.
    1. Macromolecules—Analysis. 2. Spectrum analysis.
I. Ivin, Kenneth John. [DNLM: 1. Macromolecular systems—
Congresses. 2. Spectrum analysis—Congresses.
QD380 S927 1974]

QD381.S8 1976   547'.7   75-19355
ISBN 0 471 43120 6

Set on Monophoto Filmsetter and printed in Great Britain
by J. W. Arrowsmith Ltd., Bristol.

Preface

This volume is based largely on the contributions to a meeting held at Cranfield Institute of Technology, near Bedford, England in July 1974 under the auspices of the Macromolecular Group of the Chemical Society of London. The object of the meeting, and also that of this volume, was to bring together information about the application of some of the more recently developed spectroscopic methods to the study of macromolecules.

The contributions divide into five groups, dealing with neutron scattering (1–4), far infrared and Raman spectroscopy (5–8), electron spectroscopy (9), nuclear magnetic resonance spectroscopy (10–13) and electron spin resonance spectroscopy (14–16).

These techniques allow a wide variety of information to be obtained on polymers, not only about their detailed chemical structure (9–15) but also about their conformations (7), distance between ends (1), rate of collision between ends (16), vibration frequencies (3–8), segmental motions (2, 14, 15), crystallinity (7), lamellar thickness (6), and surface composition (9).

Much of physical science may be imagined on a three-coordinate system in which the axes are labelled 'Method', 'Substance' and 'Molecular property'. In this volume are described perhaps a dozen spectroscopic methods applied to nearly a hundred macromolecular substances in order to obtain about a dozen types of molecular property. It is hoped that the coordinate points thereby established on this universal frame may help to provide the reader with a guide into unknown regions.

K. J. Ivin

Contributing Authors

ALLEN, G.   Department of Chemistry, University of Manchester, England.

ANDO, D. J.   Department of Physics, Queen Mary College, London, England.

BATCHELDER, D. N.   Department of Physics, Queen Mary College, London, England.

BLOOR, D.   Department of Physics, Queen Mary College, London, England.

BOVEY, F. A.   Bell Laboratories, Murray Hill, New Jersey, U.S.A.

BULLOCK, A. T.   Department of Chemistry, University of Aberdeen, Scotland.

CAMERON, G. G.   Department of Chemistry, University of Aberdeen, Scotland.

CLARK, D. T.   Department of Chemistry, University of Durham, England.

CUDBY, M. E. A.   I.C.I. Ltd., Plastics Division, Welwyn Garden City, England.

CUNLIFFE, A. V.   E.R.D.E., Waltham Abbey, Essex, England.

EBDON, J. R.   Department of Chemistry, University of Lancaster, England.

FULLER, P. E.   E.R.D.E., Waltham Abbey, Essex, England.

HENDRA, P. J.   Department of Chemistry, University of Southampton, England.

HIGGINS, JULIA S.   Institut Laue-Langevin, Grenoble Cedex, France.

LINDBERG, J. J.   Department of Wood & Polymer Chemistry, University of Helsinki, Finland.

PETHRICK, R. A.   Department of Pure & Applied Chemistry, University of Strathclyde, Glasgow, Scotland.

PRESTON, F. H.   Department of Physics, Queen Mary College, London, England.

SCHAEFER, J.   Monsanto Company, St. Louis, Missouri, U.S.A.

SHIMADA, K.   Department of Chemistry, SUNY College of Environmental Science and Forestry, Syracuse, N.Y., U.S.A.

SHIMANOUCHI, T.   Department of Chemistry, University of Tokyo, Japan.

SZWARC, M.   *Department of Chemistry, SUNY College of Environmental Science and Forestry, Syracuse, N.Y., U.S.A.*

TÖRMÄLÄ, P.   *Department of Wood and Polymer Chemistry, University of Helsinki, Finland.*

WHITE, J. W.   *St. John's College, Oxford, England.*

WILLIS, H. A.   *I.C.I. Ltd., Plastics Division, Welwyn Garden City, England.*

WRIGHT, C. J.   *A.E.R.E. Harwell, Didcot, Berkshire, England.*

Contents

10. High-Resolution Carbon-13 Studies of Polymer Structure

F. A. BOVEY

11. The Analysis of ^{13}C n.m.r. Relaxation Experiments on Polymers

J. SCHAEFER

12. ^{13}C n.m.r. Studies of α-Methylstyrene-Alkane Copolymers

A. V. CUNLIFFE, P. E. FULLER *and* R. A. PETHRICK

13. The Characterization of Diene Polymers by High Resolution Proton Magnetic Resonance

J. R. EBDON

14. Spin Labels and Probes in Dynamic and Structural Studies of Synthetic and Modified Polymers

P. TÖRMÄLÄ *and* J. J. LINDBERG

15. E.s.r. Studies of Spin-Labelled Synthetic Polymers

A. T. BULLOCK *and* G. G. CAMERON

16. E.s.r. Studies of Dynamic Flexibility of Molecular Chains

K. SHIMADA *and* M. SZWARC

1

A Review of Neutron Scattering with Special Reference to the Measurement of the Unperturbed Dimensions in Macromolecules

G. Allen

University of Manchester

1.1 INTRODUCTION

A few years ago a group of British chemists and physicists presented a case for the construction of a High Flux Beam Reactor (HFBR) to be devoted to neutron scattering studies of the properties of matter. The case included a programme of work concerned with the structure and dynamics of polymeric materials. In the event the Science Research Council (U.K.) obtained[1] a one-third share in the Franco-German HFBR, at that time under construction at Grenoble. The first four contributions to this volume are presented by workers who pioneered the use of the low-flux facilities at Harwell for studies of polymeric materials and who are now using the new facilities at Grenoble. Our aim is to present an overall view of the status of neutron scattering work on polymers. In this first contribution the general principles of neutron scattering from molecular systems are cursorily reviewed and then the results are presented of measurements of the radii of gyration of macromolecules by small-angle neutron scattering.

1.2 NEUTRON SCATTERING[2]

The neutron has a mass of 1 a.u., it is uncharged and has a spin $I = \frac{1}{2}$. The wavelength distribution of thermal neutrons produced from reactors has a Maxwellian peak at 1·8 Å. In most neutron scattering experiments longer wavelength neutrons are used and so cold neutron sources have been developed to enhance the flux at 5–10 Å. Liquid hydrogen refrigeration is usually used to cool the neutrons and the flux at 5 Å is enhanced by an order of magnitude relative to the corresponding flux of thermal neutrons.

At a wavelength of 5 Å, neutrons have velocities of ~ 1000 m s^{-1} and kinetic energy of 300 J mol^{-1} ($\equiv 23$ cm^{-1} $\equiv 4$ meV). For a given wavelength the kinetic energy of a neutron beam is very much lower than the photon energy of electromagnetic radiation. Furthermore much larger momentum transfers can be studied because of the large mass of the particle. Neutron scattering is distinguished by the fact that it is the only scattering phenomenon for which the energy and momentum transfers are simultaneously of the required orders of magnitude for the study of molecular systems—and hence polymers.

1.2.1 Energy and momentum transfer

When a neutron of incident wavelength λ_0 and velocity v_0 is scattered through an angle θ in an inelastic process (Figure 1.1) which changes the scattered

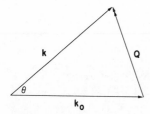

Figure 1.1 Momentum transfer in a neutron scattering event. k_0 is the initial wave vector, k the scattering wave vector and Q the momentum transfer

wavelength to λ and the velocity to v, then the energy transfer is:

$$\Delta E = \tfrac{1}{2}m(v^2 - v_0^2) = \frac{\hbar^2}{2m}(k^2 - k_0^2)$$

where m is the mass of the neutron and $k_0 = 2\pi/\lambda_0$ and $k = 2\pi/\lambda$ are the wave vectors.

From Figure 1.1, the momentum transfer is:

$$\hbar Q = \hbar(k - k_0)$$

For an inelastic process

$$Q = 2\left[\left(k_0^2 + \frac{m\,\Delta E}{\hbar^2}\right) - k_0\left(k_0^2 + \frac{2m\,\Delta E}{\hbar^2}\right)^{\frac{1}{2}}\cos\theta\right]^{\frac{1}{2}}$$

and for an elastic process, or a quasi-elastic process for which $\Delta E \sim 0$

$$Q = \frac{4\pi \sin(\theta/2)}{\lambda_0}$$

Elastic ($\Delta E = 0$), quasi-elastic ($\Delta E \sim 0$) and inelastic ($\Delta E \neq 0$) scattering events are observed in neutron experiments and these events are studied as a function of momentum transfer $\mathbf{Q}$. We shall see shortly that each type of scattering process has a specific application in the study of polymeric materials.

1.2.2 Scattering cross-sections; coherent and incoherent

In one popular type of neutron scattering spectrometer, the time-of-flight instrument, energy transfer is measured by the difference in velocity of the incident neutrons and of the neutrons scattered in a fixed direction. The momentum transfer is defined by the incident wavelength λ_0 and the angle of scatter θ for elastic and quasi-elastic events; for inelastic events ΔE is required in addition in order to measure $\mathbf{Q}$.

Being uncharged, neutrons are scattered by the *nuclei* in the sample under investigation. Consequently the optical selection rules which govern the scattering and absorption of electromagnetic radiation do not apply. In principle *all* possible energy and momentum transfers are observable in neutron scattering experiments.

The scattering cross-section σ (or alternatively the scattering length b, since $\sigma = 4\pi\langle b^2\rangle$) is different for each kind of nucleus—i.e. for each isotope, and is independent of energy for the low-energy neutrons usually used. For example $\sigma^{1H} \neq \sigma^{2H} \neq \sigma^{12C}$ etc. However, because the neutron has a spin of $\frac{1}{2}$, the scattering cross-section also depends on the total spin angular momentum of the neutron and the scattering nucleus. For a nucleus with spin I the net spin can take on values $I + \frac{1}{2}, I - \frac{1}{2}$. Thus if we consider an array of nuclei containing only one isotopic species of spin I, provided the nuclear spins are uncorrelated, two scattering components are generated:

(i) Spin-coherent scattering, with associated interference effects; its intensity is proportional to the square of the mean scattering length averaged over the array of nuclei, i.e.

$$\sigma_{coh} = 4\pi\langle b\rangle^2$$

(ii) Spin-incoherent scattering, displaying no interference effects; its intensity is proportional to the mean square of the deviation from the average over the array,

$$\sigma_{incoh} = 4\pi[\langle b^2\rangle - \langle b\rangle^2]$$

Table 1.1 Nuclear scattering cross-sections (in barns) for low-energy neutrons

	I	σ_{coh}	σ_{incoh}
^{1}H	$\frac{1}{2}$	1·8	79·7
^{2}H	1	5·6	2·0
^{12}C	0	5·6	—
^{14}N	1	11·6	0·3
^{16}O	0	4·2	—
^{28}Si	0	2·0	—

A special case arises for nuclei for which $I = 0$, since only the coherent component is generated for these isotopes. Table 1.1 lists the coherent and incoherent scattering cross-sections for the nuclei most commonly occurring in synthetic and natural polymers; note the wide range of values for both components.

1.2.3 Selection rules and molecular spectroscopy

Finally in this brief review of neutron scattering from molecular systems we must consider the consequences in molecular spectroscopy of the absence of optical selection rules. In infrared spectroscopy only the normal modes for which the element of the dipole moment tensor is finite are active (i.e. $|\partial\mu/\partial q|^2 > 0$) and correspondingly in Raman spectroscopy the element of the polarizability tensor must be finite (i.e. $|\partial\alpha/\partial q|^2 > 0$). Indeed the intensities of the observed bands depend on the magnitude of these elements. If we perform a 'Raman' experiment using neutrons rather than visible light to excite the spectrum (and noting that because $E \equiv 23\,\mathrm{cm}^{-1}$ only the anti-Stokes region will be observed) then *all* normal modes will be active, because optical selection rules do not apply for the scattering of neutrons by nuclei. Further, the intensity of each band will be related to the mean-square-displacement of the nuclei in its normal mode. Thus certain torsional intramolecular modes, especially of side groups, in polymer chains will be more intense than stretching and bending modes.

Another consequence of the relaxed selection rules is that in periodic systems such as crystals or polymer chains the dispersion of the modes has to be considered. In optical molecular spectroscopy only the phase difference $\delta = 0$ needs to be considered; in neutron molecular spectroscopy all phase differences are allowed. Thus the computation of the density of vibrational states for a neutron spectrum is in this sense a much more formidable problem than for an infrared or Raman spectrum.

There is, fortunately, an effect peculiar to neutron spectroscopy which greatly aids vibrational assignment. The large difference in cross-section between ^1H and ^2H noted in Table 1.1 means that selective deuteration of, say, a CH_3 group greatly reduces its intensity in the neutron molecular spectrum not only because of the change in cross-section but also because of the reduction in the amplitudes of nuclear displacement in the normal mode.

1.3 BASIC NEUTRON SCATTERING EXPERIMENTS IN POLYMERS

From the point of view of the neutron scattering technique there are six basic types of experiment, which can be classified as elastic, quasi-elastic and inelastic, each category having a spin-coherent and spin-incoherent sub-division. In this section we will simply list the applications to polymer problems which are already being studied or are imminent.

Elastic		
Coherent	Crystallography	Structure of crystalline polymers
	Small-angle scattering	Molecular dimensions (R_g) in bulk polymers
Incoherent	—	—
Quasi-elastic		
Coherent	Doppler broadening of elastic peak as $f(\mathbf{Q})$	Molecular dimensions and molecular dynamics in solutions and rubbers
Incoherent	Dopper broadening of elastic peak as $f(\mathbf{Q})$	Molecular dynamics in solutions and rubbers
Inelastic		
Coherent	Dispersion curves of intra- and intermolecular vibrations	Elastic constants of crystals Intermolecular potential functions
Incoherent	Molecular spectra Lattice vibrations	Molecular vibrations Intramolecular potential functions

In general, deuterated samples must be used for the study of coherent scattering phenomena to reduce the incoherent contribution from protons, and selective deuteration is useful in neutron inelastic incoherent spectroscopy. The list is not exhaustive and the remainder of this contribution will deal only with small-angle neutron scattering. The following three contributions will deal with, respectively, quasi-elastic studies of self diffusion (p. 13), molecular spectroscopy (p. 29), and the study of the dispersion curves of intra- and intermolecular vibrations in crystalline polymers (p. 41).

1.4 MEASUREMENT OF MOLECULAR DIMENSIONS IN BULK BY SMALL-ANGLE NEUTRON COHERENT SCATTERING

Some 20 years ago Flory[3] put forward the hypothesis that in the amorphous states, rubber and glass, a polymer chain obeys random flight statistics, that is to say that the molecules have unperturbed dimensions as found in θ-solvents at the θ-temperature. Although unperturbed dimensions of polymer chains are readily measured in θ-solvents by Rayleigh light scattering or by small-angle X-ray studies, hitherto no measurements have been possible in the bulk rubber or glass. These techniques are applicable to solutions because of the contrast provided by the difference in refractive index between solute and solvent in light-scattering studies and the difference in electron densities in the case of X-ray scattering.

In the bulk undeuterated polymer there is, of course, no contrast between individual polymer molecules and the matrix. However, if we consider 1% of a perdeuterated polymer dissolved in a matrix of the corresponding protonated polymer we can see by inspection of Table 1.1 that there is now a contrast arising from the difference in σ_{coh}^{1H} and σ_{coh}^{2H}. Thus we can use the intensity of elastic coherent neutron scattering from the sample to estimate the mean-square-radius of gyration of the deuterated polymer coils just as we do from Rayleigh scattering and X-ray scattering measurements. As in the case of X-rays, small-angle scattering is involved because of the short wavelength of the neutron beam. The method is more versatile than the other two because it is applicable to solutions (e.g. protonated polymer in perdeuterated solvent) as well as to bulk materials. Furthermore, since the force fields of perdeutero- and protonated-polymer molecules are very similar, their molecular conformations should not be very different in the isotopic mixture from those in the 'pure' polymers.

The analysis of results is precisely analogous to the well-established method[4] of Rayleigh light scattering. A 'Rayleigh' ratio R_θ is defined for a polymer molecule in solution which scatters an incident beam of neutrons of intensity I_0 through an angle θ with scattered intensity I measured at a distance r from the molecules, as

$$R_\theta = \frac{I}{I_0} r^2$$

For a deuterated polymer dissolved in a protonated matrix

$$R_\theta = \left(\sum_i b_i^{2H} - \frac{V_d}{V_h} \sum_i b_i^{1H} \right)^2 f\left(\frac{4\pi \sin (\theta/2)}{\lambda_0} \right)$$

where b_i^{2H} is the coherent scattering length of the ith nucleus in the deuterated polymer and b_i^{1H} is the coherent scattering length of the ith nucleus in the matrix. The molar volumes of polymer and matrix are V_d and V_h respectively. In the case of a deuterated polymer dispersed in its protonated equivalent of the same molecular weight M containing N protons or deuterons per molecule

$$R_\theta = N^2 (b_{coh}^{2H} - b_{coh}^{1H})^2 f\left(\frac{4\pi \sin (\theta/2)}{\lambda_0} \right)$$

and if the concentration of deuterated polymer is c g cm^{-3} then the value of R_θ per unit scattering volume is

$$R_\theta = N^2 (b_{coh}^{2H} - b_{coh}^{1H})^2 f\left(\frac{4\pi \sin (\theta/2)}{\lambda_0} \right) c N_0/M$$

$$= KcMf\left(\frac{4\pi \sin (\theta/2)}{\lambda_0} \right)$$

where $K = N^2 (b_{coh}^{2H} - b_{coh}^{1H})^2 N_0/M^2$.

Since the thermodynamic environment is not necessarily ideal, in practice R_θ would be expressed in the form of a Zimm equation[4]

$$\frac{Kc}{R_\theta} = \frac{1}{M} f^{-1}\left(\frac{4\pi \sin (\theta/2)}{\lambda_0} \right) + 2A_2 c + \cdots$$

where A_2 is the second virial coefficient of the system. Thus M and A_2 are obtained by plotting $\mathscr{L}_{\theta\to0}(Kc/R_\theta)$ versus concentration and since

$$\underset{c\to0}{\mathscr{L}} \; f^{-1}\left(\frac{4\pi \sin{(\theta/2)}}{\lambda_0}\right) = 1 + \frac{R_g^2}{3}\left(\frac{4\pi \sin{(\theta/2)}}{\lambda_0}\right)^2 + \cdots$$

R_g, the radius of gyration, can be obtained from a plot of $\mathscr{L}_{c\to0}(Kc/R_\theta)$ versus $\sin^2{(\theta/2)}$. Thus the Zimm plot is used, as in light-scattering studies, to obtain R_g^2, M and A_2. Finally we must note that in fact the Z-average value, $\langle R_g^2 \rangle_z$, will be obtained and of course the weight average molecular weight, M_w.

1.4.1 Apparatus and experimental technique

Typical of small-angle neutron scattering instruments is D-11 which is installed at the high-flux beam reactor at Grenoble. A schematic diagram is given in Figure 1.2. Cold neutrons from the reactor are collimated by means of

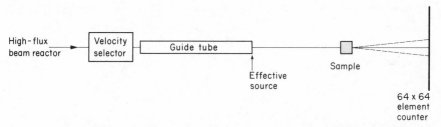

Figure 1.2 Schematic diagram of a low-angle neutron scattering spectrometer

a guide tube and then passed through a helical slot rotor which acts as a velocity selector. A further length of guide tube brings the monochromated neutrons to a convenient point in the laboratory where the diffractometer is located. The sample, about 1×1 cm and a few mm in thickness, is held in the beam in a cell with quartz windows at a point some 2–20 m from the effective source. The detector is located a similar distance D away from the sample.

The detector is specially designed for small-angle scattering work. It consists of an array of 64×64 BF_3 counters arranged in a square grid. The unit size is 1 cm $\times 1$ cm and each cell is separately linked to a computer. Since the scattering is isotropic, the cells are grouped for data processing in terms of their distance from the centre of the undeflected part of the neutron beam. The intensity at distances between $(l + 0.5)$ and $(l - 0.5)$ cm is summed, averaged and evaluated as $I(l)$ per cm^2, and thus $I(l)$ can be determined in increments of 1 cm from $l \sim 5$ to 32 cm. The scattering angle for each value of l is given by

$$\theta = \frac{l}{D}$$

8

In a set of experiments it is necessary to allow for incoherent scatter from the sample. Therefore in addition to measuring $I(\theta)$ as $f(\theta)$ for samples of deuterated homopolymer dispersed in the protonated host, it is necessary to make similar measurements on a blank which contains the same relative proportions of D and H atoms now *randomly distributed* over all molecules. This sample provides the correction for the total incoherent background and any coherent scatter which may arise from any superstructure present in the sample. Subtraction of this background gives $I_{coh}(\theta)$ which is used to compute R_θ and hence the Zimm plot can be constructed.

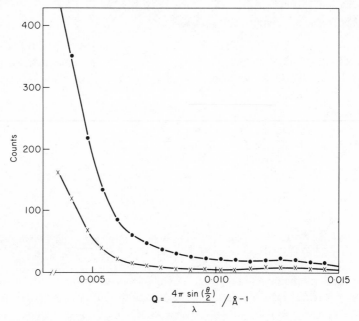

Figure 1.3 Low-angle scattering curves for deuterated poly(ethylene oxide) ($\times$) and a deuterated matrix containing 1% protonated poly-(ethylene oxide) ($\bullet$)

Typical data are shown in Figures 1.3 and 1.4. Figure 1.3 gives an idea of the magnitude of the background correction (since a randomly distributed sample gives results identical with the deuterated sample), and also of the shape of the scattering envelope. In fact the envelope follows closely the Debye equation

$$I(\mathbf{Q}) = KcM\frac{2}{x^2}(x - 1 + e^{-x})$$

where $x = Q^2 R_g^2$. A Zimm plot typical of neutron scattering results is displayed in Figure 1.4.

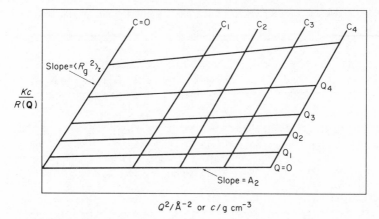

$Q^2/\text{Å}^{-2}$ or $c/\text{g cm}^{-3}$

Figure 1.4 Schematic 'Zimm' plot for neutron scattering results

1.4.2 Results of small-angle neutron scattering measurements

1.4.2.1. Poly(methyl methacrylate)

The first results on a polymer sample were reported by Kirste et al.[5] in 1973. About 1 % monodisperse poly(methyl methacrylate) was dispersed in deuterated monomer $CD_2{=}C(CD_3)COOCD_3$ and the latter polymerized to a similar molecular weight. In this way complete dispersion of the protonated matrix in the deuterated host was assured. For the blank sample monomer and perdeutero-monomer were mixed in the same proportions and polymerized so that the protonated monomer units were now dispersed at random over the whole sample.

The results obtained showed that within experimental error of about $\pm 5\%$ the dimensions of the chain in the glass was identical with that in θ-solvents, that is to say the dimensions of the polymer chain in the glass are unperturbed. In keeping with this observation the second virial coefficient A_2 is zero. For poly(methyl methacrylate) of $\overline{M}_w = 2{\cdot}5 \times 10^5$ at $25\,^\circ\text{C}$, the following values were found:

In the glass $\qquad\qquad\qquad\qquad \langle R_g^2 \rangle_z^{\frac{1}{2}} = 125\,\text{Å}, \quad A_2 = 0$
In n-butyl chloride (a θ-solvent) $\quad \langle R_g^2 \rangle_w^{\frac{1}{2}} = 110\,\text{Å}, \quad A_2 = 0$
In dioxan (an athermal solvent) $\quad \langle R_g^2 \rangle_w^{\frac{1}{2}} = 170\,\text{Å}, \quad A_2 = 5 \times 10^{-4}\,\text{cm}^3\,\text{g}^{-2}$

1.4.2.2. Polystyrene

Much more comprehensive studies of the polystyrene system have been made by Benoit et al.[6] and Wignall et al.[7] A wide range of molecular weights has been studied and dimensions have been measured in the glass and in cyclo-hexane, a θ-solvent, at $35\,^\circ\text{C}$. The results in Table 1.2 again show that in the glass the chains adopt unperturbed dimensions and that A_2 is zero.

Table 1.2 Unperturbed dimensions of polystyrene in bulk and in
solution[6]

$\overline{M}_w$	$\langle R_g^2 \rangle_w^{\frac{1}{2}}/\text{Å}$		
	In bulk[a]	In C_6H_{12}[a]	In C_6H_{12}[b]
21,000	38	42	—
57,000	59	70	—
90,000	78	88	—
97,000[c]	90	84	—
160,000	107	117	108
325,000	143	150	—
890,000	280	260	—
1,100,000	297	293	303

[a] By neutron scattering.
[b] By light scattering.
[c] Ref. 7.

These results are consistent with the general relationship

$$\langle R_g^2 \rangle_w = 0.056 \overline{M}_w \, (\text{Å})^2$$

1.4.2.3. Polyethylene

Work on polyethylene at first suggested that the dimensions in the melt
were actually larger than θ-dimensions. However, it is now clear from the
work of Wignall and Ballard[8] that $+CD_2+_n$ tends to aggregate when in
dispersion in $+CH_2+_n$ polymer. Unpublished results show that when a
molecular dispersion is achieved then the chains do display unperturbed
dimensions.

Thus the hypothesis originally advanced by Flory[3] that polymer chains in
their own matrix display unperturbed dimensions is now finally based on
experiment.

1.4.2.4. Rubber elasticity.

The kinetic theory of rubber elasticity is based on assumptions that the
lengths of chains between crosslink points in a rubber network behave as
individual Gaussian chains and that in deformation the chain vectors are
displaced in the same ratios as the bulk dimensions of the sample (i.e. affine
deformation). Thus the stress supported by a rubber network in simple extension
is:

$$f = NkT \frac{\langle r_i^2 \rangle}{\langle r_0^2 \rangle} (\alpha - \alpha^{-2})$$

where α is the extension ratio, r_i is the end-to-end distance of a chain between
crosslinks in the network and r_0 is the unperturbed dimension of the same length
of chain in the free state. In the theory developed by James and Guth,

$\langle r_i^2 \rangle / \langle r_0^2 \rangle = \frac{1}{2}$; in other theories it is not specifically evaluated but is implicitly considered to be about unity.

In Manchester[9] we have made measurements on poly(ethylene oxide) diols containing 1–2% deuterated polymer in the melt and hence have established values for $\langle r_0^2 \rangle$ in a manner exactly analogous to that described above. The polymer has then been converted into a urethane network and $\langle r_i^1 \rangle$ measured in the rubber. The preliminary results give $\langle r_i^2 \rangle / \langle r_0^2 \rangle \sim 1$ and not $\frac{1}{2}$. We are now studying the effect of extension of the sample on the anisotropy of the dimensions of the chain and also the effect of extension on the small-angle scattering from a deuterated sample containing protonated crosslinks. Ultimately therefore the basic tenets of the kinetic theory of rubber elasticity will also be tested.

1.5   REFERENCES

1. E. W. J. Mitchell, *New Scientist*, 1974.
2. See for example *Chemical Applications of Thermal Neutron Scattering*, ed. B. T. M. Willis, Oxford University Press, 1973.
3. P. J. Flory, *Principles of Polymer Chemistry*, Cornell University Press, Ithaca, 1953, ch. 12.
4. C. Tanford, *Physical Chemistry of Macromolecules*, J. Wiley, London, 1961, ch. 5, p. 275.
5. R. G. Kirste, W. A. Kruse and J. Schelten, *Makromol. Chem.*, **162,** 299, (1973).
6. H. Benoit, D. Decker, J. S. Higgins, C. Picot, J. P. Cotton, B. Farnoux and G. Jannink, *Nature*, **245,** 13, (1973).
7. D. G. H. Ballard, G. D. Wignall and J. Schelten, *European Polymer J.*, **9,** 965, (1973).
8. G. D. Wignall, private communication.
9. G. Allen, A. Maconnachie and R. H. Mobbs, to be published.

2
Diffusional Motion of Rubbers

J. S. Higgins

Institut Laue–Langevin, Grenoble

2.1 INTRODUCTION

Neutrons scattered by nuclei undergoing diffusive motion gain or lose very small amounts of energy from the scattering system. It is therefore the quasi-elastic region of the neutron scattering spectra which is of interest (see p. 4). The scattering cross-section is measured as a function of energy and momentum transfer of the neutron and is related to the space-time correlation function of the scattering nuclei (see section 2.1.1 below). With the resolution now available in neutron spectroscopy low-energy motions can be observed over distances up to about 10 Å and on a time-scale between 10^{-12} and 10^{-9} s. The technique is therefore particularly well suited to the investigation of segmental motion in polymers.

The glass-to-rubber transition in polymers is generally accepted to be due to the onset of long-range conformational motion in the main chain brought about by relative rotations of successive segments. These long-range motions are responsible for rubber elasticity. In n.m.r. measurements on polymers a minimum in the spin-lattice relaxation time T_1 is associated with the freezing out of these motions at the rubber-to-glass transition (though it usually occurs[1,2] at a temperature above T_g) and similar effects are observed in dielectric or mechanical relaxation experiments. Analysis of the temperature dependence leads to values for activation energies of the processes involved and an attempt at correlation with the molecular structure of the polymer. There is no satisfactory statistical mechanical theory of entanglement effects on diffusive motions of polymers[3,4] and thus at present no model calculations of the neutron scattering spectra of amorphous polymers. In the ideal case of infinitely long chains in dilute solution detailed calculation for long-range conformational changes have been made. Until recently experimental limitations have prevented the testing of these models in the energy and momentum transfer ranges where they are applicable. Meanwhile a considerable body of information about the nature of segmental motion in polymer chains is being built up from analysis of data obtained from amorphous polymers in the experimentally accessible ranges.

Although in principle there is much information to be gained about correlated segmental motion from the coherent scattering of neutrons, all the results here described have been obtained from polymers containing hydrogen nuclei and for which, therefore, the incoherent scattering is overwhelmingly dominant.

2.1.1 Correlation functions for polymeric systems

In any neutron scattering experiment the scattered intensity is measured as a function of energy and angle of scatter. Removal of factors arising from the particular technique chosen leads to the differential cross-section $d^2\sigma/d\Omega\,dE$. This is the probability that an incident neutron will be scattered by the sample with a known energy change dE into a given solid angle $d\Omega$.

This may be written

$$\frac{d^2\sigma}{d\Omega\,dE} = \frac{N\sigma}{4\pi}\frac{\mathbf{k}}{\mathbf{k}_0}S(\mathbf{Q},\omega) \tag{1}$$

$\mathbf{k}_0$ is the initial and $\mathbf{k}$ the final wave vector of the neutron; $\hbar\omega$ is the energy transfer $= (\hbar^2/2m)(k^2 - k_0^2)$, $\hbar\mathbf{Q}$ is the momentum transfer $= \hbar(\mathbf{k} - \mathbf{k}_0)$, Nb^2 is the cross-section of the sample nuclei for neutrons assuming N is the number of nuclei each of scattering cross section σ. In general (see Chapter 1) the cross-section comprises both coherent and incoherent terms. However, for the hydrogenous system considered here only the incoherent scattering is important. $S(\mathbf{Q},\omega)$ is the incoherent scattering law (more normally $S_{inc}(\mathbf{Q},\omega)$. For a detailed discussion of coherent and incoherent scattering see, for example, Allen and Higgins.[11] $S(\mathbf{Q},\omega)$ is proper to the scattering system, being the double Fourier transform of the space-time self-correlation function $G_s(\mathbf{r},t)$

$$S(\mathbf{Q},\omega) = \frac{1}{2\pi\hbar N}\iint \exp - i(\omega t - \mathbf{Q}.\mathbf{r})G_s(\mathbf{r},t)\,d\mathbf{r}\,dt \tag{2}$$

In the case of polymers the time correlations are interesting—the intermediate scattering function $I(\mathbf{Q},t)$, in which the spatial Fourier transform is already performed, demonstrates clearly the form of this time dependence for different models,

$$S(\mathbf{Q},\omega) = \frac{1}{2\pi}\int I(\mathbf{Q},t)e^{-i\omega t}\,dt \tag{3}$$

and for example, for a system undergoing Fick's law diffusion

$$I(\mathbf{Q},t) = e^{-DQ^2 t} \tag{4}$$

giving rise to a Lorentzian form of the scattering law (at constant Q) with a half-width varying as DQ^2.

The slow motion of long molecules in solution has been discussed by Rouse[5] in the so-called free-draining limit where hydrodynamic coupling of chain segments is neglected. In this limit successive units of the molecule equalize their average orientation by diffusive processes along the chain—so-called Rouse modes. These effects dominate the neutron spectra in the limit of low

Q and ω (i.e. long correlation times and distances)[6] giving rise to a scattering function depending on $t^{\frac{1}{2}}$

$$I(\mathbf{Q}, t) = \exp\left[-\frac{Q^2\rho^2}{3}\sqrt{\frac{W|t|}{\pi}} \right] \tag{5}$$

where ρ is the statistical step length and W^{-1} essentially the correlation time of the system.

The scattering law has an unusual Q^4 half-width dependence arising from the fact that in time t a signal travels a distance $d \propto t^{\frac{1}{2}}$ along the polymer chain; but because the chain is coiled the distance in space is proportial to $d^{\frac{1}{2}}$.

The hydrodynamic coupling effect of the solvent, discussed by Zimm,[7] gives rise to a $t^{\frac{2}{3}}$ dependence.[8]

$$I(\mathbf{Q}, t) = \exp\left[-\frac{Q^2\rho^2}{6\pi}\Gamma\left(\frac{1}{3}\right)|Wt|^{\frac{2}{3}} \right] \tag{6}$$

The scattering law then has a Q^3 half-width dependence.

These latter two models are the only examples of calculations for polymer systems, and although they only apply to strictly limited conditions of dilute solution, and low **Q**, they have been included to show the type of correlation function that may occur when nuclei are bound into polymer chains.

2.2 EXPERIMENTAL

2.2.1 Measurement

In the investigation of low-energy motions of polymer chains the most frequently used neutron technique has been the time-of-flight spectrometer, an example of which is described in detail on p. 30 (Figure 3.1). Figures 2.1 shows the time-of-flight spectrum of a polymer obtained at two different angles of scatter and illustrates the salient features of this type of spectroscopy. In this experiment the incident neutrons are monochromated at a wavelength of 8·2 Å (2080 μs m^{-1} time of flight). Most of the scattered neutrons arrive in the intense peak centred on this time of flight. The corresponding energy is only 1·2 meV which is much less than thermal energies so that inelastic processes almost invariably involve transfer of energy from the system to the neutrons. The peak at about 500 μs m^{-1} corresponds to the 1–0 transition of the torsion vibration of the methyl group (see p. 30 and Allen et al.[9]).

The spectra at the two scattering angles show distinct differences caused by the large neutron mass and hence the large momentum transfer involved in a scattering process. Apart from changing the intensity of the inelastic scattering (see p. 29) this also has an effect on the elastic scattering. Inset in the figure and on the same time-of-flight scale is the resolution function of the machine for elastic scattering. At 25° the elastic processes reproduce this incident distribution, but at 134° it is considerably broadened by quasi-elastic processes.

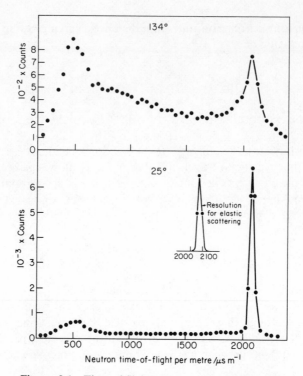

Figure 2.1 Time-of-flight spectra of polydimethyl-siloxane at scattering angles of 25° and 134°. Incident wavelength 8·2 Å. Neutron counts *vs.* time of flight with (inset) the elastic scattering from a vanadium sample

Analysis of this quasi-elastic broadening as a function of the momentum transfer gives information about the low-energy motion of the polymer chain.

2.2.2 Analysis

The ordinates in Figure 2.1 represent the double differential scattering cross-section $d^2\sigma/d\Omega\,d\tau$ where τ is the neutron time-of-flight per metre. This cross-section is related to the energy cross-section

$$\frac{d^2\sigma}{d\Omega\,dE} = -\frac{\tau^3}{m}\frac{d^2\sigma}{d\Omega\,d\tau} \tag{7}$$

where m is the neutron mass, and combining with equation (1), the scattering law introduced in section 2.1.1 is given by

$$\frac{d^2\sigma}{d\Omega\,d\tau} = -m\frac{\tau_0}{\tau^4}\frac{N\sigma}{\hbar 4\pi}S(\mathbf{Q},\omega) \tag{8}$$

where τ_0 is the initial and τ the final flight time (related to $\mathbf{k}_0$, $\mathbf{k}$ by $|\mathbf{k}| = m/\hbar\tau$).

When comparing $S(\mathbf{Q}, \omega)$ obtained from time-of-flight experiments with that calculated for model systems, allowance must be made for the fact that measurements are usually at constant θ (scattering angle) rather than constant $\mathbf{Q}$, and for the multiple scattering effects which are inevitable in finite samples.[10]

Experimentally the $S(\mathbf{Q}, \omega)$ is convoluted with the resolution function of the machine, $R(\omega)$

$$S_{exp} = \int_{-\infty}^{+\infty} S(\mathbf{Q}, \omega)R(\omega - \omega')\,d\omega' \tag{9}$$

where $S(\mathbf{Q}, \omega)$ is the theoretical function, and in practice this resolution may be of comparable half-width to the theoretical broadening. In order therefore to extract information about the scattering law it is necessary to make some assumptions about its functional form.

One example of a particularly simple scattering law, that for a liquid undergoing Fick's law diffusion, was introduced in section 2.1

$$S(\mathbf{Q}, \omega) = \frac{1}{2\pi} \frac{DQ^2}{(DQ^2)^2 + \omega^2} \tag{10}$$

where D is the classical self-diffusion constant of a liquid. At constant $\mathbf{Q}$ this function is a Lorentzian whose energy-half-width varies as DQ^2. In many experiments on liquids[11] the half-width of the quasi-elastic broadening has been deconvoluted from the resolution function and plotted against Q^2. Deviation from straight-line behaviour with slope D has been the starting point of earlier discussions about diffusive behaviour in liquids.[12]

Even for such a simple analysis in terms of half-widths it is necessary to make some assumptions about the form of the scattering function in order to remove the finite machine resolution.

2.3 BULK POLYMERS

2.3.1 Polydimethylsiloxane

This polymer (PDMS)

$$\left(\!\!\!\begin{array}{c} CH_3 \\ | \\ Si-O \\ | \\ CH_3 \end{array}\!\!\!\right)_n$$

has long been known for its high internal flexibility. Even samples of high molecular weight have low glass-transition temperatures and relatively low bulk viscosity. The availability of a series of low-molecular-weight ring and chain oligomers is another reason for the choice of PDMS as a system for investigation of segmental motion.

In an initial series of experiments quasi-elastic broadening was observed in a series of such oligomers.[9] Within the resolution of the experiments the scattering approximated to Lorentzian broadening of the Gaussian resolution function and using this functional form it was possible to extract the half-widths of the broadening and demonstrate the variation with momentum transfer. In Figure 2.2 the half-widths extracted in this way for the scattering of a 4 monomer-unit ring and chain are compared. Over the somewhat limited **Q** range,

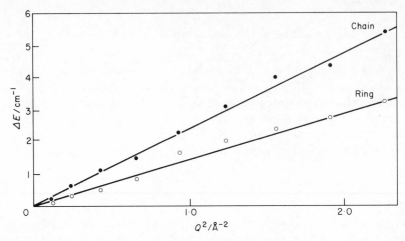

Figure 2.2 Variation of ΔE (full width at half maximum of $S(\mathbf{Q}, \omega)$) with momentum transfer, **Q**

for the ring compound
$$\left[\begin{array}{c} CH_3 \\ | \\ -Si-O- \\ | \\ CH_3 \end{array} \right]_4$$

and for the chain compound
$$CH_3 \left[\begin{array}{c} CH_3 \\ | \\ Si-O \\ | \\ CH_3 \end{array} \right]_3 \begin{array}{c} CH_3 \\ | \\ Si-CH_3 \\ | \\ CH_3 \end{array}$$

$\Delta E \propto Q^2$ and hence the slope of this linear dependence was described by an effective diffusion coefficient, D_{eff}, by analogy with equation (10). Figure 2.2 shows that D_{eff} is about twice as large for a chain than for the corresponding ring compound.

In Figure 2.3 the D_{eff} values measured analogously for the series of oligomers are plotted against molecular weight. Shown in the same figure are the diffusion coefficients obtained from spin-echo n.m.r. measurements.[13] In the spin-echo measurements changes are observed in the echo signal due to displacements of the protons in a magnetic field gradient with a characteristic observation time of about 10^{-3} s. The motion of the protons is followed over relatively long distances and for these short chain molecules it is certainly the centre-of-mass diffusion coefficient which is measured. On the other hand the characteristic observation time for neutron scattering is 10^{-10} to 10^{-12} s. Thus any higher

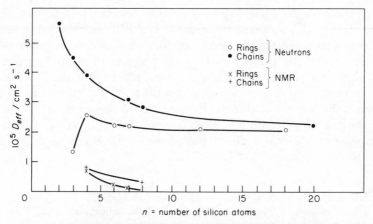

Figure 2.3 Variation of the D_{eff} values (see text) with n where n is the number of silicon atoms for cyclic and linear oligomers of PDMS. Included also are values of diffusion coefficients measured by spin-echo n.m.r.[13] The value of D_{eff} for $n = 2000$ is $1\cdot8 \times 10^5 \, cm^2 \, s^{-1}$

frequency components such as local rotations of segments of the chains (which would average out in spin-echo measurements) will be evident in the neutron spectra. These effects can be seen in Figure 2.3. The proton mobility as seen by the neutrons is much higher than the centre-of-mass diffusion. The D_{eff} values decrease with increasing molecular weight in approximate analogy with the D values but then reach a constant value which persists up to high molecular weights ($M_w > 10^5$). The highest M_w used for spin-echo measurements was 3×10^4 and for this sample[14] $D = 5 \times 10^{-9} \, cm^2 \, s^{-1}$.

The initial neutron results thus indicate a proton motion involving local mobility much greater than the centre-of-mass diffusion. This motion is insensitive to molecular weight for $n \gtrsim 15$ and effectively independent of chain conformation (rings and chains). From these first observations therefore the broadening could be associated with internal motions of the polymer chains.

2.3.2 Other polymer systems

The unusual mobility of PDMS, which made it a natural choice for the initial experiments, raises questions about the uniqueness of the above phenomena. In particular it has been suggested that the methyl side groups have a very free rotation about the Si—C bonds and it is possible that the observed broadening is caused by this motion.[15]

Experiments on poly(propylene oxide) discount this possibility.[16,17] This polymer (PPO)

$$\begin{array}{ccc} CH_3 & H & \\ | & | & \\ \text{+}C & - C - O\text{+}_n \\ | & | & \\ H & H & \end{array}$$

has one methyl side group and it is possible selectively to replace H by D (see p. 32). The D_{eff} values of three samples of PPO (substituted in this way) were compared at different temperatures. Not only was appreciable broadening observed in this chemically very different polymer (though at temperatures above room temperature) but it was almost identical in the three samples. This experiment indicated that the motion observed was equivalent for all the protons in the polymer confirming its allocation as segmental rather than side-chain motion. Broadening was also observed and D_{eff} values extracted for polymethylphenylsiloxane, polyisobutene and poly(ethylene oxide). The observations are summarized in Table 2.1.

Table 2.1 Diffusion coefficients D_{eff} and associated Arrhenius parameters D_0 and E_a of various polymers measured by inelastic neutron scattering

	$10^5 D_{eff}$ [a] $/\text{cm}^2 \text{ s}^{-1}$	$10^3 D_0/\text{cm}^2 \text{ s}^{-1}$	$E_a/\text{kJ mol}^{-1}$
Polydimethylsiloxane	2·09	0·5 ± 0·2	7·5
Polymethylphenylsiloxane	0·62	0·65 ± 0·25	11·4
Poly(propylene oxide)			
$\left.\begin{array}{l} H_6 \\ D_3HH_2 \\ H_3DD_2 \end{array}\right\}$	0·27[b]	7·0 ± 0·3	19·0
Poly(ethylene oxide)	0·20[b]	20·0 ± 10·0	17·0
Polyisobutene	0·04[b]	4·5 ± 2·5	22·0

[a] At 20 °C.   [b] Extrapolated value.

2.3.3 Temperature dependence

The temperature dependence of D_{eff} was measured for each of the polymers discussed above. A simple Arrhenius behaviour of the type $D_{eff} = D_0 e^{-E_a/RT}$ was found sufficient to describe this initial data (see, for example, Figure 2.4). The values of D_0 and E_a are listed in Table 2.1.

The D_{eff} values at room temperature are an order of magnitude larger for PDMS than for other polymers—the addition of the bulky phenyl side group is sufficient to reduce the mobility noticeably. If E_a is tentatively interpreted as an activation energy for segmental rotation, then the results in Table 2.1 indicate that this activation energy is a function of the type of bond along the chain back bone—the C—C—C bond being much 'stiffer' than the O—Si—O bond. Correspondingly, the pre-exponential factor varies with the nature of the side groups attached to the chains.

Relaxation experiments,[1,2,18-22] whether n.m.r., dielectric or mechanical, measure a relaxation frequency associated with the internal segmental motion of the chains. The activation energy associated with this relaxation process is consistently larger than the E_a values from neutron measurements.[16]

The activation energies for bulk viscous flow—also associated with this type of motion—are correspondingly larger still; values are summarized in

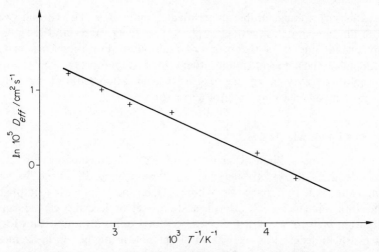

Figure 2.4 Variation of $\ln D_{eff}$ with inverse absolute temperature for a high molecular weight PDMS sample

Table 2.2 Activation energies for segmental motion in rubbers determined by various techniques

| | $E_a/\text{kJ mol}^{-1}$ | | |
Sample	Incoherent neutron scattering	Relaxation data	Viscosity
Polydimethylsiloxane	7·5 ($-35 \rightarrow 100\,°C$)	9 [2] ($> -60\,°C$)	15·27 [19] ($-21 \rightarrow 141\,°C$)
Polymethylphenylsiloxane	11·4 ($15 \rightarrow 94\,°C$)	18 [2] ($>60\,°C$)	
Polyisobutene	22 ($60 \rightarrow 200\,°C$)	~ 70 [1] ($40\,°C$)	67 [20] ($60 \rightarrow 100\,°C$)
Poly(propylene oxide)	19 ($22 \rightarrow 104\,°C$)	40·5 [18] ($\sim 0\,°C$)	
Poly(ethylene oxide)	22 ($70 \rightarrow 170\,°C$)	29·3 [22] ($20\,°C$)	48·9 [21] ($68 \rightarrow 120\,°C$)

Table 2.2. These effects are a manifestation of the different observation time scales of the measurement techniques, a difference already mentioned, to be discussed in section 2.5.

2.4 POLYMERS IN SOLUTION

The slow development of this field arises from two experimental difficulties. Firstly, if experiment is to be compared with theory the measurements must be made at concentrations of 1 % or less where the effect of overlap of chains can be allowed for. This means, however, the use of very high incident neutron fluxes, this poses considerable experimental difficulties. The second problem is more fundamental: the model applies at small Q values and the predicted energy broadening is of the order of microvolts. For these reasons initial experiments in the field were made either on relatively short molecules for which dilute solutions may be up to 10% in concentration without serious chain overlap, or involved a phenomenological approach.

2.4.1 Small molecules in CS_2[23]

In neutron time-of-flight experiments on two low-molecular-weight ethylene oxide oligomers (ethyl ether and 1,2-diethoxyethane) dissolved in CS_2 an attempt was made to describe the whole $S(Q, \omega)$ spectrum with a simple model for the vibrational inelastic scattering and a model for the quasi-elastic scattering which attempted to account for centre-of-mass diffusion (important for these short chain molecules) as well as random motion of the chain segments as described by de Gennes calculations.[6] In so far as separation of these various motions was successful, values of a centre-of-mass diffusion coefficient were found to be larger than those measured by n.m.r., possibly because of fluctuations in the end-to-end distances of the chains (a motion not included in the model for the segmental motion), and possibly because of overall rotation of the molecules which had not been considered as a separate contribution. The Q^4 half-width dependence of the quasi-elastic broadening predicted by the de Gennes model was not found, but this is hardly surprising for these very short molecules.

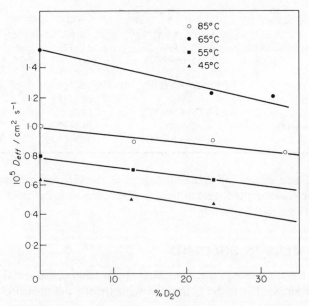

Figure 2.5 Variation of D_{eff} values with concentration for a series of solutions of poly(ethylene oxide) and D_2O at four temperatures

2.4.2 Poly(ethylene oxide) in solution[24]

In a series of experiments on poly(ethylene oxide) dissolved in D_2O the solutions were very concentrated. D_{eff} values were extracted in the same way as for the bulk polymers. In Figure 2.5 the D_{eff} values measured for the polymer at several different temperatures are plotted against the solution concentration. At these concentrations the scattering is dominated by the incoherent scattering from hydrogen in the polymers. The molecular weight was 7.8×10^3 so that the centre-of-mass diffusion should not be important compared to the segmental motion. The D_{eff} values increase as the concentration increases indicating that the local segmental motion is hindered by the solvent molecules. One possible explanation is the solvation of the chains by the D_2O molecules. This idea is supported by the increase in the (Arrhenius) activation energy from 17 kJ mol^{-1} for the polymer to 25 kJ mol^{-1} for the sample 23% diluted with D_2O.

2.5 HIGHER RESOLUTION STUDIES OF BULK POLYMERS

2.5.1 The functional form of $S(Q, \omega)$

Differences of orders of magnitude between the diffusion coefficients observed by spin-echo measurements and the D_{eff} values by neutron spectroscopy may not be surprising in view of the very different characteristic time-scales of the two types of measurement. However it has been suggested[25] that the neutron results may consist of a central very narrow component corresponding to the centre-of-mass diffusion superimposed on a broad spectrum resulting from segmental motion. Such a two-component spectrum would certainly not be resolved with the resolution (about 0.5 meV) of the initial experiments.[26]

The first experiments on bulk polymers had already indicated that the Lorentzian form for the scattering law which was assumed in order to extract the half-width data was not a very good fit for the higher molecular weight compounds especially in the wings. It was thus of great interest to perform experiments at very high resolution to try to gain further information about the actual shape of the scattering signal.

The availability of the high flux facility (p. 1) has recently made such measurements possible. Two series of experiments were undertaken, one involving improved resolution on a time-of-flight spectrometer and the other the use of ultra-high resolution from the so-called 'back-scattering' spectrometer.[27,28]

2.5.2 High resolution experiments[17]

The spectrometer shown in Figure 2.6 is essentially a triple-axis machine (p. 41) with a fixed-energy analyser. The monochromator and analyser crystals have Bragg angles very near 90°, giving extremely high resolution (about 1 μeV). Doppler motion of the monochromator crystal along the direction of the primary beam is used to vary the incident energy on the sample. In Figure 2.7 the quasi-elastic scattering from a high-molecular weight PDMS sample at

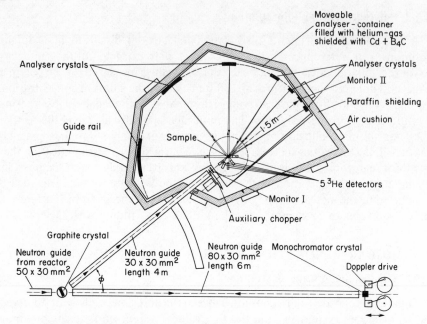

Figure 2.6 The back-scattering spectrometer (IN10) at HFR Grenoble

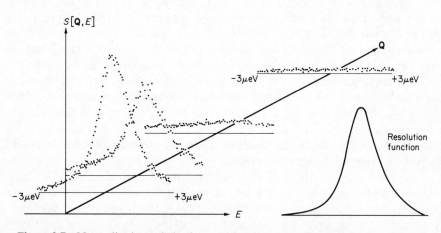

Figure 2.7 Normalized quasi-elastic scattering from polydimethylsiloxane shown on a 3-coordinate plot at four values of $\mathbf{Q}$ (0·15, 0·27, 0·57, and 0·97 Å^{-1}). At the right-hand side the resolution function of the machine is shown on the same scale

four scattering angles is compared with the Lorentzian resolution function of this machine. At the lowest angle, broadening is barely visible; at angles where $\Delta E_{\frac{1}{2}}$ can be measured it corresponds to $D_{eff} \approx 3 \times 10^{-7}\,\text{cm}^2\,\text{s}^{-1}$. Careful examination of the intensity of the quasi-elastic scattering eliminated the possibility that a broad underlying component extended beyond the range of the experiment. At high angles of scatter a centre-of-mass diffusion component would be clearly resolved from the broad distribution in fact observed.

Two important points arise from these results. Firstly it is not possible, even at this high resolution, to distinguish a single component of the quasi-elastic scattering corresponding to centre-of-mass diffusion, and it seems clear that such a component does not exist with any measurable intensity. Secondly, however, the D_{eff} values are two orders of magnitude smaller than those measured in the previous experiments.

The range of momentum transfers (corresponding to $Q = 0.15$ to 1Å^{-1}) is lower than for earlier experiments ($Q = 0.6$ to 2Å^{-1}) for which there were already indications[9] of deviations from linearity of ΔE with Q^2 for the lower Q values. Further experiments were therefore carried out at medium momentum transfer and a resolution about ten times better than the first bulk measurements: $\Delta E_{res} = 0.05$ meV. Spectroscopy using this resolution has lately become possible using a high-resolution time-of-flight spectrometer at the HFBR.

2.5.3 The broadening function and its half-width variation

Figure 2.8 shows all the presently available measurements of half-widths against Q^2 for a high-molecular-weight PDMS sample;[30] the resolution is also indicated. The resolution becomes worse for the higher Q measurements

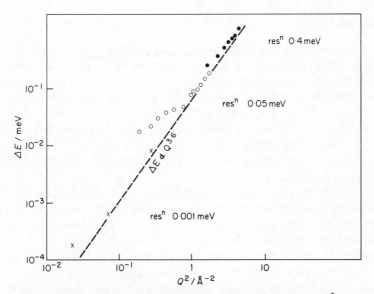

Figure 2.8 Variation of ΔE (full width, half maximum) with Q^2 on a log-log scale for PDMS in three different experiments: × 6·2 Å incident wavelength, back-scattering experiment; ○ 8·2 Å incident wavelength, time-of-flight experiment; ● 4·8 Å incident wavelength, time-of-flight experiment

because these are made with a higher incident energy for which monochromation is worse. However at these higher Q values the broadening is so large that the resolution can reasonably be sacrificed.

Various points may be made from Figure 2.8.

(i) The half-width measurements are compatible with a continuous Q dependence but the exponent is nearer 4 than 2.

(ii) Each *series* of experiments shows deviations from the continuous line at the lower Q values where the measured broadening is large compared with the resolution function.

(iii) These errors, caused by the difficulty of removing the resolution function, permit small sections of the curve to be approximated by the Q^2 dependence found in the initial measurements. Effectively the measured diffusion coefficients obtained from plots of ΔE against Q^2 are themselves a function of momentum transfer. This does not invalidate the initial results obtained for the bulk polymers because comparisons were always made in the same momentum transfer range.[9,16]

Reduction of the momentum transfer in an experiment corresponds to observation of the protons as they move over longer and longer distances. If the D_{eff} values are changing with Q it is hardly surprising that the spin-echo results, which involve distances of the order of 0·01 cm, are orders of magnitude smaller than the neutron results.

2.6 OUTLOOK

Neutron scattering is still a developing technique. The availability of the high-flux source at Grenoble has considerably increased the energy- and momentum-transfer ranges available for neutron spectroscopic measurements. It is now possible to consider experiments in which the quasi-elastic broadening is less than 1 μeV and the momentum transfer less than $0.5 \, \text{Å}^{-1}$. This is the range in which the long-range conformational changes described by de Gennes[6,8] should be observed. Experiments of this type on polymers in dilute solution are in preparation at Grenoble and results will probably be available by the time this review is published.

In the case of bulk polymers some high-resolution data is already available but difficulty arises from the theoretical problem of describing entangled polymer chains in the bulk.[3,4] Only one comparison of a theoretical prediction with experimental data has so far been made for bulk polymers. It can be shown that polymerization removes one degree of freedom per point in the chain and this should be observable as a change in the second moment of the scattering function $S(Q, \omega)$ on going from a monomeric to a polymeric system. Some evidence of this change was obtained from early experiments on bulk PDMS.[3] The problem of calculating the form of $S(Q, \omega)$ itself is not yet solved. It may be that since the neutron experiments are sensitive only to short segments of the chain, the entanglements can be ignored or at least introduced as a second-order effect. In this case the effect of the surrounding polymer chains reduces to a microscopic viscosity, similar to that for ordinary solvent molecules, and the problem is considerably simplified.

Two models for polymers in dilute solution were described in section 2.1.1. They produced correlation functions varying as $t^{\frac{1}{2}}$ and $t^{\frac{2}{3}}$ and scattering law functions with Q^4 and Q^3 half-width variation. In Figure 2.9 an experimental

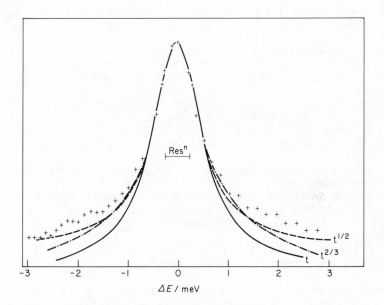

Figure 2.9 Comparison of experimental $S(\mathbf{Q}, \omega)$ as a function of ΔE with calculations from $I(\mathbf{Q}, t)$ as a function of t (full line), $t^{1/2}$ (dashed line) and $t^{2/3}$ (dotted line)

scattering law is compared to three model calculations—the two for polymers in dilute solution and the conventional Lorentzian for diffusion processes. In each case the calculation has been folded with the experimental resolution. The higher-frequency components arising when the time dependence is less than first power give rise to intensity in the wings of the curves and there is some evidence from the figure that this is already closer to the experimental situation than the Lorentzian.[29]

The next year will certainly see considerable advances in high-resolution neutron spectroscopy from polymers in the bulk and, it is to be hoped, corresponding successes in comparing this data with well-based theoretical calculations. Then it should be possible to extract from the measurements information about the statistical step length, the local viscosity and other microscopic chain parameters.

2.7 REFERENCES

1. J. G. Powles and K. Luszczynski, *Physics*, **25**, 455, (1959).
2. J. A. Barrie, M. J. Fredrickson and R. Sheppard, *Polymer*, **13**, 431, (1972).
3. S. F. Edwards and A. G. Goodyear, *J. Phys. A: Math. Nucl. Gen.*, **6**, L31, (1973).
4. S. F. Edwards and G. Grant, *J. Phys. A: Math. Nucl. Gen.*, **6**, 1169, (1973).
5. P. E. Rouse, *J. Chem. Phys.*, **21**, 1272, (1953).
6. P. G. de Gennes, *Physics*, **3**, 37, (1967).
7. B. H. Zimm, *J. Chem. Phys.*, **24**, 269, (1956).
8. P. G. de Gennes and E. Dubois-Violette, *Physics*, **3**, 181, (1967).
9. G. Allen, P. N. Brier, A. G. Goodyear and J. S. Higgins, *J. Chem. Soc. Faraday Discussion*, **6**, 169, (1972).
10. M. W. Johnson, *A.E.R.E. Report No. R7682*, (1974).
11. G. Allen and J. S. Higgins, *Reports on Progress in Physics*, **36**, 1073, (1973).
12. K. E. Larrson, *Inelastic Scattering of Neutrons*, Proc. I.A.E.A. Copenhagen, **1**, 397, (1968).
13. D. W. McCall and E. W. Anderson, *J. Chem. Phys.*, **34**, 804, (1961).
14. D. W. McCall and C. M. Huggins, *Appl. Phys: Letters*, **7**, 153, (1965).
15. A. W. Henry and G. J. Safford, *J. Polymer Sci.*, **A2**, 7, 433, (1969).
16. G. Allen, J. S. Higgins and C. J. Wright, *J. Chem. Soc. Faraday Trans. II*, **70**, 348, (1974).
17. G. Allen, R. E. Ghosh, A. Heidemann, J. S. Higgins and W. S. Howells, *Chem. Phys. Letters*, **27**, 308, (1974).
18. T. M. Connor, D. J. Blears and G. Allen, *Trans. Faraday Soc.*, **61**, 1097, (1965).
19. D. J. Plazek, V. Dannhauser and J. D. Ferry, *J. Coll. Sci.*, **16**, 101, (1961).
20. J. D. Ferry, L. D. Grandine, J. R. Fitzgerald and E. R. Fitzgerald, *J. Appl. Phys.*, **24**, 911, (1953).
21. T. P. Yin, S. E. Lovell and J. D. Ferry, *J. Phys. Chem.*, **65**, 534, (1961).
22. G. Allen, T. M. Connor and H. Pursey, *Trans. Faraday Soc.*, **59**, 1525, (1963).
23. G. Jannink, U. K. Deniz, P. Taupin, *J. Chem. Phys.*, **55**, 2384 (1970).
24. A. Macconachie and G. Allen, unpublished results.
25. K. E. Larrson, *J. Chem. Soc. Faraday Discussion*, **6**, 167, (1972).
26. T. Springer, *Springer Tracts in Modern Physics*, **64** (1972), Springer Verlag.
27. B. Alefeld, M. Birr and A. Heidemann, *Nucl. Instr. Methods*, **95**, 435, (1971).
28. B. Alefeld, *Kerntechnik*, **14**, 15, (1972).
29. G. Allen, R. E. Ghosh, J. S. Higgins and R. E. Lechner, to be published.

3
Torsion Vibrations in Polymers

C. J. Wright

A.E.R.E. Harwell

3.1 INTRODUCTION

The torsion vibrations of polymers can be ideally studied with inelastic incoherent neutron scattering (i.n.s.), the particular spectroscopic method of this contribution. Interest in torsion vibrations derives from their intimate connection with relaxation processes in polymers and their sensitivity to polymer microstructure which can be used on both a qualitative and a quantitative level.

The double differential cross-section for incoherent scattering of neutrons by a single atom undergoing displacement in normal mode j is given by

$$\frac{\partial^2 \sigma_{inc}}{\partial \Omega \, \partial E} = \hbar \frac{\mathbf{k}}{\mathbf{k}_0} b_{inc}^2 \frac{(\mathbf{Q} \cdot \mathbf{U}_j)^2}{2\omega_j m} \frac{\exp\left(-\hbar\omega_j/kT\right) e^{-2W}}{1 - \exp\left(-\hbar\omega_j/kT\right)} \delta(E - \hbar\omega_j)$$

where $\mathbf{k}$ and $\mathbf{k}_0$ are the scattered and initial neutron wave vectors, $\mathbf{Q}$ is the momentum transfer on scattering defined as $\hbar\mathbf{Q} = \hbar\mathbf{k} - \hbar\mathbf{k}_0$ and ω is the energy transfer. $\mathbf{U}$ is the mass-weighted displacement vector of the atom in the normal mode and e^{-2W} is the Debye–Waller factor for the atom where

$$W = \frac{\hbar}{4m} \sum_j \frac{(\mathbf{Q} \cdot \mathbf{U}_j)^2}{\omega_j} \coth\left(\hbar\omega_j/2kT\right)$$

This equation enables us to understand why neutron spectroscopy is so valuable for studying torsion vibrations.

Firstly there is the atomic incoherent cross-section factor $4\pi b_{inc}^2$ which for protons is a factor of 10 greater than that of other commonly occurring atoms in polymers (see p. 3).

Secondly the differential cross-section is proportional to the square of the displacement of an atom involved in a scattering process, which for light atoms will be high because of the inverse relationship between U^2 and mass. Consider, for example, the scattering associated with lower energy vibrations, such as methyl rocking and methyl torsion. For the former, the square of the displacement for a single hydrogen atom will be approximately inversely proportional to the mass of the whole methyl group, i.e. $\propto 1/15$; for the latter, the square of the displacement will be proportional to r^2/I (where I is the moment of inertia about the axis of oscillation and r is the distance of the H atoms from this axis), i.e. $\propto 1/3$. Scattering associated with torsion vibrations of methyl groups is thus expected to be five times as intense as that associated with rocking vibrations.

A final factor, about which there is little direct evidence, is the degree of dispersion, i.e. the variation in energy transfer with wave vector, of torsion modes. It is generally assumed that a torsion vibration suffers little dispersion, which in turn leads to a sharper distribution of the energy transfer on scattering.

All these factors combine to increase the scattering from torsion vibrations so that, in contrast with optical spectroscopy, these are usually very intense in a neutron spectrum. The deuterium cross-section is much less than that of the proton, and this, together with its increased mass, allows the further identification of a torsion vibration by isotopic substitution of the suspected rotor, which greatly reduces its scattering intensity.

3.2 EXPERIMENTAL

Figure 3.1 shows the 6H time-of-flight spectrometer at A.E.R.E. Harwell which was used to record all the spectra shown here. The neutrons leave the reactor core to be monochromated by a pair of phased rotating discs through which only neutrons with preselected velocities may pass. After scattering from the sample, which may readily be cooled or heated, the neutrons are detected at a bank of counters covering the angles 18° to 90°. The energy transfers suffered by the neutrons on scattering can be determined from their flight times in travelling between the sample and the detectors.

3.3 RESULTS

3.3.1 Poly(propylene oxide) and polypropylene: torsion frequencies and energy barriers

Figure 3.2 shows the double differential scattering cross-section of poly-(propylene oxide)

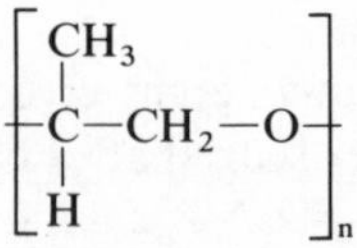

$$
\left[
\begin{array}{c}
CH_3 \\
| \\
-C-CH_2-O- \\
| \\
H
\end{array}
\right]_n
$$

measured in this way[1] at a scattering angle of 90°. There is high intensity due to the elastically scattered neutrons at a time of flight of 1066 μs m^{-1}. The inelastically scattered neutrons which have gained energy from the sample arrive more quickly and Figure 3.2 shows a peak in their distribution at 330 μs m^{-1}.

Inelastic neutron scattering spectra from polycrystalline materials are frequently converted to an amplitude-weighted density of phonon states $\rho(\omega)$

31

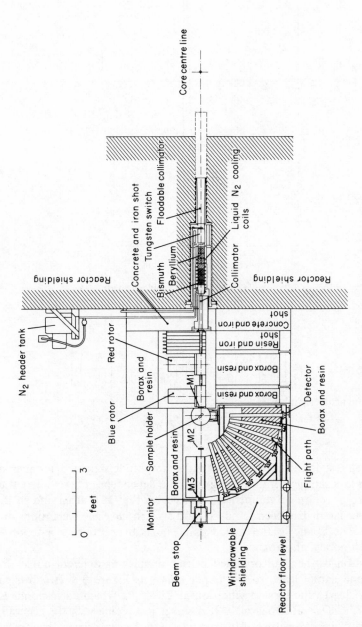

Figure 3.1 The 6H time-of-flight spectrometer

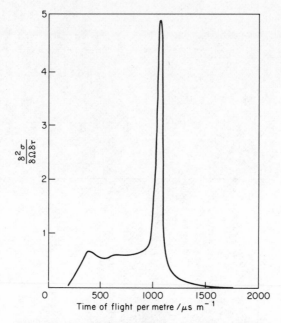

Figure 3.2 $\partial^2\sigma/\partial\Omega\,\partial\tau$ for poly(propylene oxide) at 90° of scatter

where

$$\rho(\omega) = \underset{Q\to 0}{\mathscr{L}}\left[\frac{S(\mathbf{Q},\omega)}{Q^2}\left(\frac{2\hbar\omega}{kT}\right)\exp\left(\frac{\hbar\omega}{2kT}\right)\sinh\left(\frac{\hbar\omega}{2kT}\right)\right]$$

and where the scattering law is

$$S(\mathbf{Q},\omega) = \frac{\mathbf{k}_0}{\mathbf{k}}\frac{1}{Nb_{inc}^2}\frac{\partial^2\sigma_{inc}}{\partial\Omega\,\partial E}$$

$\rho(\omega)$ is of interest for it corresponds to the amplitude-weighted number of energy levels in a sample per unit energy. All the dependence of the differential scattering cross-section on momentum transfer in the $\mathbf{Q}\,.\,\mathbf{U}$ term and in the Debye–Waller factor has been removed, and so also have the weightings due to the thermal occupancy of the excited state. $\rho(\omega)$ is an appropriate function to compare with optical absorbance.

Extrapolation of the data obtained in this manner at different angles for poly(propylene oxide) leads to the upper curve in Figure 3.3. It shows an intense peak due to the torsion vibration at $228\ cm^{-1}$ which corresponds to the peak in $\partial^2\sigma/\partial\Omega\,\partial\tau$ at $330\ \mu s\ m^{-1}$. It is sharp, in keeping with the generally assumed non-dispersive nature of these modes. On the same figure there is the curve obtained in a similar way for poly(propylene oxide)-CD_3, and the lack of significant intensity at $228\ cm^{-1}$ supports the assignment of the scattering at this energy in the undeuterated polymer to the torsion vibration.

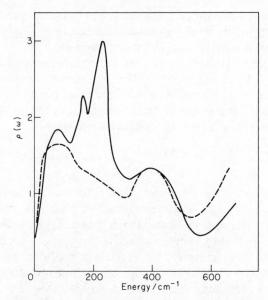

Figure 3.3 $\rho(\omega)$ for poly(propylene oxide)—CH_3 (full line) and poly(propylene oxide)—CD_3 (dashed line)

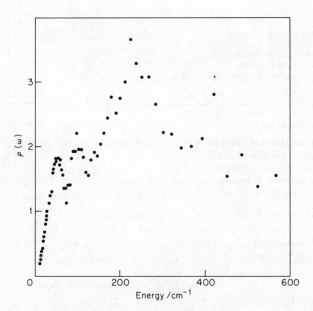

Figure 3.4 $\rho(\omega)$ for polypropylene

Figure 3.4 shows $\rho(\omega)$ for polypropylene, and its overall similarity in shape to that of poly(propylene oxide) indicates that the intense peak in its neutron spectrum is also due to the methyl torsion vibration.

From the torsion frequencies so obtained it is possible to derive rotational energy barriers. Thus, assuming that the torsion is a vibration which is intra-molecular in character and unmixed with any other vibrations of its own chain, and if the potential energy can be described with only the first term in a Fourier expansion, then the rotational barrier $\Delta E\dagger$ is given by

$$\Delta E\dagger = \frac{8\pi^2\omega^2 I}{9} - \text{(zero-point energy)}$$

Some evidence for the intra-molecular nature of the torsion comes from experiments on swollen poly(methyl methacrylate) in which the torsion frequency is unchanged from that in the dry state. The correctness of the other assumptions is shown by comparing activation energies obtained by neutron scattering with those measured by relaxation methods, such as n.m.r. (see Table 3.1). This verification allows the calculation of barriers to rotation in other polymers containing methyl groups which cannot be easily studied by other means, e.g. poly(α-methylstyrene); (see below).

Table 3.1 Barriers to rotation of methyl side groups in polymers, from neutron scattering and n.m.r. experiments

	Barrier/kJ mol^{-1}	
	Neutron scattering	n.m.r.
Poly(propylene oxide)	13·8	18·4
Poly(methyl methacrylate) (syndiotactic)	33	30
Poly(methyl methacrylate) (isotactic)	23	22·6
Polypropylene	13·8	16·7

3.3.2 Poly(methyl methacrylate) and poly(α-methylstyrene): effects of microstructure

In poly(methyl methacrylate) there are two sizeable but dissimilar side-groups, $-CH_3$ and $-COOCH_3$, and it was found that the measured torsion frequency of the α-methyl group depended on the tacticity of the polymer. Thus in Figure 3.5 it may be seen that the 'isotactic' and 'syndiotactic' forms peak at 300 and 360 cm^{-1} respectively.[2] In both cases the methyl group in the ester side-chain was deuterated and the distribution of triads for the two samples was

	Isotactic	Heterotactic	Syndiotactic
'Isotactic'	75%	12%	13%
'Syndiotactic'	8%	31%	61%

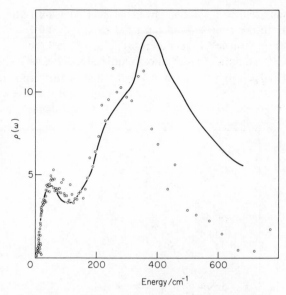

Figure 3.5 $\rho(\omega)$ for isotactic (circles) and syndiotactic (line) poly(methyl methacrylate)

The question arises as to whether the difference in measured torsion frequencies of the derived barriers (Table 3.1) can be explained in terms of the structure of the polymers. The results of some potential energy calculations based on pure tactic forms and an assumed chain conformation are compared with experiment in Table 3.2. It can be seen that although the calculated values are about 3 to 4 times too high the syndiotactic form is found to have a higher barrier than the isotactic, as observed.

Table 3.2 Calculated barriers/kJ mol^{-1} for rotation of α-methyl groups in poly(methyl methacrylate) and comparison with experiment[2]

	Isotactic	Syndiotactic	
Assumed main-chain conformation	tgtg	tt	ttgg
Calculated	54	88	138
Experimental	23	33	

Measurements on the syndiotactic/isotactic poly(methyl methacrylate) stereocomplex, in which syndiotactic molecules crystallize in cylindrical holes in an expanded isotactic polymer lattice,[3] should enable the conformations of the two polymers in this form to be determined.

Poly(methyl methacrylate) has two side-groups which are bulky but differ in volume. Poly(α-methylstyrene) on the other hand has bulky side-groups, but with similar van der Waals' radii, the methyl group and phenyl group (perpendicular to the ring) having radii of 0·20 nm and 0·185 nm respectively. This is possibly the reason that 'heterotactic' (51% heterotactic, 29% syndiotactic and 20% isotactic triads) and 'syndiotactic' (90% syndiotactic, 10% heterotactic triads) polymers show no difference in their spectra (Figure 3.6).

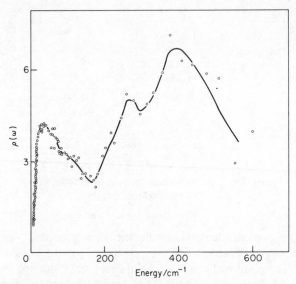

Figure 3.6 $\rho(\omega)$ for heterotactic and syndiotactic head-to-tail poly(α-methylstyrene)

This polymer is also obtainable in head-to-head, tail-to-tail form and its spectrum

$$\left[CH_2-\underset{\underset{Ph}{|}}{\overset{\overset{Me}{|}}{C}}-\underset{\underset{Ph}{|}}{\overset{\overset{Me}{|}}{C}}-CH_2 \right]_n \qquad \left[CH_2-\underset{\underset{Ph}{|}}{\overset{\overset{Me}{|}}{C}}-CH_2-\underset{\underset{Ph}{|}}{\overset{\overset{Me}{|}}{C}} \right]_n$$

head-to-head, tail-to-tail   head-to-tail

(Figure 3.7) shows a peak at 300 cm^{-1} instead of 380 cm^{-1} as found for the normal head-to-tail form (Figure 3.6). The respective barriers to rotation are 23 and 37 kJ mol^{-1}. An explanation of this is that steric hindrance in the head-to-head structure forces the central C—C bond (see above) into a *trans* conformation so reducing the hindrance to methyl group rotation compared with that in the head-to-tail form.

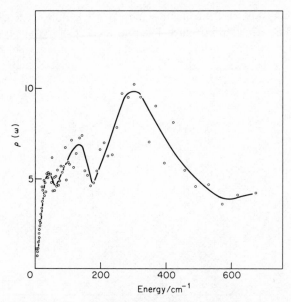

Figure 3.7 $\rho(\omega)$ for head-to-head, tail-to-tail poly-(α-methylstyrene)

These examples show how torsion frequencies can give insight into the local conformations of polymer chains; they may also be expected to provide useful information in the conformational analysis of stereoblock copolymers.

3.3.3 Polystyrene: torsion frequency of the phenyl group

Figure 3.8 shows that isotactic polystyrene gives a strong peak at 58 cm^{-1} which has been assigned to the torsion vibration frequency of the phenyl group

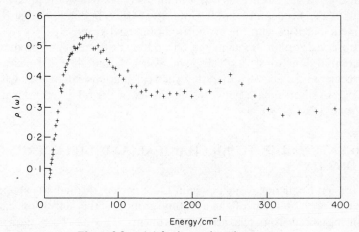

Figure 3.8 $\rho(\omega)$ for isotactic polystyrene

38

on the basis of its intensity, the increase in its relative intensity to other bands on deuterating the main chain (Figure 3.9), and its lack of frequency- or intensity-

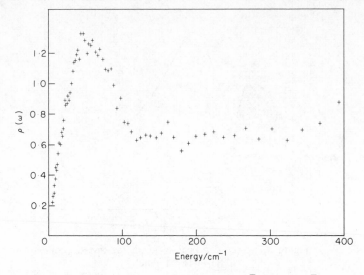

Figure 3.9 $\rho(\omega)$ for isotactic polystyrene $\left[\begin{array}{c} CD-CD_2 \\ | \\ C_6H_5 \end{array}\right]_n$

sensitivity to para-substitution by heavy halogen atoms.[4] (Para-substitution has no effect on the moment of inertia of the phenyl group for rotation about the bond attaching it to the main chain.) Comparing Figure 3.9 with Figure 3.8 it can be seen that the peak at $250\ \mathrm{cm^{-1}}$, a chain frequency, has disappeared and that the peak at $26\ \mathrm{cm^{-1}}$ is lower in intensity relative to the phenyl torsion band. It has also been found that the intensity of this band in phenyl-substituted copolymers increases with the proportion of phenyl groups.[5]

The observed torsion frequency of $58\ \mathrm{cm^{-1}}$ is bridged by the predictions of two potential functions[6,7] which give $89\ \mathrm{cm^{-1}}$ and $48\ \mathrm{cm^{-1}}$ respectively. The agreement is sufficiently good that such calculations can be used with some confidence to discriminate between the causes of the observed relaxations in polystyrene; (see below).

3.4 RELATIONSHIP TO MECHANICAL AND DIELECTRIC LOSS DATA

The torsion frequency of a side-chain is connected, through the shape of the potential energy function of the rotating group, to the activation energy of rotation measured from relaxation experiments. This has been shown above for the simple case of the threefold periodic methyl group. The low temperature relaxations of polymers currently being studied are associated with defects of the chain, inclusions, and torsion vibrations of side-groups, but it is not usually possible to decide between alternative mechanisms on the basis of relaxation evidence alone. In polystyrene there are a pair of relaxations, the γ and the δ, which have resisted assignment and recent work has turned to calculating the barrier to rotation of the phenyl group and comparing this with the relaxation activation energies to decide whether it could be associated with either of the two peaks.

Such calculations are only weakly supported by experiment, and are easy to fault, because of neglect of interchain interactions and poor potential functions. However this approach can be supported by using such calculations to predict the torsion frequency and comparing this with experiment. This is the background to the work on polystyrene described above.

Finally we may contrast the various spectroscopic methods that have been mentioned, and further delineate the bounds within which we may connect activation energies and torsion frequencies. It is important to remember that a rotor's near neighbours will only be stationary at low temperatures. In the rubbery state the fluctuating environment about a single rotor will blur the shape of its potential barrier and the measurement of a torsion frequency will only qualitatively be relatable to an activation energy which itself will vary with temperature.

The higher activation energies of segmental rotations (α processes), compared with side-chain rotations, in polymers such as polystyrene and poly(methyl methacrylate), means that relaxations measured at low temperatures will be less strongly intermixing. Consequently low-temperature inelastic neutron scattering measurements should be more comparable with mechanical relaxation data than with nuclear magnetic resonance data recorded at higher temperatures and frequencies.

3.5 CONCLUSION

These results show the range of uses to which torsion spectra can be applied and it is to be expected that they will be a useful tool in the further understanding of polymer structure and dynamics.

3.6 REFERENCES

1. G. Allen, P. N. Brier and J. S. Higgins, *Polymer*, **15**, 319, (1974).
2. G. Allen, J. S. Higgins and C. J. Wright, *Polymer*, **13**, 157, (1972).
3. A. M. Liquori, G. Anzuino, V. M. Coiro, M. d'Alagni, P. de Santis and M. Savino, *Nature*, **206**, 358, (1965).
4. G. Allen and C. J. Wright, *Polymer*, to be published.
5. J. W. White and D. A. Peace, *Ricerca Sci.*, **N84**, 56, (1973).
6. S. Reich and A. Eisenberg, *J. Polymer Sci.*, **A2, 10,** 1397, (1972).
7. J. L. De Caen, G. Elefante, A. M. Liquori and A. Damiani, *Nature*, **216,** 910, (1967).

4

Studies of Phonons in Polymer Crystals

J. W. White

University of Oxford

4.1 INTRODUCTION

In 1968 the first measurements by inelastic neutron scattering of phonons travelling along the chain axis of polyethylene were reported by Feldkamp, Venkataraman and King.[1] The chain dynamics of polytetrafluoroethylene have also been studied in stretch-oriented specimens[2,3] but no evidence was available from any of these measurements, of phonons* perpendicular to the chain axis. Since the binding within the chains can be described quite well by valence forces, it is highly desirable to measure phonon dispersion curves perpendicular to the chain axis to find a correct model for the intermolecular part of the crystal potential.

Ways of doing this have been developed at Oxford in the last five years[4,5,6] and collective excitations of the crystalline parts of bulk polymer specimens can be observed even in unoriented material. In addition to this technique, we have prepared highly oriented polyethylene and deuteropolyethylene specimens by high pressure annealing[7] and have produced large, fully deuterated specimens of polyoxymethylene[8] which have single crystal texture and allow for the first time the measurement of transverse acoustic modes of vibration in a crystal.

The neutron method allows us to measure the elastic properties of polymer single crystals and, in the case of polyoxymethylene, most of the elements of the compliance matrix have been found. This information is uniquely valuable when coupled with bulk tensile measurements to understand the properties of normal, composite polymers. In contrast to this rather practical application of neutron scattering is the eventual hope that, by this method and allied studies, we may be able to develop a microscopic understanding of the forces which lead to the conformations, tertiary and quaternary structures and crystal packing in biological macromolecules. At the end of this contribution a description is given of the recent use of neutron diffraction to study the structure of single

---

* Phonons are high-frequency sound waves, the superposition of which in crystalline materials can be used to describe the atomic thermal displacements at all temperatures above absolute zero. For a simple discussion see reference 11.

collagen fibrils, perhaps one of the most highly oriented biological macro-molecules.

4.2 POLYMER VIBRATIONS AND FORCES

One of the simplest vibrational modes available to a polymer chain is illustrated by the 'accordion' mode of the normal paraffin hydrocarbons and polyethylene. These modes are shown diagrammatically in Figure 4.1 and

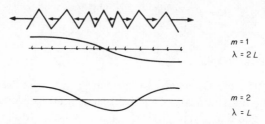

$m = 1$
$\lambda = 2L$

$m = 2$
$\lambda = L$

Figure 4.1 The accordion mode of vibration for hydrocarbon and polyethylene chains, showing the fundamental and first overtone vibrations

correspond to the symmetrical elongation and contraction of the molecular chain parallel to its length. The frequencies of these modes have been observed in elegant Raman scattering experiments by Schaufele and Shimanouchi.[9] We may construct the dispersion curve for this accordion mode by plotting the frequency of these vibrations as a function of the reciprocal of the wavelength for a given mode. This is in effect what Schaufele and Shimanouchi did in their classic paper wherein they showed that the data from a very large number of hydrocarbons whose length was different could be superimposed on the same curve. A sketch of their curve is shown in Figure 4.2. It can be seen that the mode frequency is linear in the reciprocal of the mode wavelength for the longest waves in the molecules. As the mode wavelength becomes shorter and approaches twice the repeat distance in the polymer chain, the mode frequency becomes gradually less dependent upon the inverse mode wavelength.

The great importance of Schaufele and Shimanouchi's plot (Figure 4.2) is that it shows that the vibrations of fairly short hydrocarbon chains in the solid phase depend mainly upon the internal forces between carbon atoms and only to a minor extent upon the crystal structure and molecular packing.

In addition to these intra-chain modes there will also exist modes of vibration of one polymer chain or molecule with respect to its neighbours packed in the crystal. Such modes are formally similar to the accordion mode and may be represented as shown in Figure 4.3.

The arrows in this figure indicate the amplitude of vibration of the whole molecules relative to a nodal point. We can see that, whereas for the accordion mode it was chiefly the intra-molecular forces which determined the molecular

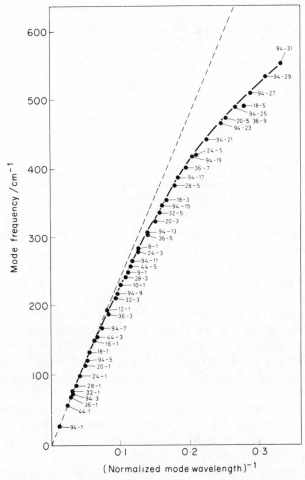

Figure 4.2 The dispersion curve for the accordion mode of vibration in hydrocarbon chains.[9] The mode frequency from Raman scattering has been plotted as a function of the reciprocal of the mode wavelength

mode frequencies, for these so-called 'external' modes it is the forces between the molecules in the crystal that determine the frequencies of vibration. Inelastic neutron scattering spectroscopy is able to determine the frequencies of these external modes of vibration as a function of the wave vector of the excitation (proportional to the reciprocal wavelength) and thereby to allow us to construct dispersion curves for the modes of vibration in which the inter-molecular forces play a dominant role.

We may now consider the importance of making such studies. A primary reason, and the long-term goal of all of these measurements, is to obtain experimental information on the inter-molecular forces which lead to the fascinating

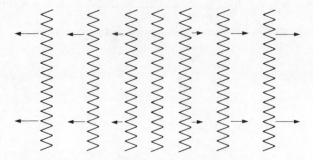

Figure 4.3 Longitudinal acoustic intermolecular mode analogous to the accordion mode in the chain (the fundamental vibration is shown)

variety of polymer conformations, tertiary and quaternary structures, and crystal structures found in synthetic polymeric and biological polymer systems. There is also a more immediate goal. The problem of what determines the macroscopic elastic properties of synthetic polymers is one of long standing and arises chiefly because polymers are composite materials. In general we may take as a model for a synthetic polymer a system of intermingled crystalline blocks and amorphous regions shared by one or many individual polymer chains. When one stresses the polymer, strains develop both in the crystalline regions and in the amorphous fraction. To understand the observed properties one must know all about the elastic properties of both regions as well as their physical distribution.

At the outset one might hope that the amorphous region could be described by an isotropic modulus; for the crystalline regions it is certain that, in most polymers, many elastic constants will be needed, since the crystal structures are, in general, much less symmetrical than cubic.

A two-fold strategy was adopted. Firstly we tried to see whether unique properties of the neutron scattering cross-section could allow information to be extracted preferentially for the crystalline regions by special adaptation of standard techniques to polymers. The second element in our strategy was to try to produce better and larger single crystal specimens of polymers suitable for coherent neutron scattering measurements. To this end we also had to develop techniques for deuterating common polymeric molecules.

4.3 TECHNIQUES OF NEUTRON SCATTERING SPECTROSCOPY

The experimental details of spectrometers for making inelastic neutron scattering measurements have been extensively reviewed recently.[10,11] It may suffice to say that the spectrum of the neutrons scattered at a certain scattering angle, θ, may be analysed by determining either the velocity of the scattered neutrons (for example by measuring their time of flight over a given distance) or by determining their wavelength by selection using a crystal monochromator and Bragg's law. Using these techniques it is possible to measure the energy spectrum of neutrons scattered from a sample at any scattering angle and, if the incident neutron beam is mono-energetic, this energy spectrum will be the excitation spectrum of the sample.

Coherent neutron scattering spectroscopy differs from incoherent neutron scattering[11] in that the intensity of inelastic scattering events is controlled, not only by the density of excitations in the sample, but also by the coupling together of the inelastic event with crystal diffraction. Thus, in general, for longitudinal acoustic vibrations in a solid, the intensity of the excitations will be strong when the incident neutron wavelength and the angle of scattering are almost correct for Bragg diffraction within the sample. It is this coupling together of Bragg diffraction and inelastic excitation that allows us to use the neutron method to extract information about the crystalline regions of the polymer specimens without interference from excitations which occur in the amorphous regions.

Mathematically we may show that the scattered neutron spectrum from an incident monochromatic beam suffering fractional angular frequency change $\delta\omega$ after scattering into a fractional solid angle $\delta\Omega$ is given by a double differential scattering cross-section

$$\frac{d^2\sigma}{d\Omega\, dE_{coh}} = (2\pi)^3 \sum_{j\theta} \frac{\mathbf{k}}{\mathbf{k}_0} \delta(\hbar\omega - \hbar\omega_j)(\mathbf{q})_\theta F_j(\theta) \tag{1}$$

where

$$F_j(\theta) = G(\theta)|g_j(\mathbf{Q})|^2 (n_{\mathbf{q}j} + 1)\, \delta(\mathbf{Q} + \mathbf{q} - 2\pi\mathbf{\tau})_\theta \tag{2}$$

and

$$g_j(\mathbf{Q}) = \sum_v \left(\frac{\hbar}{2M_v\omega}\right)^{\frac{1}{2}} b_v \cdot \mathbf{Q} \cdot \zeta_{vj}(\mathbf{q}) \cdot e^{-W_v}\, e^{i\mathbf{Q}\alpha_v} \tag{3}$$

is the dynamical structure factor. In this expression the momentum transfer, $\hbar\mathbf{Q}$, is defined in terms of the incident and outgoing neutron wave vectors, $\mathbf{k}_0$, $\mathbf{k}$, by equation (4).

$$\mathbf{Q} = \mathbf{k} - \mathbf{k}_0 \tag{4}$$

$\mathbf{q}$ is the wave vector of the phonon excited by the neutron scattering event in the crystal and $\mathbf{\tau}$ is the reciprocal lattice vector for a particular set of planes in the crystal. The quantity $n_{\mathbf{q}j}$ is the Bose population factor for the mode whose wave vector is $\mathbf{q}$ and M_v is the mass of the scattering atom v; $\zeta_{vj}(\mathbf{q})$ is the cartesian amplitude of vibration associated with normal mode j, atom v being displaced with the wave vector $\mathbf{q}$.

When coupled with the use of slow neutrons which control the maximum value for momentum transfer, $\mathbf{Q}$, thereby limiting excitations to such low momentum transfers that only modes in the first or possibly second Brillouin zones are excited, this formula allows selective measurements of the inter-molecular vibrations of the polymer crystal to be made.

4.4 PHONONS IN POLYCRYSTALLINE POLYTETRAFLUOROETHYLENE

Figure 4.4 shows the powder diffraction pattern from polycrystalline fluon and from a teflon fibre sample taken at room temperature. A striking feature of

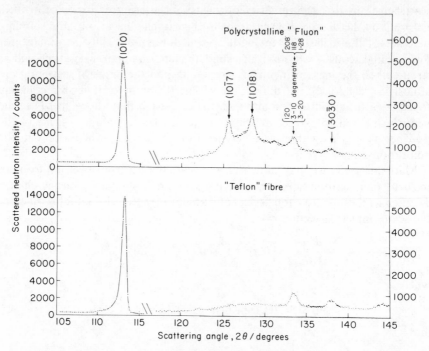

Figure 4.4 Diffraction pattern at 296 K for incident 1·035 Å neutrons on highly crystalline fluon powder and for teflon fibres oriented perpendicular to the beam direction (zero angle = 101·05°)

this diffraction pattern is the very large Bragg peak at scattering angles (2θ) of about 113°. This Bragg peak comes from the prismatic reflection $(10\bar{1}0)$ associated with the sideways-stacking of the almost cylindrical polymer molecules in this substance. By adjusting the neutron spectrometer to work with momentum transfers in the region of this very strong Bragg peak we can choose to look at excitations whose eigenvectors are directed perpendicular to the axis of the rod. In particular, longitudinal acoustic modes might possibly be seen and, since the crystal is hexagonal, there will be no anisotropy in the basal plane. Thus the phonon frequency is independent of the angle of the scattering vector with respect to the a-axis in this plane.

Figure 4.5 shows the neutron time-of-flight spectrum for hexagonal polytetrafluoroethylene at 296 K taken at four scattering angles $\theta = 27°$, 36°, 45° and 54° to the incident beam direction.[4] A remarkably sharp peak which arises from energy transfer to the neutrons from excitations in the crystal can be readily seen. By contrast with Raman scattering measurements, the frequency of this peak (determined by its energy separation from the elastic peak which falls near $1200\,\mu s\,m^{-1}$) can be seen to get smaller as the angle of scattering is increased and indeed if $\theta = 54°$ the peak is almost merged with the elastic scattering. In fact, at a slightly higher angle than the 54°, the elastic scattering is intense and has superimposed upon the incoherent scattering the large $10\bar{1}0$ Bragg peak seen in Figure 4.4.

This strong singularity which moves as a function of the scattering angle is characteristic of phonon excitations in a molecular crystal. The reason that its frequency changes as the scattering angle increases is that by changing the

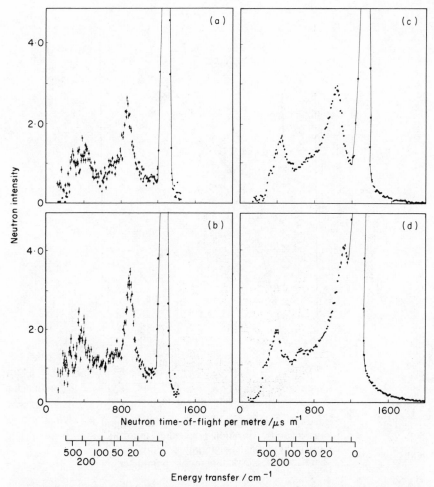

Figure 4.5 Neutron time-of-flight spectra for incident 5·0 Å neutrons on poly-crystalline hexagonal polytetrafluoroethylene at 296 K. The spectra were recorded at scattering angles θ to the incident beam direction of (a) 27, (b) 36, (c) 45, (d) 54°

48

scattering angle we change the amount of momentum delivered to the crystal and hence, by momentum conservation, we change the momentum of the lattice waves. It can be seen by the application of de Broglie's relationship

$$mv = \frac{h}{\lambda}$$

that by delivering more or less momentum to the crystal we change the wavelength of the crystal excitation produced.

This can also be seen by reference to the dispersion curve obtained by plotting out data such as those from Figure 4.5. This is shown in Figure 4.6 where the

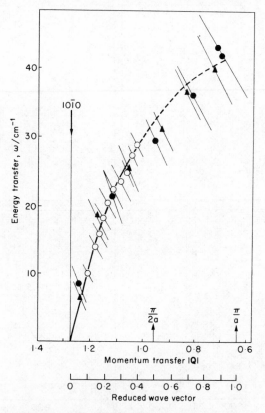

Figure 4.6 The dispersion curve for the longitudinal acoustic vibration perpendicular to the chains in hexagonal polytetrafluoroethylene at 296 K. a is the crystallographic parameter

frequencies of a large number of phonons determined in different runs are plotted as a function of the momentum transfer and also as a function of the reduced wave vector for the mode excited in the inelastic collision. It can be seen that the frequency becomes zero when the momentum transfer is exactly that associated with excitation of Bragg reflection for the $10\bar{1}0$ planes. As in the case of the dispersion curve for the accordion mode of the hydrocarbons (Figure 4.2), the slope of this curve at low values of the reduced vector (long wavelength excitations in the crystal) gives the velocity of sound waves propagating in the longitudinal compressional mode perpendicular to the chains. Since we know the density of crystalline polytetrafluoroethylene, we may calculate the compliance constant for this mode. It comes to be 1.8×10^{10} N m^{-2}. This number is not available from any other measurements.

4.5 POLYETHYLENE

The crystal structure of polyethylene is rather more complicated than that of teflon at room temperature. It has an orthorhombic unit cell containing two molecules and this means that the Bragg scattering pattern is less simple than for a hexagonal structure. Nevertheless it has been possible to extract acoustic phonon data from polycrystalline and partly ordered deuteropolyethylene samples.[5,6] These results will not be described in detail but they illustrate how in a more complicated crystal one must combine the neutron scattering selection rule with calculations based on an assumed force field for the intermolecular forces in the crystal. By this joint approach one is able to interpret even 'complicated' neutron spectra from only partly ordered materials.

A much better approach in the case of polyethylene is to try to produce highly oriented material. We were led to do this by some experiments of Dr. P. Bowden and Dr. D. Young of the Department of Metallurgy, Cambridge University, who had succeeded in obtaining three-dimensional orientation in small pieces of polyethylene. Their method was to anneal the polyethylene, after one-dimensional stretching at a temperature rather above its normal melting point, under high pressure. This method was further developed in Oxford by Mr. J. F. Twisleton; the final technique consisted of taking the stretched material (usually eight times the normal extension), rolling it to produce some orientation perpendicular to the stretch direction, and then high pressure annealing of discs of the stretched and rolled strips.[7] Figure 4.7 shows the pole figures for deuteropolyethylene samples produced in this way at the three stages of production. The pole figures measure the strength of Bragg reflection for a given direction with respect to the original drawing direction.

It can be seen that the originally drawn material (Figure 4.7a) has a well-developed pole along the draw direction z which is the chain-axis direction and the c-axis of the orthorhombic crystal. After rolling in the x (a) direction, some orientation perpendicular to the c-axis is produced (Figure 4.7b). Clear 110 poles can be seen but it is also noticeable that the sample is twinned. By the process of rolling no deterioration of the c-axis pole strength has occurred. After high-pressure annealing it can be seen that the twins are removed (Figure

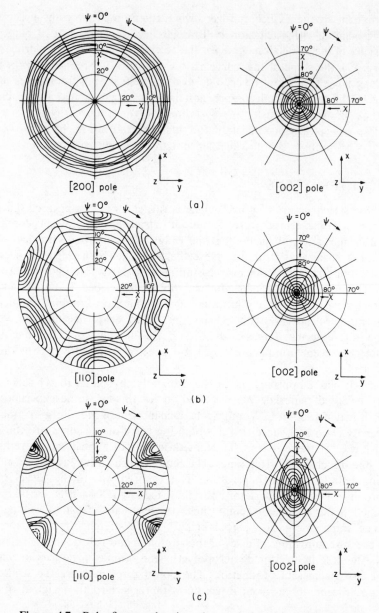

Figure 4.7 Pole figures showing the evolution of 3-dimensional crystalline order in an initially uniaxial, stretch-oriented deuteropoly-ethylene strip (a), as a function of rolling (b), and annealing at high pressure (c)

4.7c) without appreciable deterioration of orientation in the c-axis direction. This final sample has a mosaic about the c-direction of approximately $\pm 5°$ and about the a- and b-directions of the orthorhombic crystal of ± 8 to $\pm 10°$.

Such a specimen can be used to measure phonon dispersion curves for longitudinal acoustic vibrations and these will be reported elsewhere. Suffice it to say that the measurements along the 110 direction are in good agreement with those obtained from polycrystalline and partly ordered specimens using time-of-flight techniques.

4.6 POLYOXYMETHYLENE

Efforts to obtain large single crystal specimens of polymers for coherent neutron scattering measurements have been much more successful for polyoxymethylene than for polyethylene. This is because it is possible to obtain single crystal texture specimens of this polymer by the topotactic polymerization of monomer single crystals. In particular, single crystals of trioxane, $(CH_2O)_3$, and tetroxane, $(CH_2O)_4$, can be polymerized almost completely to the polymer by irradiation with X- or gamma-rays followed by curing at temperatures between 45 °C and 62°C.[8,12] By this method we have produced single crystals of the hydrogenous polyoxymethylene of large dimensions (Figure 4.8) and a 6·4 g fully deuterated single crystal for coherent neutron scattering studies.

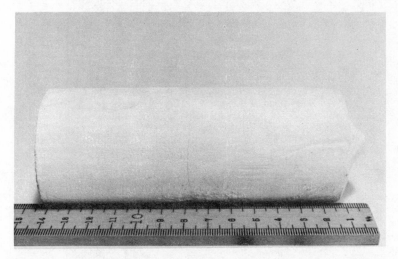

Figure 4.8 Photograph of a large single crystal texture specimen of polyoxymethylene produced by gamma-ray induced polymerization of a single crystal of trioxane

These 'single crystals' when freshly produced are, in fact, composites of twins each of which has some disorder about the c-axis. Annealing[12] can eliminate most of the twinned material and gives a specimen very suitable for making neutron scattering measurements.

Figure 4.9 shows rotation photographs for a small single crystal of the polymerized trioxane. In Figure 4.9a the short exposure shows the well-

52

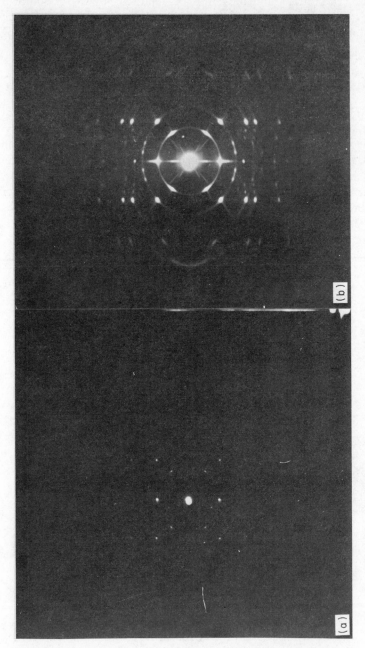

Figure 4.9 X-ray rotation photographs for a small single crystal of gamma-irradiated trioxane after some heat treatment. The two X-ray rotation photographs were taken at different exposures and the long exposure photograph (b) shows some of the disorder and twinning present in an incompletely annealed specimen

developed crystalline reflections in this material and the sharpness of the spots is much greater than that for the ordered polyethylene specimens discussed above. The longer exposure shown in Figure 4.9b reveals the presence of some twinned material in this unannealed sample; the streaking also indicates the presence of some disorder.

The polyoxymethylene molecule in such single crystals is in a 9/5 helix conformation and so is rather more complicated than the simple planar zig-zag structure found in polyethylene. Figure 4.10 shows the dimensions of one repeat unit of the polymer and its projection perpendicular to the chain axis. In fact, the projection is almost circular and the molecules pack on a pseudo-hexagonal basis. The disorder in these crystals is not fully understood and its

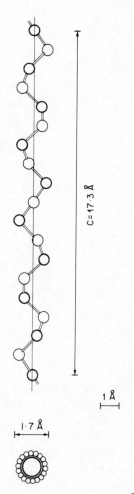

C = 17·3 Å

1 Å

1·7 Å

Reduced wave vector

Figure 4.10 Molecular conformation and dimensions of the repeat unit in polyoxymethylene crystals

54

presence somewhat limits the neutron scattering measurements that can be made. Nevertheless, the high crystalline order illustrated in Figure 4.9 allows, for the first time, an attempt to be made to measure transverse acoustic excitations in a polymer crystal. The low-frequency (long-wavelength) slopes of the dispersion curves for the transverse acoustic modes give information about the shear moduli (or compliance constants) which are even more difficult to obtain than the compressibility constants for polymer crystals.

Neutron inelastic scattering measurements were made on the 6·4 g fully deuterated single crystal of polyoxymethylene using the IN2 3-axis neutron scattering spectrometer at the Institut Laue-Langevin in Grenoble in February 1973. Transverse acoustic excitations were observed there for the first time and Figure 4.11 shows the Brillouin triplet for the *c*-axis transverse acoustic phonon

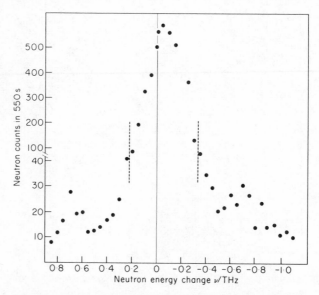

Figure 4.11 Brillouin triplet for the *c*-axis transverse acoustic phonon in deuteropolyoxymethylene single crystal observed with a constant **Q** scan (1, 0, 0·6)

in polyoxymethylene. This figure shows a plot of scattered neutron counts as a function of the neutron energy change. At zero neutron energy change we have a large incoherent neutron scattering central peak. In light scattering this would be the Rayleigh component. On either side of this peak there is a single small peak corresponding to the transverse acoustic phonon frequency at the particular wavevector of the constant **Q** scan taken through the dispersion surface. On the side of positive neutron-energy-gain, the peak is sharp and well-focussed by the spectrometer conditions. By contrast the spectrometer conditions engender a poorly-focussed peak on the negative-energy-gain side. The background counts due to stray neutrons on the IN2 spectrometer are so low that these peaks represent good signal-to-noise. Many such excitations were observed and it was possible to construct the transverse acoustic branch of the dispersion curves for some distance into the Brillouin zone of this crystal.

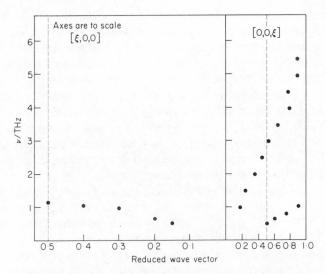

Figure 4.12 Phonon dispersion curves for two chain-axis phonons and for the transverse acoustic mode perpendicular to the chain direction in deuteropolyoxymethylene at 296 K

Figure 4.12 summarizes some of the experimental phonon dispersion curves for deuteropolyoxymethylene. Two branches have been observed along the c-axis (chain axis) corresponding to the longitudinal acoustic branch and the transverse acoustic branch respectively. Perpendicular to the chain, a mode which corresponds to the transverse acoustic mode along the $\zeta00$ direction has been observed with a velocity of sound comparable to the transverse acoustic branch along the chain direction. A very much higher velocity of sound was found for the chain-axis, longitudinal acoustic branch and shows the strong anisotropy of the compliance constants in this polymer crystal.

The velocity of sound for the longitudinal acoustic branch along the chain is 1×10^6 cm s^{-1} which corresponds to an elastic constant $c_{33} = 1.49 \times 10^{11}$ N m^{-2}. For the transverse acoustic mode polarized in this direction the velocity of sound is 2×10^5 cm s^{-1} which gives a transverse, shear compliance constant $c_{44} = 6 \times 10^9$ N m^{-2}.

Measurements such as those on polyoxymethylene are really only just beginning and very much more work, firstly to unravel the disorder in this polymer crystal, and secondly to generate suitable 3-dimensional force fields for it, has to be done. Experiments on polyethylene can undoubtedly be extended further since new types of neutron spectrometer adapted for studying polymers are at present being constructed. We may then ask what is the possibility of doing experiments on biological macromolecules?

4.7 NEUTRON DIFFRACTION FROM COLLAGEN

Collagen molecules, for example the fibrils from the tail of a rat, when suitably stretched, rank amongst those biological materials which give the strongest diffraction patterns for X-rays. With neutron scattering there exists a unique possibility for analysing the diffraction pattern from aqueous biological samples. This arises because the coherent scattering length for protons is negative while the scattering length for deuterons is positive. Using mixtures of H_2O and D_2O, the medium surrounding the biological specimen may have any desired mean scattering length between a negative and a positive value. This fact has already been exploited to good effect in the study of dispersed biological materials[13] but can also be used in the understanding and analysis of Bragg diffraction from regular structures.

For collagen the amino-acid sequence is already known[14] and so it is possible to calculate the neutron scattering length density for individual amino-acids in the chain. By varying the H_2O/D_2O ratio in the surrounding medium and allowing for proton/deuteron exchange with the amino-acids, it is possible to arrange that the scattering length contrast between different amino-acid groupings and the surrounding medium is zero. By this means we are able to vary the contribution of individual groups to the scattering pattern and hence test models for the spatial arrangement of these groups in the molecule.

This idea was tried out for the first time in October 1973.[15] The diffraction pattern from collagen immersed in H_2O and D_2O was measured using the D11 low-angle neutron diffractometer at the Institut Laue-Langevin in Grenoble. This machine has remarkably high sensitivity and it was possible to obtain useful diffraction patterns from a single, 1 cm long, 0.5 mm diameter, collagen fibril in about 1000 s exposure time. A very interesting difference was observed between the two specimens. In the actual experiment the same fibre was measured first with the H_2O surrounding and then, without being shifted at all, after equilibration overnight in a D_2O bath.

In Figure 4.13 it can be seen that, after an exposure of 4800 s, the collagen/H_2O system gave a strong diffraction in the first and third orders with the fifth order appearing at the bottom of the picture. The diffraction peaks indexed on a repeat distance of 670 Å. In the shorter exposure, collagen/D_2O system, it is clear that not only can the first and the third orders of diffraction be seen with a trace of the fifth order, but that the fourth order of diffraction is beginning to appear. We suspect that this is a real effect due to the change in contrast between the D_2O surroundings and the collagen molecule. More recent experiments have shown that at least 15 orders of diffraction can be observed[16] and it is expected that this isotopic contrast method may be of direct value for interpreting the 3-dimensional structure of collagen and also for finding other molecules present in the biological material.

57

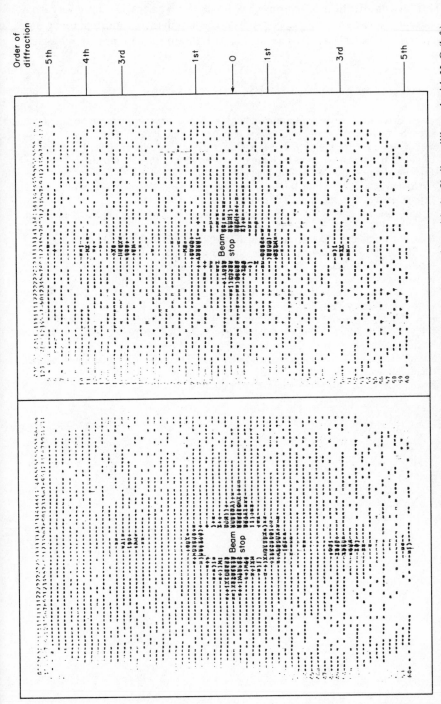

Figure 4.13 The low-angle neutron diffraction from a single collagen fibril, 1 cm long, taken with the fibril equilibrated with H_2O (left) and with the same fibril equilibrated with D_2O (right). Only the fibril was in the beam for the diffraction although it was wet with the equilibrated liquid. $\lambda_0 = 17.9$ Å. Exposures = 4800 s (left), 1500 s (right)

58

This final illustration of the power of neutron scattering is a long way from the inelastic scattering experiments on synthetic polymers reported above but, nevertheless, it is well worth making a start on even rather complex biological materials.

4.8 REFERENCES

1. L. A. Feldkamp, G. Venkataraman and J. S. King, *Neutron Inelastic Scattering*, **2,** Proc. Symposium Copenhagen, 1968. I.A.E.A., Vienna, 1968, p. 165.
2. V. LaGarde, H. Prask and S. Trevino, *Disc. Faraday Soc.*, **48,** 15, (1969).
3. L. Piseri, B. M. Powell and G. Dolling, *J. Chem. Phys.*, **58,** 158, (1973).
4. J. F. Twisleton and J. W. White, *Polymer*, **13,** 40, (1972).
5. L. Holliday and J. W. White, International Symposium on Macromolecules I.U.P.A.C., Leiden, 1970. *Pure and Applied Chemistry*, **26,** 545, (1971).
6. J. F. Twisleton and J. W. White, *Neutron Inelastic Scattering*. Proc. Symposium, Grenoble 1972. I.A.E.A., Vienna, 1972, p. 301.
7. P. A. Reynolds, J. F. Twisleton and J. W. White, *J. Polymer Sci.*, submitted.
8. R. Currat, D. K. Steinmann, M. B. M. Harryman and J. W. White. Report to the Neutron Beam Research Committee of the Science Research Council, 1973, p. 152.
9. R. F. Shaufele and T. Shimanouchi, *J. Chem. Phys.*, **47,** 3605, (1967).
10. G. C. Stirling, *Chemical Applications of Neutron Scattering*, ed. B. T. M. Willis, Oxford University Press, 1973.
11. J. W. White, *Polymer Science*, Chapter 27, ed. A. D. Jenkins, North Holland, 1972.
12. J. P. Colson and D. H. Reneker, *J. Appl. Phys.*, **41,** 4296, (1970).
13. H. B. Stuhrmann, *J. Appl. Crystallography*, **7,** 173, (1974).
14. See for example, D. J. S. Hulmes, A. Miller, D. A. D. Parry, K. A. Piez and J. Woodhead-Galloway, *J. Mol. Biol.*, **79,** 137, (1973).
15. K. Ibel, A. Miller and J. W. White, to be published.
16. G. T. Jenkin, unpublished results.

5

A New Aspect of Vibration Spectroscopy of High Polymers

T. Shimanouchi

University of Tokyo

5.1 INTRODUCTION

Infrared and Raman spectra of synthetic and biological polymers have been studied extensively these twenty years. Normal coordinate treatments have been made. The chemical structure, configuration, conformation and orientation have been revealed for many polymers.

The procedures of these treatments have been mainly limited to the local vibrations like the CH_2, CH_3, $C=O$, $C-Cl$ stretching and deformation vibrations. The author wishes to emphasize that the vibrations of polymer chains have another aspect and that the overall vibrations like the stretching, bending, and twisting vibrations of the whole molecule are important.[1]

The frequencies of these overall vibrations are roughly proportional to the reciprocal dimensions. For high polymer molecules these frequencies are usually small and, hence, the amplitudes due to thermal motions are large. These frequencies are related to Young's modulus and other elastic constants. For biopolymer molecules these vibrations are expected to be closely related with biological functions. In the present article the recent advances in this field will be surveyed.

5.2 n-PARAFFIN MOLECULES

The first basic polymer is polyethylene $(-CH_2-)_n$. Fortunately in this case we have n-paraffin molecules which have the chains with definite length. For the study of the overall vibrations it is convenient to have such molecules and n-paraffins have been studied in detail.

The polyethylene chain takes the extended zig-zag conformation in the crystalline state. This is a kind of one-dimensional crystal and we have optical and acoustical modes. The normal coordinate treatment for this one-dimensional crystal has been made[2-5] and Figure 5.1 shows the dispersion curve obtained.

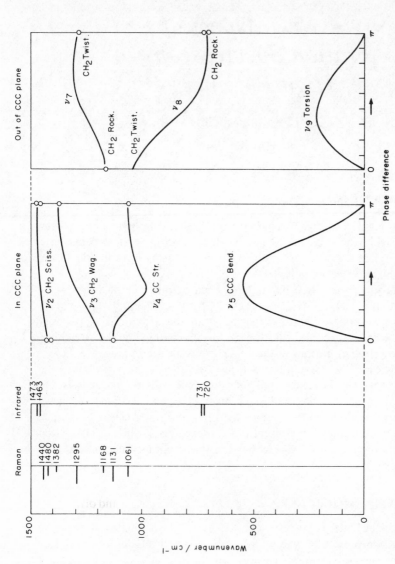

Figure 5.1 The dispersion curves for the polyethylene chain $(CH_2)_n$ taking the extended zig-zag conformation. The frequencies are given as functions of the phase difference between the neighbouring CH_2 groups. The observed Raman and infrared bands given on the left side correspond to the circles given in the dispersion curves. The CH_2 symmetric (ν_1) and antisymmetric (ν_6) vibrations are in the $3000\,cm^{-1}$ region and are omitted. The CH_2 wagging vibration ν_3 with zero phase difference is infrared active, but is too weak to be observed. The CC stretching and CCC bending vibrations are coupled with each other and appear as ν_4 and ν_5. The CH_2 rocking and twisting vibrations are also strongly coupled in a certain region of the phase difference. The CH_2 wagging vibrations with zero and π phase differences and the CH_2 rocking vibration with π phase differences are split into doublets in the crystalline state

The ν_5 and ν_9 branches are acoustical and the others are optical. The overall vibrations belong to the former and the local vibrations to the latter.

For the infinite chain the ν_5 and ν_9 branches give only zero frequency for the infrared and Raman-active modes. The finite chain consisting of n identical units has more modes that are active. The frequencies are given as a function of the phase difference, $m/(n + 1)$ (for the free-end case), where m is the order of the vibrational modes. Figures 5.2 and 5.3 give the modes for various m.

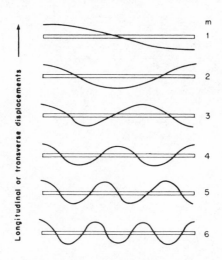

Figure 5.2 Modes of harmonics of the overall vibrations

The calculations for the polyethylene chain give the relation between the acoustical branches and the overall vibrations as shown in Figure 5.4. Each of them are dealt with in the following sections.

The n-paraffin molecule has a few points to be mentioned. First the terminal CH_3 group is slightly different from the CH_2 unit of the chain. However, the approximation of treating the former as identical to the latter is less serious than we first thought, although the local vibrations of the CH_3 group like the CH_3 torsion or rocking should be treated separately.

The second problem is what is the dynamical unit. If it is the CH_2 group, n is equal to N_c, the number of carbon atoms. If it is the C—C bond, the CCC angle, or the CC—CC torsional axis, n is $N_c - 1$, $N_c - 2$, or $N_c - 3$, respectively. There are ambiguities to some extent for the interpretation of the phase difference $m/(n + 1)$.[6] Moreover, the corresponding modes are coupled with each other if they belong to the same symmetry species.

Third is the coupling of these vibrations with the translational and rotational lattice vibrations in the crystal. These lattice vibrations correspond to the case $m = 0$ and two of $m = 1$. Fortunately the coupling is not serious for the case of

62

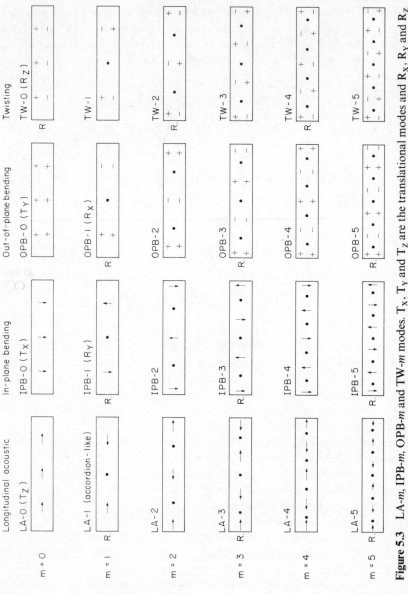

Figure 5.3 LA-*m*, IPB-*m*, OPB-*m* and TW-*m* modes. T_X, T_Y and T_Z are the translational modes and R_X, R_Y and R_Z are the rotational modes. The displacements perpendicular to the paper plane are denoted by + or −. R denotes the Raman-active modes

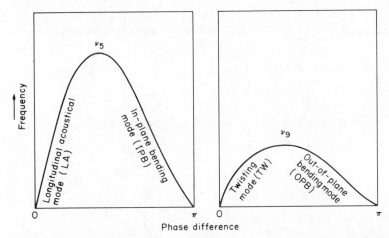

Figure 5.4 Frequency versus phase difference curves for acoustical vibrations of the polyethylene chain and their vibrational modes as overall vibrations

the longitudinal acoustic modes to be discussed in the next section. However it is serious for the other modes. The case for the triclinic modification will be discussed in section 5.4.

5.3 LONGITUDINAL ACOUSTICAL MODES OF n-PARAFFIN MOLECULES

The mode $m = 1$ was first observed by Mizushima *et al.*[7,8] in 1941 in the Raman spectra of crystalline n-paraffins including butane, pentane, hexane, heptane, octane, decane, dodecane and cetane. This was the first reported observation of the overall vibration. They found that for most of the Raman bands of n-paraffins the frequencies are independent of the number of carbon atoms,[9] with the exception of the lowest frequency band appearing in the skeletal deformation region. The frequencies are proportional to the reciprocal length of the molecule and are assigned to the accordion-like vibration which corresponds to the case $m = 1$ of the longitudinal acoustic vibration LA-1 in Figure 5.3.

For a homogeneous rod the frequency of LA-1 is given by

$$v = \frac{1}{2l}\sqrt{\frac{E}{\rho}} \tag{1}$$

where l, E and ρ are the length, Young's modulus, and the density of the rod, respectively. The observed Raman frequency is expressed by

$$v = \frac{2400}{N_c}\,(\text{cm}^{-1}) \tag{2}$$

Table 5.1 The frequencies observed for the LA-1 (accordion-like) vibrations of n-paraffins and those calculated by equation (2)

Number of carbon atoms	Observed (cm^{-1})	Calculated (cm^{-1})	Reference
8	283	300	8, 16
9	249	267	8
10	231	240	8, 16
12	194	200	8, 16
14	168	171	16
16	150	150	8, 16
18	133	133	10, 16
20	120	120	16
22	112	109	16
24	103	100	16
28	85	86	10
32	76	75	10
36	67	67	10
44	56	55	10
94	26	26	10

(see Table 5.1). By comparing these two equations, they obtained Young's modulus of the polyethylene chain: $E = 34 \times 10^{10}$ N m^{-2}.

The development of the laser-Raman technique has made the situation clearer. Schaufele and the author[10] observed the LA-1 band for longer n-paraffins in 1967. The observed frequencies are in excellent agreement with those calculated from equation (2) for the longer paraffins as shown in Table 5.1.

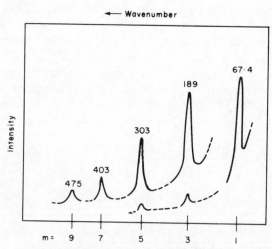

Figure 5.5 Low-frequency Raman spectrum of n-$C_{36}H_{74}$ showing the accordion-like vibration and its harmonics (LA-m) with the observed wavenumber/cm^{-1} and the value of the band order m

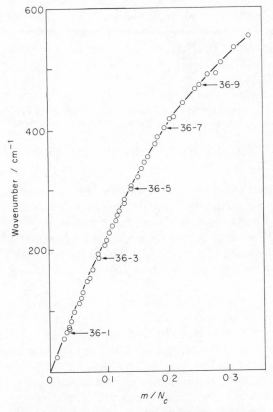

Figure 5.6 Frequencies of observed LA-m Raman bands as a function of m/N_c. The assignment N_c-m beside the circles correspond to the frequency observed for n-$C_{36}H_{74}$. The full line indicates part of the ν_5 dispersion curve, the corresponding phase difference being given by $\pi(m/N_c)$

In addition they observed many LA-m bands with odd m. One example is shown in Figure 5.5. The frequencies are given by the curve shown in Figure 5.6 which is identical to that calculated for ν_5 of Figure 5.1.

The accordion-like modes are little affected by the arrangements of neighbouring molecules. This is the reason why we obtain the smooth curve given in Figure 5.6 for all the crystal modifications. Even in the liquid state the frequency does not shift appreciably, provided the extended zig-zag molecule is present.

When the molecule does not take the extended all-*trans* conformation, it does not have the frequency given by equation (2); but if it does, the existence of the all-*trans* conformation is proved. This relation was used for the conformational study of the n-paraffin molecules in the liquid state.[8,10,11] The octane molecule takes the all-*trans* form at least partly. For the cetane molecule this form cannot be found in the liquid state. The fact that we observed the LA vibrations for crystalline $C_{94}H_{190}$ means that this molecule takes the completely extended zig-zag form in crystals. The possibility of estimating the thickness of the folded polyethylene chains in crystals by the LA Raman frequency was suggested by Schaufele and the author.[10] Peticolas *et al.*,[12] Peterlin *et al.*[13] and Koenig *et al.*[14] proved that this method can be used practically.

5.4 LOW-FREQUENCY RAMAN BANDS OF CRYSTALLINE n-PARAFFINS

The next problem in the overall vibrations of the n-paraffin molecules is to find the in-plane bending vibrations IPB-*m*, out-of-plane bending vibrations OPB-*m*, and twisting vibrations TW-*m* shown in Figure 5.4.

The bands IPB-*m* and OPB-*m* with odd values of *m* and TW-*m* with even values of *m* are expected to appear in the Raman spectra at the lower frequency region. Recently the low-frequency bands were studied in detail by numerous authors.[15-20]

The n-paraffin crystals take the triclinic, monoclinic or orthorhombic modification depending on the number of carbon atoms.[21] The author will describe first the interpretation given for the triclinic series by his collaborators[16] and discuss afterward the experimental results obtained by other authors.

The spectra of the triclinic crystals of n-paraffins including C_{24} to C_8 are shown in Figure 5.7a. As shown in this figure we have six series of Raman bands, the frequencies of which decrease as the carbon number increases. The series 6 of Figure 5.7a is the strongest in intensity and is assigned to the accordion-like LA-1 vibrations. The assignments of the other series are not simple. For the triclinic crystal the Bravais unit cell has only one molecule. Accordingly, only one set of the vibrations given in Figure 5.3 should appear, although these vibrations may be coupled with each other to some extent.

The problem is how to assign the remaining five series of Figure 5.7a to vibrations of Figure 5.3. When the number of carbon atoms increases, series 1 becomes close to series 2 and series 3 to series 4. This fact means that series 1 and 2 are similar in nature to each other and series 3 and 4 are also similar.

In order to make reasonable assignments, we calculated the normal frequencies and modes of these triclinic crystals, assuming adequate intra- and inter-molecular force constants and also molecular packings when they are not available. The result is shown in Figure 5.7b. It is of interest that in the calculated frequencies we also have two sets of series, R_X (OPB-1) and R_Y (IPB-1), and OPB-3 and IPB-3.

When the molecule is in a free state, R_X and R_Y are pure rotations and their frequencies are zero. OPB-3 and IPB-3 have different frequencies, but they are very small in the free state. The reason that these two sets of vibrations have the frequencies given in Figure 5.7b is that they have force constants due to intermolecular interactions. As for these interactions, R_X is similar to R_Y,

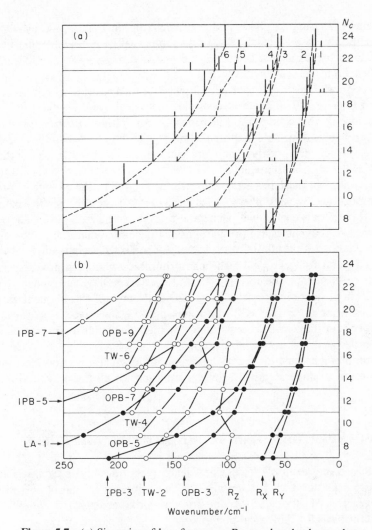

Figure 5.7 (a) Six series of low-frequency Raman bands observed for triclinic n-paraffins from C_8H_{18} to $C_{24}H_{50}$. (b) Frequencies calculated for the Raman-active overall vibrations of triclinic n-paraffins from C_8H_{18} to $C_{24}H_{50}$. The filled circles correspond to the six series given above. The values for $C_{20}H_{42}$, $C_{22}H_{46}$ and $C_{24}H_{50}$ are obtained from the dispersion curves which are given by the values calculated for the lower paraffins (see ref. 16)

and OPB-3 is similar to IPB-3. This is the reason that their frequencies are respectively close to each other for the longer n-paraffin molecules.

The above results show that series 1 and 2 are assigned to R_Y and R_X, and series 3 and 4 to OPB-3 and IPB-3. Similarly series 5 is assigned to OPB-5. As shown in Figure 5.7b the calculated frequencies for R_Z (TW-0) and TW-2 are somewhat irregular. This is due to the couplings among these vibrations and OPB vibrations. Figure 5.7a shows some weakly observed Raman lines in addition to the six series. They may be assigned to R_Z, TW-2, IPB-5, OPB-7 or other modes.

Fanconi and his collaborators observed similar series of low frequency Raman lines.[17,18] The procedure similar to that above gives the final assignments for all the crystal modifications of n-paraffins. Brunel and Dows[20] observed a broad and weak Raman line at $47.5 \, cm^{-1}$ in crystalline n-octane at 20 K, which disappeared at higher temperatures. They assigned this line to R_Z. However, the frequency is too low to be assigned to this vibration.

The change of these low-frequency vibrations due to the change of pressure is of importance. Wu and Nicol[19] measured the pressure effect of the frequencies of the Raman bands of polyethylene and $n\text{-}C_{23}H_{48}$ and $n\text{-}C_{44}H_{90}$. They found that the LA frequencies are pressure-insensitive and some others are sensitive. These data are potentially very useful for making assignments. However, $n\text{-}C_{23}H_{48}$ and $n\text{-}C_{44}H_{90}$ take the orthorhombic or monoclinic modification and the results cannot be correlated with the above assignments for the triclinic modification. Moreover, it is to be noted that some of the frequencies of the intramolecular OPB and IPB vibrations may be sensitive to pressure, since their force constants are largely due to the intermolecular forces and also they are coupled with the lattice vibrations.

In summary, at least for the triclinic n-paraffin crystals, we now have enough information to understand the general situation for all the overall vibrations.

5.5 POLYPEPTIDE CHAINS AND PROTEIN MOLECULES

As the second example let us consider polypeptides and proteins. The unit structure $-CHR-CO-NH-$ is far more complex than that of polyethylene. However, the vibrations of the unit have already been studied in detail as an important basis for biological phenomena.

The example of polyethylene given above shows that for the study of the overall vibrations the chain with a definite length is very useful. Substances corresponding to n-paraffins are the poly(amino-acid)s like polyglycine, polyalanine or polyglutamate, consisting of definite number of amino-acid residues.

At present such a set of standard samples is not available, although the technique of separating the heterodisperse mixture into their components is developing very rapidly. However, an important point is that the natural protein is homodisperse. The molecular weight is definite, the substance crystallizes and the X-ray analysis shows that the molecule has α-helix or other conformations with definite lengths. This is the reason why we chose them as the second example for the study of the overall vibrations, although this is at present only in a preliminary stage.

5.6 LOCAL AND NON-LOCAL VIBRATIONS OF POLYPEPTIDE CHAINS

Many of the vibrations of the polypeptide chain appearing in the infrared and Raman spectra are local in nature.[22] However, they are not so definite as the case of the polyethylene chain. The amide I and II bands appearing in the region $1700 \sim 1500 \, cm^{-1}$ are clearly localized in the $-CO-NH-$ group. Their frequencies do not depend on the neighbouring amino-acid residue side-chains, but only depend on the conformations such as the α-helix, the antiparallel β, or the random coil as was first pointed out by Miyazawa and Blout.[23]

The bands due to the less localized vibrations appear in the region $500 \sim 200 \, cm^{-1}$. They are characteristic of the amino-acid side-chain and also of its conformation like the right-handed or left-handed α-helix or the β type.[24-26] The frequencies are not affected by the kind of the neighbouring amino-acid residues. (See Table 5.2.) This fact means that these vibrations are localized in part of the $(CO-NH)-CHR-(CO-NH)$ structure.

Table 5.2 Far infrared bands characteristic of amino-acid residues with various main-chain conformations (cm^{-1})[22]

Amino-acid residue	Right-handed α-helix (cm^{-1})	Left-handed α-helix (cm^{-1})	β-Form (cm^{-1})
L-Ala	527–523, 375–371	478, 420	446–440
L-Val	415–409, 541		
L-Leu	472–467, 396–394		
L-Phe	484–481		502–501

In the lower-frequency region, a band due to a far less localized vibration appears. An example is the $264 \, cm^{-1}$ Raman band of poly-L-alanine. The intensity of this band is not appreciably affected by the inclusion of glycine or L-valine residues in the α-helix of poly-L-alanine. However, the inclusion of D-alanine in the poly-L-alanine α-helix appreciably decreases its intensity.[27] This fact shows that the band is not localized in a part of the chain and is sensitively affected by a slight deformation in the α-helix. The calculation of the normal vibrations of the polyalanine α-helix shows that this band is assigned to the breathing vibration of the α-helix.[28]

The dispersion curve calculated by Itoh and the author[29] for the right-handed poly-L-alanine is shown in Figure 5.8. The complexity of the curve, especially in the low-frequency region, shows that the optical and acoustical vibrations are not clearly separated from each other and that local vibrations occur to various degrees. A dispersion curve similar to those given in Figure 5.8 was also given by Fanconi et al.[30] However, the $CO\cdots HN$ hydrogen-bond stretching force constant used for the calculation is too large and gives too large frequencies in the low-frequency region.[22]

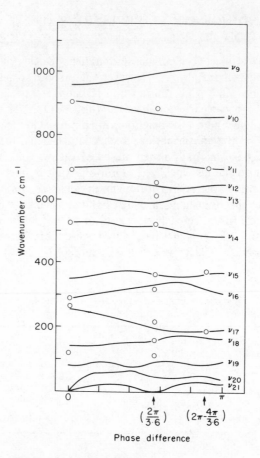

Figure 5.8 Dispersion curves of the vibrations for the poly-L-alanine α-helix. The vibrations with the phase differences of zero, $2\pi/3.6$ and $4\pi/3.6$ belong to the species A_1, E_1 and E_2, respectively. All of them are Raman-active and the former two are infrared-active. The circles give the frequencies of the infrared and Raman bands observed. The given assignments are supported by their infrared dichroism and Raman anisotropy (see ref. 22, 28 and 29)

5.7 OVERALL VIBRATIONS OF POLYPEPTIDE CHAINS AND PROTEIN MOLECULES

The overall vibrations of a polypeptide chain with definite length are similar to those of n-paraffin chain. The accordion-like vibration (LA-1) has the frequency given in equation (1), when the chain is long. The values estimated from the dispersion curve given in Figure 5.8 are shown in Table 5.3.

Table 5.3 Frequencies calculated[a] for the overall vibrations of
α-helical poly-L-alanine[22]

Number of residues	Accordion-like vibn. (cm^{-1})	Twisting vibn. (cm^{-1})	Bending vibn. (cm^{-1})
5	56	16	25
10	40	9	10
20	23	5	4
30	15	1	1
40	12	0	0
50	10	0	0

[a] There have been discussions about the phase conditions. These values were obtained by $\pi/(N + 1)$. Fanconi et al.[30] favour other conditions. We are dealing with the case with free ends, assuming the number of units is equal to the number of amino-acid residues. See also ref. 6.

The calculated value of the Young's modulus is

$$E = 2.31 \times 10^{10}\,\text{N m}^{-2}$$

These values may be used for the estimation of the length of polyalanine α-helices as in the case of polyethylene chains.[29]

The next problem is the bending vibrations. The symmetry properties of the α-helix give the same frequency for IPB-m and OPB-m. They may be called TA-m (transverse acoustical). The values of these frequencies are estimated from the curve given in Figure 5.8, where the phase difference is near 0.56π for v_{21}, and are given in Table 5.3.[22] The value for the twisting vibrations can also be estimated from the curve v_{21} of Figure 5.8 where the phase difference is nearly zero. They are far lower than the LA vibrations and are given in Table 5.3.[22]

The protein molecule having definite lengths of α-helix is expected to have such LA, TA and TW low-frequency vibrations. In addition the molecule may have various β-structures and random parts. A random part with only non-polar side-chains is flexible. The part with the α-helix, and polar or cystine S—S side-chains, may be more rigid. When the molecule has the heme group with the central heavy metal atom, this part is especially rigid. Accordingly, the rigid part may vibrate translationally or rotationally in the medium consisting of the soft parts of the protein molecule. This may give low-frequency modes. In addition, the protein molecule often consists of two, four or more subunits. The inter-subunit vibrations may also give low frequencies.

The measurements of the low-frequency region of the infrared, Raman and neutron inelastic scattering spectra are expected to give useful information about these vibrations. However, the interpretation of the spectra should be made cautiously, taking into account the above factors.

5.8 REFERENCES

1. T. Shimanouchi, *Pure Appl. Chem.*, **36**, 93, (1973).
2. M. Tasumi, T. Shimanouchi and T. Miyazawa, *J. Mol. Spectroscopy*, **9**, 261, (1962); **11**, 422, (1963).
3. R. G. Snyder and J. H. Schachtschneider, *Spectrochim. Acta*, **19**, 85, 117, (1963).
4. M. Tasumi and T. Shimanouchi, *J. Chem. Phys.*, **43**, 1245, (1965).
5. M. Tasumi and S. Krimm, *J. Chem. Phys.*, **46**, 755, (1967).
6. T. Shimanouchi and M. Tasumi, *Indian J. Pure Appl. Phys.*, **9**, 958, (1971).
7. S. Mizushima and T. Shimanouchi, *Proc. Imp. Acad. Tokyo*, **20**, 86, (1944).
8. S. Mizushima and T. Shimanouchi, *J. Amer. Chem. Soc.*, **71**, 1320, (1949).
9. T. Shimanouchi and S. Mizushima, *J. Chem. Phys.*, **17**, 1102, (1949).
10. R. F. Schaufele and T. Shimanouchi, *J. Chem. Phys.*, **47**, 3605, (1967).
11. R. F. Schaufele, *J. Chem. Phys.*, **49**, 4168, (1968).
12. W. L. Peticolas, G. W. Hibler, J. L. Lippert, A. Peterlin and H. G. Olf, *Appl. Phys. Letters*, **18**, 87, (1971).
13. A. Peterlin, H. G. Olf, W. L. Peticolas, G. W. Hibler and J. L. Lippert, *Polymer Letters*, **9**, 583, (1971).
14. J. L. Koenig and D. L. Tabb, *J. Macromol. Sci.-Phys.*, **B9**, 141, (1974).
15. G. Vergoten, G. Fleury, M. Tasumi and T. Shimanouchi, *Chem. Phys. Letters*, **19**, 191, (1973).
16. H. Takeuchi, T. Shimanouchi, M. Tasumi, G. Vergoten and G. Fleury, *Chem. Phys. Letters*, **28**, 449, (1974).
17. H. G. Olf and B. Fanconi, *J. Chem. Phys.*, **59**, 534, (1973).
18. F. Khouy, B. Fanconi, J. D. Barnes and L. H. Bolz, *J. Chem. Phys.*, **59**, 5849, (1973).
19. C. K. Wu and M. Nicol, *J. Chem. Phys.*, **58**, 5150, (1973).
20. L. C. Brunel and D. A. Dows, *Spectrochim. Acta*, **30A**, 929, (1974).
21. A. Müller and K. Lonsdale, *Acta Crystallogr.*, **1**, 129, (1948) and papers cited there.
22. T. Shimanouchi, Y. Koyama and K. Itoh, *Progress in Polymer Science Japan (Kodansha)*, **7**, 273, (1974).
23. T. Miyazawa and E. Blout, *J. Amer. Chem. Soc.*, **83**, 712, (1961).
24. K. Itoh, H. Katabuchi and T. Shimanouchi, *Nature New Biology*, **239**, 42, (1972).
25. K. Itoh and H. Katabuchi, *Biopolymers*, **11**, 1593, (1972).
26. K. Itoh and H. Katabuchi, *Biopolymers*, **12**, 921, (1973).
27. K. Itoh, H. Hinomoto and T. Shimanouchi, *Biopolymers*, **13**, 307, (1974).
28. K. Itoh and T. Shimanouchi, *Biopolymers*, **10**, 1419, (1971).
29. K. Itoh and T. Shimanouchi, *Biopolymers*, **9**, 383, 1413, (1970).
30. B. Fanconi, E. W. Small and W. L. Peticolas, *Biopolymers*, **10**, 1277, (1971).

6

The Measurement of
Lamellar Thickness by Raman Methods

P. J. Hendra

University of Southampton

6.1 METHODS FOR THE INVESTIGATION OF LAMELLAE IN POLYMERS

Many polymers solidify from the melt in spherulites composed of lamellar structural units, and the mechanical stability and ultimate strength of these materials are essentially limited by the interlamellar forces. However, our knowledge of the detailed structure of polymer lamellae and the thermodynamic and kinetic criteria which govern their formation is by no means perfect.

Methods available for determining the structure of lamellae include:

(a) High-angle X-ray diffraction—which enables us to identify the repeat structure in the crystalline lamellar core and also the degree of crystallinity.
(b) Vibrational spectroscopy—a much more limited technique in the structural sense but one capable of giving some information on interchain association forces.
(c) Low-angle X-ray diffraction—(LAXD)—the best method of measuring lamellar thickness.
(d) Electron microscopy of fracture surfaces—a method of measuring lamellar thickness which is normally used to support LAXD.
(e) Chemical methods—usually involving oxidative destruction of the chain folds at the lamellar surface followed by analysis of the molecular fragments.

None of these techniques allows us to state within a given lamellar unit the magnitude of l_c or l_f as defined in Figure 6.1. The 'long spacing' from LAXD measurements give us $l_c + 2l_f$ and some of the chemical data can indicate a rough value of l_c but the situation is far from satisfactory. Thus, when Raman methods were proposed as a technique for measuring lamellar thickness their potential was appreciated quickly in this field.

As summarized in Professor Shimanouchi's contribution to this volume (p. 61) the Raman spectrum of polyethylene contains bands with low-frequency shifts characteristic of an accordion-like acoustic vibration of the planar zig-zag

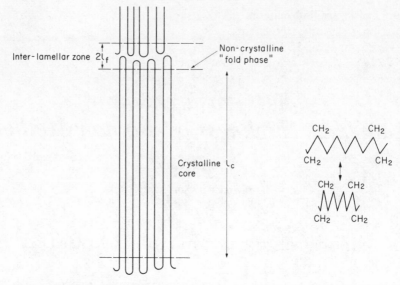

Figure 6.1 Definitions of 'core' and 'fold' zone thicknesses and (right) the longitudinal acoustic (LA) mode of vibration in polyethylene

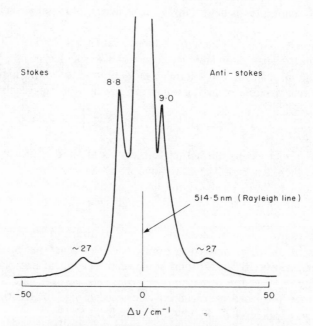

Figure 6.2 An example of a longitudinal acoustic mode Raman spectrum in polyethylene. Cary 82 spectrometer, 100 mW Ar^+ laser radiation, spectral bandwidth $2\,cm^{-1}$

sectors of crystalline material (the LA mode). Further, the frequency of the vibration is sensitive to the length of the planar system; see Figure 6.2.

Naively, one could assume that the planar zig-zag polymethylenic sectors of the crystalline core vibrate exactly as they do in crystalline paraffins of related morphology and that the vibration is restricted by the first *gauche* CH_2-CH_2 unit encountered at the interface between the crystalline and fold phases. As a consequence, the measurement of acoustic vibrational frequency could be used to give a precise value of l_c. If combined with good LAXD data a value of l_f would then be calculable. Early promise was however not realized; it became clear that the thickness determined by the Raman method agreed roughly with that from LAXD but was not strictly comparable with it.

Recently, Peterlin and co-workers[1] have pointed out that the assumption regarding the lack of involvement of the fold phase in the acoustic vibration may be incorrect. They have analysed the problem of the three-zone vibrator comprising an upper fold phase, an intermediate crystalline core and a lower layer also consisting of a fold phase, and have allowed for probable density differences between the phases (see Figure 6.1). They conclude that the vibrational anti-node assumed to lie at the fold-core interface may, in fact, lie within the fold zone, i.e. the length estimated from the measurement of the acoustic mode may be significantly greater than l_c. Experimental and theoretical evidence has, however, been accumulating which suggests that the anti-node lies close to the fold-core interface but that the vibrational force constant may not be exactly the same as for paraffins. The evidence may be summarized as follows.

(a) The LA mode frequency in polyethylene is not temperature-sensitive. At low temperatures the modulus of elasticity and the density of the fold phase will be very different from that at higher temperatures. If this phase contributed significantly to the vibration one would expect to see a shift in frequency on cooling.[2]
(b) Removal of the fold zone by 'etching' with fuming nitric acid does not alter the frequency of the LA mode.[2]
(c) Tilting the chains with respect to the lamellar surfaces does not alter their acoustic frequency.[3]
and
(d) Calculations by Fanconi[4] suggest that a *gauche* CH_2-CH_2 unit or similar structural discontinuity will confine the LA mode, but a methyl side-group will affect it very little.

Thus, at this time, we may conclude that the LA mode frequency is related to the core thickness l_c but the relationship is not known precisely. Even with this limitation the technique has value and this will be demonstrated.

6.2. EXPERIMENTAL MÉTHODS

The Raman technique for identifying and measuring the Raman shift associated with lamellae in polyethylene is experimentally fairly easy. The

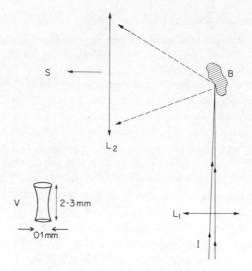

Figure 6.3 Typical illumination and collection geometry in a contemporary laser-Raman spectrometer. I = incident laser beam, ~ 100 mW; L_1 = focussing lens; L_2 = collection lens, aperture $\sim f = 1.5$; B = bulk sample of polymer; V represents typical size of illuminated and viewed volume; S = spectrometer

Raman experiment is normally carried out by illuminating the sample with monochromatic laser radiation and then viewing the scattered light normal to the illuminating direction—see Figure 6.3. The main experimental problems are as follows.

(a) The extremely low efficiency of Raman scattering requiring the use of sophisticated optical detectors and powerful sources. It is typical for Raman bands to have intensities of $< 10^4$ photons per second when the sample is illuminated with 0.2 W of radiation.

(b) Most crystalline polymers are turbid and the amount of Raman-scattered light can be as low as 10^{-10} of the total scattered light, requiring the use of high quality multiple monochromators. For the study of low energy modes, e.g. the LA mode, one usually resorts to using triple monochromators.

(c) Interference with the Raman effect from fluorescence; this is fortunately absent in most polyethylene samples.

The main experimental advantage is versatility: the sample can be heated, cooled, mechanically manipulated or even exposed to high pressures, all with comparative ease. Further, the cross-section of the sample need only be a little bigger than the focussed laser spot, say $100\,\mu$ diameter, and the length need only be 3 mm. If the sample is large, Raman methods can thus be used to give data with excellent spatial resolution; see Figure 6.4. Many aspects of the subject have been reviewed recently.[5]

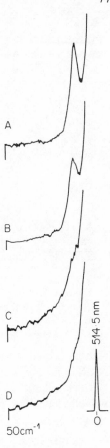

Figure 6.4 Low-frequency Raman spectra of drawn polyethylene at various points through the necking region; A, in the
undrawn part of the sample; B, close to the neck in the undrawn
part; C, in the neck; D, in the fibre side of the neck

6.3 APPLICATIONS OF THE RAMAN TECHNIQUE

The method has been applied to polyethylene at temperatures between
$-180\,^{\circ}$C and $+140\,^{\circ}$C. One advantage in studying specimens at low temperatures is that the Rayleigh band narrows, allowing us to approach closer to the
laser wavelength. The data strongly support the defect model of melting due to
Reneker[6] rather than the alternative, pre-melting proposal.[2] It is also clear that
as the temperature is raised, there is lamellar thickening.[7,8] Rolled and drawn
polyethylene have also been studied, with particular reference to the breakdown
of extended polymethylenic chains of regular length on passing through the
deformation neck or as a function of extension ratio in rolling.[8]

In order to elucidate the method of formation of lamellae from the melt it is
essential to monitor the kinetics of the process. One method of carrying this out
is to prepare a glass from the melt and allow it to crystallize at low temperatures
at a controlled rate. Although glasses have been made for most synthetic
polymers and in some cases are relatively stable e.g. poly(ethylene terephthalate),
polyethylene has not previously been made in the glassy state. Recently, we
have succeeded in preparing polyethylene glass by cooling the melt from

140 °C to − 150 °C rapidly. We find that material so produced has a vibrational spectrum similar to that of the melt but that on warming to − 90 °C crystallization sets in. As this occurs, lamellae form, of thickness ∼ 180 Å; see Figure 6.5. We therefore conclude that T_g lies below − 90 °C and that any quenching

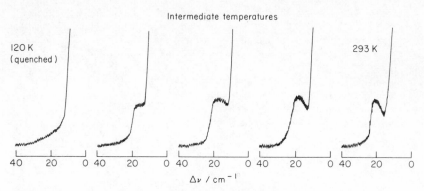

Figure 6.5 Longitudinal acoustic mode Raman spectra of polyethylene glass—as quenched, and then when slowly warmed. The observed band indicates a lamellar thickness near 180 Å

routine which allows the specimen to exceed − 90 °C before analysis will result in it containing lamellae of thickness ∼ 180 Å.[9]

The applications of Raman spectroscopy discussed above have all referred to polyethylene. LA mode spectra have also been recorded for the following systems:

(a) cis-1,4-polybutadiene[10]: band between Δν 4·5 and 6·5 cm^{-1};
(b) poly(decamethylene sebacate)[3];
(c) polytetrahydrofuran and some related polyethers[11];
(d) poly(ethylene sulphide) and higher poly(methylenic sulphides).[12]

In poly(decamethylene sebacate) the bands occur between 36 and 27 cm^{-1} and originate in acoustic motions of the polymethylenic sectors. Using the modulus normally accepted for polymethylene the agreement between lamellar thickness derived from Raman and LAXD data was excellent. Examples of the results from a poly(methylenic sulphide) are shown in Figure 6.6. It is thought that the prominent low-frequency band near 15 cm^{-1} is due to the third order acoustic mode, the fundamental being 'buried' in the Rayleigh band except in the cases involving the thinnest lamellae. Unfortunately, the precise relationship between LA mode frequency and lamellar thickness cannot yet be defined in this case as the LAXD pattern for this material has proved elusive. It is clear already though that lamellar thickening occurs over a wide temperature range and that the lamellae can double in thickness if treated appropriately; and this in a polymer whose propensity to form lamellae had not previously been proven.

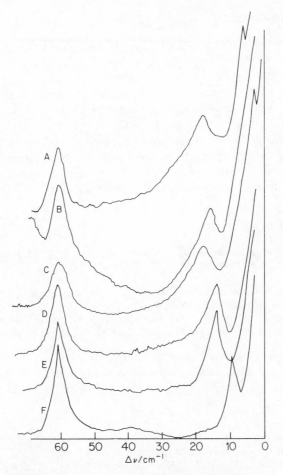

Figure 6.6 Low-frequency Raman spectra of poly-
(ethylene sulphide). A, rapidly quenched from melt;
B, slowly cooled from melt; C, D, E and F: specimen
A annealed for 2 hours at 120, 150, 170 and 190°C
respectively

Raman bands due to fundamental (i.e. $m = 1$) LA modes in melt-crystallized
polymers are relatively intense but the higher order modes give rise to only
weak bands (only the 'odd' order modes are Raman active and only $m = 1$
and $m = 3$ have been seen to-date). In crystalline normal paraffins, long series
of bands are frequently found and may well include the 21st–31st order lines in
favourable cases. Further, the Raman bands encountered in polymers tend to be
broad and it may well be that the intensity and band-width data can be made
to yield information on the distribution of lamellar thicknesses within a
specimen.

6.4 CONCLUSION

The Raman method of measuring approximate lamellar thicknesses is now an established tool in polymethylenic systems. It is likely that it will eventually prove to be of value in many other synthetic polymers. Currently, the attraction of the Raman method is that it inspects a minute sample volume and does so in simple or unlikely experimental environments with consummate ease. It is hopefully anticipated that the method will shortly lead us to a much more detailed knowledge of the structure of fold sequences and hence of interlamellar adhesion (and thus mechanical properties of the bulk material). In addition, the mechanism of lamellar formation and the parameters which govern it may well advance through the use of this novel technique.

6.5 REFERENCES

1. H. G. Olf, A. Peterlin and W. L. Peticolas, *J. Polymer Sci.*, **B12**, 359, (1974).
2. P. J. Hendra, E. P. Marsden, M. E. A. Cudby and H. A. Willis, *Makromol. Chem.*, in press.
3. M. J. Folkes, A. Keller, J. Stejny, P. L. Goggin, G. V. Fraser and P. J. Hendra, *Kolloid Z. und Z. Polymere* **253**, 354, (1975).
4. B. Fanconi and J. Crissman, *J. Polymer Sci., Polymer Letters*, **13**, 421, (1975).
5. P. J. Hendra, in *Polymer Spectroscopy*, ed. D. O. Hummel, Verlag Chemie, Berlin, 1974.
6. D. H. Reneker, *J. Polymer Sci.*, **59**, 839, (1962).
7. J. L. Koenig and D. L. Tabb, *J. Macromol. Sci., Phys., Part B*, **9**, 141, (1974).
8. G. V. Fraser, P. J. Hendra, M. E. A. Cudby and H. A. Willis, *J. Materials Sci.*, **9**, 1270, (1974).
9. P. J. Hendra, H. P. Jobic and K. Holland-Moritz, *J. Polymer Sci., Polymer Letters*, **13**, 365, (1975).
10. J. B. Downes, University of Southampton, private communication.
11. G. V. Fraser, University of Bristol, private communication.
12. P. J. Hendra and H. A. Majid, *J. Materials Sci.*, in press.

7

The Study of Crystallinity in Synthetic Polymers by Low-Frequency Vibrational Spectroscopy

H. A. Willis and M. E. A. Cudby

ICI Ltd., Welwyn Garden City

7.1 INTRODUCTION

The vibrational spectra of synthetic polymers between 4000 cm^{-1} and 400 cm^{-1}, the so-called 'mid-infrared' region, have been the subject of intensive study since the technological innovations of the early 1940's established infrared spectroscopy as a means of investigating molecular structure. There was at that time a particular need for methods to characterize the many synthetic polymers which had been produced during the preceding twenty years. The ability to observe the presence and to determine the concentration of structural units such as OH, NH, CH$_3$ and C=O in these comparatively intractable materials lead to a great upsurge of interest, and following the considerable success achieved in this chemical characterization, attention was given to the possible application of spectroscopy to the morphology of synthetic polymers.

Initial investigations[1-3] met with some apparent success, and in the 1950's a number of papers suggested that the infrared spectrum might be used to study crystallinity, because differences could be observed in the spectra of common polymers such as polyethylene, polyethylene terephthalate, and nylon which could be related to crystallinity as determined by X-ray diffraction or by density measurements.

Further investigation showed,[4] unfortunately, that this success was largely illusory. What was being observed in the vibrational spectrum was not necessarily the presence of crystalline material in a specimen, but the presence of the particular rotational conformers of a polymer molecule which can exist in the crystal. For example, only the fully extended trans conformation of the polyethylene chain will pack into the paraffin crystal, and in the case of poly(ethylene terephthalate) the glycol residue must be in the trans conformation in the crystalline regions. However, in both these polymers the trans conformer is also present in amorphous regions; hence the supposition that measuring the trans/gauche ratio is equivalent to measuring the crystalline/amorphous ratio is clearly false.

It must not be concluded, however, that spectra in the mid-infrared region have nothing to say about the crystallinity of polymers. For example, a close examination of the infrared spectrum of a crystallizable polymer will usually reveal that, with increasing crystallinity, those bands in the spectrum arising from molecular units which are in the correct conformation to enter the crystal-

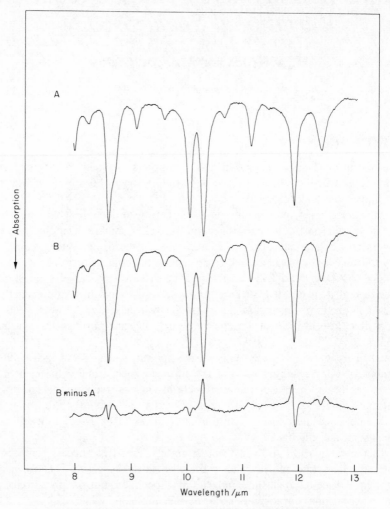

Figure 7.1 Mid-infrared spectrum of an isotactic polypropylene sample: (A) melt cast; (B) same film heat-crystallized. (B) minus (A) shows typical effects of band narrowing (narrow positive peak with broad wings) intensity reduction (negative peak) and "flyover" due to shift of band maximum on crystallization. All these effects result from increase of crystallinity on thermal treatment

line regions become considerably sharper and narrowed while bands due to non-crystallizable conformers have decreased in intensity. The effect is emphasized in the 'derivative' spectrum (Figure 7.1). Presumably change in long range order is responsible for the change in band width. The molecular unit in the 'crystallizable' conformation exists entirely within an environment of similar units in the crystal, while such a unit in an amorphous region may experience a multitude of environments, each of which has its own particular effect in modifying the resonant frequency of the vibrators. Hence broad bands are seen from the molecular unit in amorphous regions, and sharp, narrow bands from the same unit in the crystal. Interesting though this effect may be, it is not a straightforward matter to determine crystallinity from the band width.

The basic problem in applying vibrational spectra in the $4000 \, cm^{-1}$ to $400 \, cm^{-1}$ region to the observation of crystallinity in synthetic polymers is that the major features of the spectra arise from individual molecules and their conformation, while crystallinity is concerned with packing and the interrelation of molecules. The intermolecular effect classically observed in vibrational spectroscopy is hydrogen bonding. Unfortunately in nylon, where it has been presumed that hydrogen bonding plays a part in stabilizing the crystal, it has not been possible to establish any difference between the position, or band width of the absorption bands arising from the $N—H$ group in crystalline and amorphous samples. Ingenious, if not helpful, explanations have been offered for this.[5]

In the spectra of some polymers interchain interaction in crystalline forms containing more than one molecule in the unit cell is revealed by correlation-splitting of certain of the bands. The best known example is the methylene rocking mode, which appears in the infrared spectrum of polyethylene in the orthorhombic form as a sharp doublet at $730 \, cm^{-1}$ and $720 \, cm^{-1}$. In the monoclinic form (one molecule per unit cell) the same vibrational mode is observed as a single sharp band at $720 \, cm^{-1}$. Extended chains in the amorphous regions of polyethylene give rise to a broader band centred near $720 \, cm^{-1}$. Due to the overlapping of the bands from these different forms it is a dubious proposition to use them to determine crystallinity, especially as the bands of the crystalline species are highly dichroic.

7.2 THE FAR INFRARED REGION (400–10 cm^{-1})

The absorption bands arising from the vibration of neighbouring chains in a molecule are known to appear in the infrared region mainly at frequencies below $400 \, cm^{-1}$. Long range relaxational motions give rise to absorption at very low frequencies in the radio region ($< 1 \, cm^{-1}$). In the far infrared, which spans the gap between the mid-infrared and radio regions, bands due to some internal molecular deformation modes may appear, but nevertheless it would be the favoured region for the direct observation of intermolecular modes in polymers. Delay in exploring the far infrared region has been largely a matter of experimental difficulty. It has so far been necessary to use black body

84

radiation sources, and these have low intensity in this region. There are, further-more, problems with the means of dispersion. All these difficulties are overcome by using an interferometer rather than a conventional dispersive spectrometer, and recent improvements in interferometric practice have made this form of spectroscopy attractive. Now that good equipment is available it may be anticipated that the application of spectroscopy to the study of polymer crystallinity will make substantial progress. The following examples will illustrate the kind of information which can be obtained.

7.3 POLYETHYLENE

Polyethylene is the polymer in which a crystal mode was first observed.[6] It is the simplest case, because the lowest frequency internal mode has been calculated to occur at 720 cm^{-1}; hence the band observed at 72 cm^{-1} is clearly and external (crystal) mode. The behaviour of the band as a function of temperature is also as expected, in that its frequency increases as the specimen is cooled from room temperature (72 cm^{-1}) to liquid nitrogen temperature (78 cm^{-1}). The frequency increase is expected as the crystal shrinks on cooling so that intermolecular interaction becomes stronger. Two infrared active

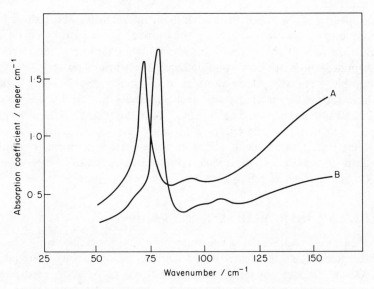

Figure 7.2 Far infrared spectrum of linear high-density highly crystalline ($\sim 90\%$) polyethylene: A, 25 °C; B, -150 °C

N.B. The absorption coefficients plotted in Figures 7.2, 7.3, 7.5 and 7.6 are defined as ln I_0/I divided by the thickness in cm. [Reproduced from ref. 8 by permission of the North Holland Publishing Co. Ltd., Amsterdam, and the authors]

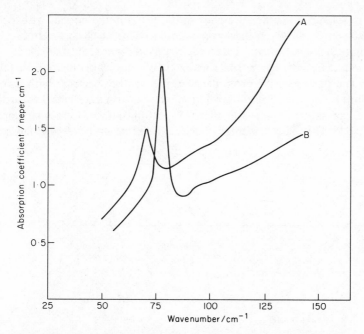

Figure 7.3 Far infrared spectrum of a low-density, chain-branched, medium crystallinity ($\sim 60\%$) polyethylene: A, 25 °C; B, -150 °C. [Reproduced from ref. 8 by permission of the North Holland Publishing Co. Ltd., Amsterdam and the authors]

crystal modes are expected in polyethylene and a weaker band observed at 109 cm^{-1} at liquid helium temperature[7] and 94 cm^{-1} at room temperature[8] is thought to be the B_{2u} crystal mode.

The careful study by Fleming *et al.*[8] shows the effects for highly crystalline polyethylene (Rigidex) and a less crystalline low-density polyethylene (Figures 7.2 and 7.3). All prominent features of the spectrum sharpen considerably as the temperature of the sample is reduced, presumably because multiphonon processes are suppressed. However, the distinct features associated with the crystalline regions are weaker, and the underlying continuum is stronger, in the less crystalline material. The continuum is thought to be a 'liquid lattice' band corresponding to the 'smearing out' of the lattice spectrum due to lack of correlation between the molecules in amorphous regions.[9]

7.4 POLYPROPYLENE

Polypropylene presents a more formidable problem, as the heavily branched structure permits the appearance of a number of very low-frequency deformation and torsional internal molecular vibrations.

On an isolated chain model, eight bands due to internal modes are expected[10] below 400 cm^{-1} of which the lowest should occur at 63 cm^{-1}. In fact, many more than eight bands are observed in this region. Goldstein *et al.*[11] have made their analysis by comparing the spectra of samples of different crystallinity, and the sharpening of the prominent features in the spectrum with increasing crystallinity (i.e. increasing conformational regularity) is well marked, especially at low temperatures where the effect of multiphonon processes is suppressed (Figure 7.4). They supplemented their analysis by study of oriented samples

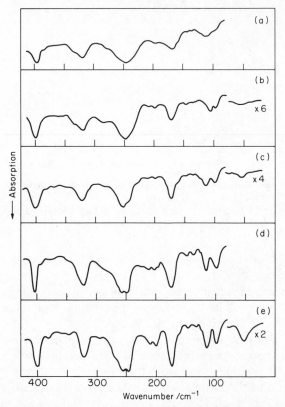

Figure 7.4 Far infrared spectra of isotactic polypropylene of different crystallinities, all recorded at about −150 °C: (a) 52%; (b) 58%; (c) 66%; (d) 70%; (e) 75%. For the region below 80 cm^{-1}, samples were increased in thickness by the factors shown. [Reproduced from ref. 11 by permission of I.P.C. Business Press Ltd., (C), London, and the authors]

with polarized radiation. They conclude that much of the complexity in the spectrum arises because almost all the bands due to internal vibrations are split as a result of intermolecular interaction in the crystalline regions of the polymer. A band near 50 cm^{-1} remains unassigned as an internal mode;

it increases in frequency from 48 cm^{-1} at room temperature to 55 cm^{-1} at ~ 100 K, while no other band in the region moves significantly in frequency as the temperature is changed. Furthermore, its intensity increases rapidly with increasing crystallinity. There seems good reason to suppose that this is a true crystal mode. No evidence was found for the remaining four infrared active crystal modes.

7.5 POLYTETRAFLUOROETHYLENE

Calculations suggest that the infrared spectrum of polytetrafluoroethylene should contain two infrared active modes due to internal molecular vibrations below 100 cm^{-1}, near 30 cm^{-1} and 15 cm^{-1}. Both of these have been observed very close to the predicted frequencies.[12,13] X-ray data suggest that polytetrafluoroethylene contains only one molecule per unit cell, both above and below the well-known room temperature transition.[14,15] This being so, there should be but one infrared active band, that corresponds to the rotation about the longitudinal molecular axis. It was with some surprise, therefore, that the far infrared spectrum of highly crystalline polytetrafluoroethylene was found[16] to consist, at low temperatures, of a series of sharp lines, although above the room-temperature transition a single, though rather broad, feature at about 50 cm^{-1} could well be the anticipated single lattice mode (Figure 7.5).

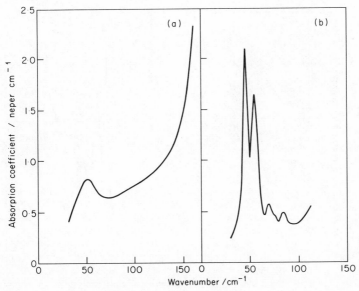

Figure 7.5 Far infrared spectrum of unsintered, highly crystalline (~90%) polytetrafluoroethylene: (a) 25 °C; (b) −150 °C

88

Chantry *et al.*[16] claim that this multi-lined spectrum appears at low temperature because the room temperature transition is accompanied by a change of crystal form in which the number of molecules per unit cell changes from one to two. Such a change should lead to the appearance of five infrared-active crystal models. Three of these are at 46, 70 and 85 cm^{-1}, corresponding to translational modes, while the most prominent band, at 55 cm^{-1}, is a closely spaced band pair[17] corresponding to the two expected rotational modes. The latter combines into the single active rotational mode above the room temperature phase transition.

Piseri *et al.*[12] reject this assignment completely, and attribute all this structure in the spectrum to the activation of internal chain modes due to conformational disorder as suggested by Zerbi.[18]

To throw some light on this controversy, Chantry *et al.*[19] have extended their work to compare the far infrared spectra of 'unsintered' and 'sintered' polytetrafluoroethylene. The former material, which is polytetrafluoroethylene as made, is of extremely high crystallinity and consists essentially of chain-extended molecules.[20] After the heat treatment known as 'sintering', the crystallinity falls significantly because the molecules chain-fold into relatively short conformationally regular segments, interrupted by conformational faults.[21]

In the interpretation of Piseri *et al.*,[12] sintering gives increased conformational disorder, and should lead to an enhancement of intensity of the group of sharp bands near 50 cm^{-1}. In the view of Chantry *et al.*[16] the reduction in crystal-

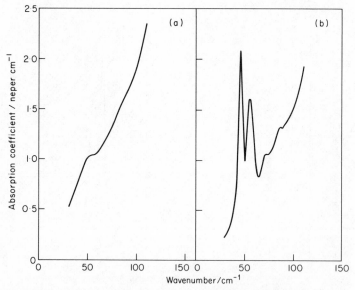

Figure 7.6 Far infrared spectrum of sintered polytetrafluoroethylene, crystallinity approx. 65%: (a) 25 °C; (b) −150 °C

linity accompanying sintering should lead to a reduction in intensity of the crystal modes, and the appearance of the familiar 'liquid lattice' continuum from the conformationally irregular material. Comparison of Figures 7.5 and 7.6 shows that sintering has the effect predicted by Chantry *et al.* and it may therefore be said that in at least one instance the complete infrared lattice spectrum of a synthetic polymer has been measured.

We would like to acknowledge the very considerable assistance we have received, both in the experimental programme and in discussion, from our co-workers at the National Physical Laboratory and at the Polytechnic of North London.

7.6 REFERENCES

1. W. H. Cobbs and R. L. Burton, *J. Polymer Sci.*, **10**, 275, (1953).
2. R. G. Quynn, J. L. Riley, D. A. Young and H. D. Noether, *J. Appl. Polymer Sci.*, **2**, 166, (1959).
3. R. G. J. Miller and H. A. Willis, *J. Polymer Sci.*, **19**, 485, (1956).
4. I. M. Ward, *Chem. and Ind.* (*London*), 905, (1956).
5. A. Elliott, *Infrared spectra and structure of organic long chain polymers*, Arnold, London, 1969, p. 85.
6. M. I. Bank and S. Krimm, *J. Chem. Phys.*, **42**, 4059, (1965).
7. G. D. Dean and D. H. Martin, *Chem. Phys. Letters*, **1**, 415, (1967).
8. J. W. Fleming, G. W. Chantry, P. A. Turner, E. A. Nicol, H. A. Willis and M. E. A. Cudby, *Chem. Phys. Letters*, **17**, 84, (1972).
9. R. W. Zwanzig, *Phys. Rev.*, **156**, 190, (1967).
10. T. Miyazawa, K. Fukushima and Y. Ideguchi, *J. Polymer Sci.*, **(B) 1**, 385, (1963).
11. M. Goldstein, M. E. Seeley, H. A. Willis and V. J. I. Zichy, *Polymer*, **14**, 530, (1973).
12. L. Piseri, B. M. Powell and G. Dolling, *J. Chem. Phys.*, **58**, 158, (1973).
13. F. J. Boerio and J. L. Koenig, *J. Chem. Phys.*, **52**, 4826, (1970).
14. S. Krimm, *Fortschr. Hochpolym. Forsch.*, **2**, 51, (1960).
15. H. G. Kilian, *Kolloid-Z. und Z. für Polymere*, **185**, 13, (1962).
16. G. W. Chantry, J. W. Fleming, E. A. Nicol, H. A. Willis, M. E. A. Cudby and F. J. Boerio, *Polymer*, **15**, 69, (1974).
17. K. W. Johnson and J. F. Rabolt, *J. Chem. Phys.*, **58**, 4536, (1973).
18. G. Zerbi and M. Sacchi, *Macromolecules*, **6**, 692, (1973).
19. G. W. Chantry, J. W. Fleming, E. A. Nicol, H. A. Willis and M. E. A. Cudby, unpublished work.
20. F. J. Rahl, M. A. Evanco, R. J. Fredericks and A. C. Reimschuessel, *J. Polymer Sci.*, **A2, 10**, 1337, (1972).
21. L. Melillo and B. Wunderlich, *Kolloid-Z. und Z. für Polymere*, **250**, 417, (1972).

8

Resonant Raman Scattering from Diacetylene Polymers

D. Bloor, F. H. Preston, D. J. Ando and D. N. Batchelder

Queen Mary College, University of London

8.1 INTRODUCTION

The diacetylene polymers are a unique class of crystalline polymers which are produced by the solid state polymerization of di-, tri- and more complex acetylenic compounds. This polymerization has been known for many years[1] but only recently has its true nature and importance been revealed by Wegner and his collaborators.[2-7] X-ray analysis of the polymers produced from the bis(phenylurethane) and bis(p-toluene sulphonate) diacetylenes[8,9] has shown that they are highly perfect crystals containing a chain-extended polymer with a conjugated backbone. The repeat unit consists of alternate double, single, triple and single bonds $\mbox{$=$}CR-C\equiv C-CR\mbox{$=$}$ in a *trans* conformation. The general arrangement of monomer molecules and the resulting polymer are shown in Figure 8.1.

The physical properties of such materials are of great interest because of the possibility that they may be one-dimensional semiconductors. Hückel π-electron molecular orbital calculations have predicted[10] a gap between the valence and conduction bands, which results from the bond alternation, of from 1 to 2 eV. The diacetylene polymers have a golden or coppery metallic appearance which is consistent with this prediction. We have reported[11] the reflection and absorption spectrum of the bis(p-toluene sulphonate) polymer. A sharp absorption edge occurs just below 2 eV and is accompanied by a distinct peak in the reflection spectrum for light polarized along the axis of the conjugated chain. For light polarized perpendicular to the chain the absorption is less intense and the reflection is constant and low across the visible region. The absorption must, therefore, result from an electronic excitation of the polymer backbone. These observations agree well with the predicted behaviour for band-gap transitions.[10]

The occurrence of this intense optical absorption has an important consequence in the Raman spectroscopy of these polymers. When the exciting radiation is close in frequency to a real electronic excitation of the polymer, strong resonant Raman scattering can be expected. This Raman spectrum

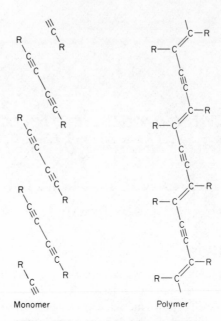

Figure 8.1 Arrangement of diacetylene monomer molecules and diacetylene polymer produced by solid state polymerization. The side groups, R, isolate the monomer units so that there is a unique crystallographic direction for growth of the polymer chains

should be dominated by those molecular motions which couple strongly with the electronic states of the backbone. The vast majority of the normal vibrational modes of the polymer will be associated with side-group motions which do not interact strongly with the backbone electrons. Thus, the resonant Raman spectrum will be much simpler than the non-resonant spectrum and we have a powerful tool for determining the structure of the polymer backbone.

We have investigated the Raman spectra of single crystal polymers obtained from the bis(p-toluene sulphonate) of 2,4-hexadiyne-1,6-diol, TSHD, and from the bis(phenylurethane) of 4,6-decadiyne-1,10-diol, PUDD. Features characteristic of resonant Raman scattering have been clearly identified. The assignment of the observed Raman lines has been made using simple calculations of the vibrational frequencies of a model diacetylene chain. The structure of the backbones of amorphous and polycrystalline diacetylene polymers have been deduced from their Raman spectra. For samples of TSHD polymer extracted from partially polymerized monomer the structure has also been investigated by X-ray diffraction to test the validity of this analysis.

8.2 RESONANT RAMAN SCATTERING

A brief summary of the theory of resonant Raman scattering[12] is presented below. This indicates the origins of the features characteristic of resonant Raman scattering that we observe for the diacetylene polymers. The Raman scattering process can be represented by the simple scheme of Figure 8.2a.

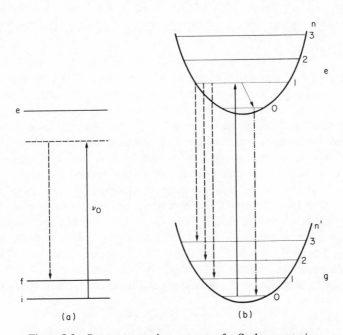

Figure 8.2 Raman scattering processes for Stokes scattering. (a) Laser frequency, v_0, less than that of excited molecular electronic states, e. The dashed level is a virtual state. (b) Laser frequency equal to that of an excited molecular state. Vibrational levels are superimposed on the electronic states. Raman transitions are shown by the heavy transition arrows; fluorescent decay by the fine arrows

The intensity of Raman scattering due to the excitation of a molecule in an initial vibrational level, i, to a final level, f, for exciting radiation of frequency v_0 is:

$$I_{if} = N_i \frac{64\pi^2}{3c^2}(v_0 - v_{if})^4 P_{if}^2 \tag{1}$$

where N_i is the population of molecules in the initial level, v_{if} is the vibrational frequency for the transition from level i to level f, P_{if} is the electric moment associated with the transition i to f induced by the exciting radiation. P_{if} is, therefore, a vector related to the electric field vector of the exciting radiation through the polarizability tensor.

$$\mathbf{P}_{if} = \alpha_{if}\mathbf{E} \tag{2}$$

α_{if} is a property of the scattering molecule and can be expressed as a sum over all the excited electronic states, e in Figure 8.2a, of the molecule.

$$\alpha_{if} = \frac{1}{h}\sum_e \left[\frac{\mu_{ef}\mu_{ie}}{v_{ei} - v_0 + i\delta_e} + \frac{\mu_{ie}\mu_{ef}}{v_{ef} + v_0 + i\delta_e} \right] \tag{3}$$

where μ_{ab} is the transition electric dipole moment for transitions from level a to level b and δ_e is the damping constant of the excited level e. For a gas or liquid an averaged value of α^2 is obtained by summing over all the spatial components of α. For a crystal, when the incident and scattered light polarizations and directions are defined, then a few, or perhaps only one, components of α are involved in determining the scattering intensity. The disappearance of certain components, i.e. selection rules, can be determined by group theory once the molecular site symmetry in the crystal is known.

The form of the energy denominator in equation (3) leads to a resonant enhancement of the scattering intensity when the exciting frequency approaches that of an excited electronic state of the system ($v_0 \Rightarrow v_{ei}$). The damping constant for the excited state prevents the intensity diverging when the frequencies are coincident. Sharp electronic states, with a low damping constant, will have the largest resonant enhancement. Near resonance, one term of the sum over excited states will dominate the scattering and the other terms can be neglected. This term will be:

$$\alpha_{if} = \frac{1}{h}\frac{\mu_{ef}\mu_{ie}}{v_{ei} - v_0 + i\delta_e} \tag{4}$$

This equation does not include the vibrational levels of the excited electronic state. These must be included since their separations are small compared with that of the electronic states.

Figure 8.2b depicts a simple case in which the molecular coordinates are the same in both electronic states and there is only one vibrational mode. The transitions occurring in resonant scattering from one of the vibronic levels are shown by the heavy full and dashed lines as in Figure 8.2a. Denoting the electronic states by subscripts (g for ground, e for excited) and the vibrational levels by superscripts (n', n for the number of vibrational quanta in the ground and excited states) we can rewrite equation (4) as:

$$\alpha_{gg}^{n_1'n_2'} = \frac{1}{h}\sum_n \frac{\mu_{eg}^{nn_2'}\mu_{ge}^{n_1'n}}{v_{eg}^{nn_1'} - v_0 + i\delta_e} \tag{5}$$

where the sum runs over all excited vibronic states. We will usually be concerned with the special case $n' = 0$. The denominator is unchanged from that of equation (4) but the numerator can be written, in the Franck–Condon approximation, as the product of the electric dipole transition moments for constant (zero) internal coordinate and vibrational overlap integrals.

$$\mu_{eg}^{nn_2'}\mu_{ge}^{n_1'n} = \mu_{eg}\mu_{ge}(e, n|g, n_2')(g, n_1'|e, n) \tag{6}$$

The occurrence of these overlap integrals means that the polarizability can be large even for large differences in n_1' and n_2'. If there are several vibrational modes for the molecule the overlap terms can lead to strong scattering of combination frequencies. The damping constant affects the intensity of overtones and combinations since if δ is small only one vibronic level need be considered in equation (5). If δ is large the sum must be carried out over the overlapping vibronic levels.

The vibrationally excited states (e, n) can relax, as shown by the light full line in Figure 8.2b, to levels with lower vibrational excitation. Fluorescent emission can then occur to the ground state giving lines with frequency shifts equal to either the vibrational excitation in the excited state or the difference between total vibrational energies in the ground and excited states. These processes often give rise to a broad fluorescent background to the Raman spectrum but in some cases they give rise to sharp resonant fluorescence lines. These can only appear when the frequency of the exciting radiation is greater than the frequency of the pure electronic transition. This limit is sometimes taken as separating the resonance Raman (or pre-resonance) regime from the resonance fluorescence (or rigorous resonance) regime.[13]

The principal results of interest are:

(i) the Raman scattering intensity is strongly enhanced when the frequency of the exciting radiation coincides with that of an electronic transition of the molecule;
(ii) overtone and combination bands occur in the resonant Raman spectrum;
(iii) the enhancement is related to the damping (i.e. to the linewidth) of the excited vibronic states;
(iv) vibrations which do not couple with the electronic states will not be resonantly enhanced; and
(v) with exciting frequencies higher than that of the pure electronic transition there will be additional features in the spectrum due to fluorescence processes.

8.3 VIBRATIONAL MODES OF THE DIACETYLENE BACKBONE

It follows from conclusion (iv) and the interpretation of the reflection spectrum that in order to assign the resonant Raman spectral lines of the diacetylene polymers we need only consider the vibrational modes of the backbone.

The repeat unit of the polymer as shown in Figure 8.1 has C_{2h} symmetry. Since we will ignore the side-groups we will use a simple model in which they are represented by a single atom R with an appropriately chosen mass. There are then six atoms in the repeat unit of the polymer chain which give rise to six A_g,

96

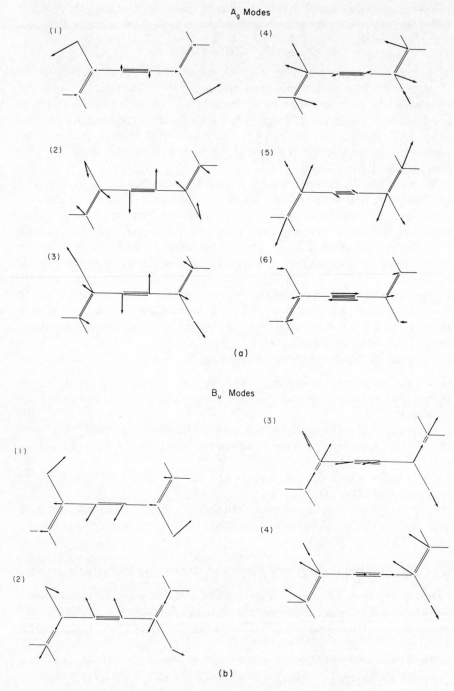

Figure 8.3 Vibrational modes of a model diacetylene chain. For details of the model see text

three B_g, three A_u and six B_u vibrational modes. One A_u and two B_u modes are zero-frequency translational modes. The remaining modes are finite frequency modes in the crystalline solid. The symmetric (g) modes are Raman-active and the asymmetric (u) modes are infrared-active. The out-of-plane modes will all be bond-bending modes. These will be relatively low in frequency and will have small changes in bond polarizability so that their Raman intensity will be low. We shall, therefore, only consider the in-plane modes since these contain the more important bond-stretching motions. There are six A_g and four B_u in-plane modes (Figure 8.3).

We use a simple force field to determine the vibrational frequencies. Nearest neighbour forces are taken to have their literature values: C—C 520 N m^{-1}, C=C 960 N m^{-1} and C≡C 1560 N m^{-1}. The bond-bending force constant was taken to be 25 N m^{-1} for all bonds. We ignore bond interaction force constants. These rather gross approximations are made in order to simplify the calculation of vibrational frequencies.

Table 8.1 Raman active (A_g) modes of the polydiacetylene trans ={CR—C≡C—CR}=$_n$ chain

Mode	Calculated frequency in units of cm^{-1}	Observed frequency[a]	
		TSHD polymer	PUDD polymer
1	170·4	—	—
2	443·1	535	705
3	595·0	955	1215
4	1261	1205	1345
5	1808	1487	1460
6	2303	2088	2072

[a] Experimental values have a precision of ± 2 cm^{-1}.

The calculated frequencies are listed in Table 8.1 and the corresponding normal modes are shown in Figure 8.3. These results are appropriate for a side-group of mass 15 (i.e. a point like CH_3 group). The substitution of either zero- or infinite mass side-groups does not greatly affect the frequencies of the five high-frequency modes since they involve relatively small amounts of side-group motion. The three high-frequency Raman-active modes are clearly seen to be mainly triple bond, double bond and single bond stretching. The latter motion also involves necessarily some bending about the double bond. All the low-frequency modes contain significant bond-bending components in the atomic motions.

The three high-frequency A_g modes will be the strongest components of the Raman spectrum since the stretching motions give rise to the largest changes in bond polarizability. The bond polarizabilities, and hence the changes for a given bond distortion, are approximately related to the order of the bond;

thus, it follows from the normal modes shown in Figure 8.3 that the Raman intensity should increase as the scattering frequency increases.

This calculation cannot be expected to give accurate values for the vibrational frequencies. We adopt this approach because fitting of force constants to vibrational frequencies is not feasible. In general the infrared-active backbone modes are obscured by side-group absorptions and the Raman-active mode frequencies are insufficient. The model is sufficiently accurate to enable us to identify the most intense Raman modes as stretching vibrations, with frequencies related to the bond lengths. Thus, even though we cannot obtain accurate force constants the resonant Raman spectrum is capable of providing information about the backbone structure.

8.4 EXPERIMENTAL TECHNIQUES

The TSHD monomer was prepared by the method due to Wegner.[14] Single crystals of monomer were grown from acetone solution. Crystals with dimensions up to 10 by 10 by 2 mm were grown by solvent evaporation and larger crystals up to 1 cm³ in volume were grown by slow cooling of a saturated solution. The best crystals were clearly-facetted prisms and were dichroic with a deep red colour due to traces of polymer produced by thermal polymerization during crystal growth. The polymerization proceeds slowly at room temperature but more rapidly when the crystals are heated above 50 °C. During this process the larger crystal facets assume a golden metallic appearance. The conversion to polymer goes to completion after prolonged heating. Excessive heating was avoided to prevent degradation of the polymer. This has been investigated fully by electron spin resonance spectroscopy and will be reported elsewhere.[15]

Single crystals of polymer give X-ray diffraction patterns with spot sizes limited by instrumental factors. The full structure has been reported by Kobelt and Paulus.[9] They find the symmetry to be $P2_1/c$ (using the second setting) with two polymer chains per unit cell extended in the b-direction. We prefer to use the equivalent symmetry $P2_1/b$ (in the first setting) in which the unique c-axis is aligned with the polymer backbone in accord with accepted use in polymer systems. The largest facets are (010) planes and the polymer cleaves easily in this plane. The cleaved faces have been examined by scanning electron microscopy and have been found, for the best samples, to be free of macroscopic defects. These samples are ideal for Raman spectroscopy since the intense specularly reflected beam can easily be directed away from the spectrometer aperture. Samples with morphologies varying from amorphous to partially crystalline were obtained by dissolving partially polymerized monomer crystals in acetone and extracting the insoluble residues.

Crystals of PUDD polymer obtained by radiation polymerization were provided by Dr. R. H. Baughman of the Allied Chemical Co. These appeared less perfect than the TSHD crystals but had flat specular faces large enough for Raman spectroscopy. 1-hydroxy-2,4-hexadiyne was prepared by the method due to Wegner.[16] This monomer sublimes readily at room temperature and loss of monomer during thermal polymerization produced amorphous polymer samples.

The region of strong optical absorption of diacetylene polymers spans the visible region as do the lines from a krypton-argon mixed-gas ion laser. These initial studies were made with such a laser used in conjunction with a Cary Raman spectrometer at the Chemistry Department, Imperial College, London. This instrument is operated as part of the University of London Intercollegiate Research Service. All spectra were obtained with the samples at room temperature. For crystals, light polarized parallel to the polymer chain will mostly be reflected, while perpendicularly polarized light will be scattered from the bulk of the crystal. Since the crystals do not decompose rapidly until heated above 200 °C little damage is expected in the laser beam. Typical input powers were from 100 to 200 milliwatt but up to 600 mW was used without significant damage to the crystal. Amorphous and partially crystalline samples were observed to decompose at high power levels. Spectra were recorded at a resolution of 5 cm^{-1} over the range from 100 to 4000 cm^{-1} from the exciting line. The intense laser lines at 647·1, 568·2, 530·8 and 520·8 nm were used to excite the Raman spectra. The input beam was polarized. The scattered light was collected using a large aperture condensing lens so that the scattering direction and polarization was not defined. A back-scattering geometry was used as this simplifies the corrections for absorption in the sample. The large collecting aperture meant, however, that such corrections could only be approximate.

8.5 RESONANT RAMAN SCATTERING FROM SINGLE CRYSTAL DIACETYLENE POLYMERS

The reflection spectrum of the TSHD polymer, reported elsewhere,[11] shows clearly resolved vibrational structure at room temperature. The damping constant, δ_e, will, therefore, be quite small and we would expect a strong resonant Raman effect. The spectra observed with 647·1 and 568·2 nm laser lines are shown in Figure 8.4. The incident light was polarized along the polymer chain. It is immediately obvious that the spectra contain only a few very intense lines as expected for a resonant Raman process.

Figure 8.5 shows the absorption spectra of a number of different TSHD polymer samples with different morphologies. The full lines show the absorption spectrum for a single crystal film, cast between microscope slides, for light polarized both parallel and perpendicular to the polymer backbone. The dotted line is the spectrum of a thin cleaved crystal with parallel polarization. The vibrational spectrum does not stand out clearly since, at the low light levels observed, stray light reduces contrast in the spectrum and limits the accuracy of the measurements. The four laser lines used to excite the Raman spectra are indicated by the arrows at the top of the figure. The 647·1 nm line is just on the absorption edge so that it occurs in the resonant Raman scattering regime. The other lines fall in the region of the vibrationally excited states and additional

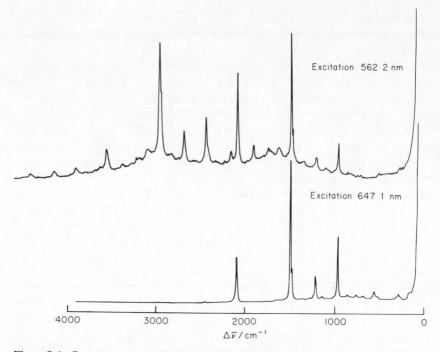

Excitation 562·2 nm

Excitation 647·1 nm

4000	3000	2000	1000	0

$\Delta \bar{\nu} / \mathrm{cm}^{-1}$

Figure 8.4   Raman spectrum of TSHD polymer single crystal for laser wavelengths of 647·1 nm, lower curve, and 568·2 nm, upper curve. Incident light polarized parallel to the polymer chain, scattered light unpolarized

effects due to fluorescence processes can occur. We can also see that for the 647·1 nm line the Stokes Raman lines fall in the region where the samples are relatively transparent. The Stokes lines for the other exciting wavelengths will fall in the region of strong sample absorption. The absorption will also affect the depth to which the laser light can penetrate the sample. Thus, to compare the scattering intensities corrections must be applied for self-absorption by the sample and changes in the scattering volume.

We will first consider the individual spectra. With 647·1 nm excitation the spectrum contains four intense components with frequencies of 955, 1205, 1487 and 2088 cm^{-1}, all with a precision of ± 2 cm^{-1}. We can clearly identify the two highest frequencies with the A_g modes 5 and 6 of Figure 8.3a. The frequencies we measure for these modes are in close agreement with the data of Baughman et al.[17,18] As can be seen from Table 8.1 the calculated frequencies are somewhat higher than the observed frequencies but this is to be expected since bond interaction and electron delocalization will both lead to a lowering in effective force constants and frequencies.[18] The absence of strong features above 2100 cm^{-1} does not conclusively prove the absence of combination lines in this case as the instrumental sensitivity falls off rapidly in this region. The assignment of the other modes is less clear. It would appear, however, that our choice of

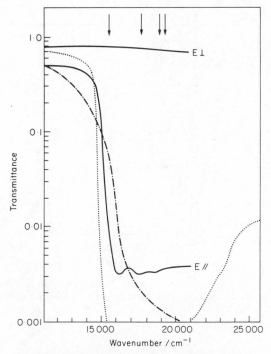

Figure 8.5 Absorption spectra of TSHD polymer samples with different morphologies. Full curves, thin single crystal between glass slides for light polarized parallel and perpendicular to the plane of the polymer backbone. Dotted curve, cleaved crystal flake for light polarized along the polymer chain. Chain curve, amorphous powder extracted from partially polymerized monomer

force constants has fortuitously given a correct frequency for A_g mode 4 although our choice of bending force constant has placed mode 3 too low, as indicated in Table 8.1. There are no obvious lines corresponding to A_g modes 1 and 2. In an attempt to distinguish resonant lines from any non-resonant lines in the spectrum spectra were recorded with the incident light polarized parallel and perpendicular to the polymer chain. The results are shown in Figure 8.6. The intensity of all the Raman lines is reduced for the perpendicular polarization but the spectrum is still a resonant Raman spectrum. This is because although the incident light was polarized perpendicular to the polymer chain it was not polarized perpendicular to the plane of the backbone, as was the case[10,11] for the sample shown in Figure 8.5. The polymer backbone lies at an angle to the (010) cleavage face[9] so this condition could not be met with normal incidence as used in this work. Thus, the absorption remains quite high and resonant scattering is still observed. The differences in intensity of the four strong lines

102

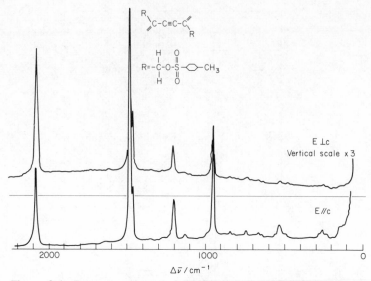

Figure 8.6 Raman spectrum of TSHD polymer single crystal with 647·1 nm excitation for incident light polarized parallel to the polymer chain, lower curve; and perpendicular to the chain, upper curve

reflect the differences in the polarizability tensors for these vibrations. However, all the lines are resonantly enhanced. This suggests that most of the weak lines result from side-group normal modes which will have small but finite displacements of the backbone atoms. These only appear in our model as A_g and B_u modes 1 in Figure 8.3. We can tentatively assign A_g mode 2 to the line at 535 cm^{-1} as this is the most intense of the weak lines and the remaining lines to modes equivalent to mode 1 for the more complex side-groups.

The 568·2 nm laser line is close to the phonon side-band produced by the double-bond stretching mode in the optical spectrum[11]; see Figure 8.5. Thus, all single-phonon Raman lines will be subject to sample absorption. Despite this they are clearly visible together with numerous multi-phonon lines, as shown in Figure 8.4. The multi-phonon lines are mainly combinations with the harmonics of the double-bond stretching mode. This is illustrated in Figure 8.7 where the spectrum is plotted using the double bond harmonics as origin. Two additional sharp lines at 2158 and 1910 cm^{-1} have the frequencies expected for a combination of the 1205 and 955 cm^{-1} lines and twice 955 cm^{-1}. Two broad fluorescence bands are clearly visible and there are a number of additional weak medium-width peaks which could originate from fluorescence processes.

The spectra recorded using the 530·8 and 520·8 nm laser emissions suffer from severe sample absorption. In general they have more fluorescent background and the differences in intensity of the major and minor features observed for longer wavelength excitation are much smaller. However, distinct changes occur in the spectra recorded at these two wavelengths even though the change

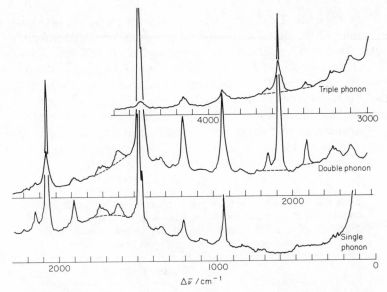

Figure 8.7 Raman spectrum of TSHD polymer single crystal with 568·2 nm excitation redrawn to show the origin of the main high-frequency components as combinations with the harmonics of the double-bond stretching vibration

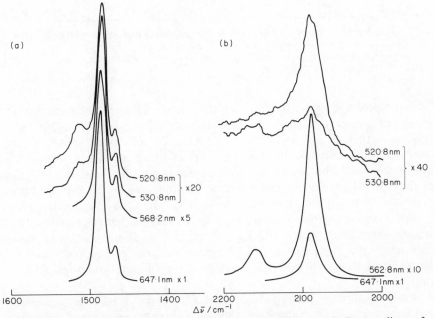

Figure 8.8 (a) Double-bond and (b) triple-bond stretching-mode Raman lines of TSHD polymer single-crystal recorded with the four laser lines at 647·1, 568·2, 530·8 and 520·8 nm. Incident light polarized parallel to the polymer chain. Excitation wavelength and scale magnification are shown by each spectrum

in wavelength is small. This is illustrated in Figure 8.8 which shows the spectra in the vicinity of the double- and triple-bond stretching modes for the four laser lines. The relative scattering intensities were obtained from the measured intensities shown in Figure 8.8. Corrections were applied for photomultiplier response, excitation level, the v^4 dependence of the scattering (see equation (1)) and the sample absorption. For a back-scattering geometry the absorption correction is just the sum of the absorption coefficients at the incident and scattered frequencies. As noted above, absolute absorption coefficients are difficult to determine so approximate relative values deduced from the absorption curves were used. The corrected intensities are listed in Table 8.2.

Table 8.2 Relative resonant Raman scattering intensities for the double and triple bond stretching vibrations of TSHD polymer single crystal as a function of incident wavelength. Incident light polarized parallel to the polymer chain.

Exciting wavelength (nm)	Relative intensities	
	1487 cm^{-1} line	2088 cm^{-1} line
647·1[a]	1·0	0·42
568·2	12·1	4·0
530·8	1·6	0·24
520·8	1·7	0·48

[a] The 647·1 nm exciting line falls on the absorption edge of the TSHD polymer; the absorption coefficient is thus liable to a large error. The relative intensities are liable to a similar error with respect to the values for the other exciting lines.

These spectra show all the features characteristic of resonant Raman scattering as outlined in section 8.2. Similar results are observed for PUDD polymer crystals as indicated in Figures 8.9 and 8.10. The observed frequencies and assignments are listed in Table 8.1. The crystal structure of this polymer has not been reported but we can see that the spectrum correlates well with that of the TSHD polymer. Modes 5 and 6, which are predominantly backbone modes, have nearly the same frequencies while the remaining modes, which involve some side-group motion, are all shifted to higher frequencies. Thus, we can conclude that the PUDD polymer has the same backbone bond-sequence as the TSHD polymer. Changes in frequency of the backbone bond-stretching modes can be correlated with bond length using established relationships.[18]

Before considering the application of this method to less perfect diacetylene polymers we wish to comment further on the scattering process in crystals. The model we have employed to discuss this effect was a simple molecular one. The polymer spectra fit this mode, but it is likely that their true explanation is more complex. In general a system with delocalized π-electrons should have a

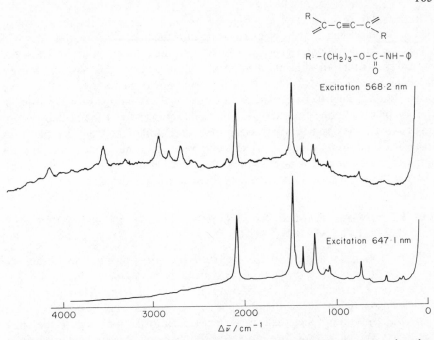

Figure 8.9 Raman spectrum of PUDD polymer single crystal for laser wavelengths of 647·1 nm, lower curve; and 568·2 nm, upper curve. Incident light polarized parallel to the polymer chain, scattered light unpolarized

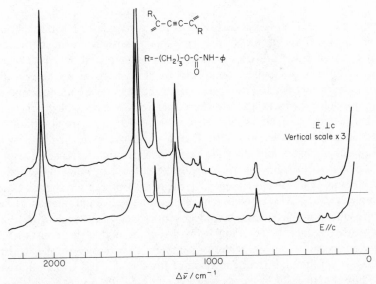

Figure 8.10 Raman spectrum of PUDD polymer single crystal with 647·1 nm excitation for incident light polarized parallel to the polymer chain, lower curve; and perpendicular to the chain, upper curve

featureless absorption edge as observed in most semi-conductors. However, the poly-diacetylenes are essentially one-dimensional systems and this will affect the band edge and give rise to distinct phonon side-bands.[10] Thus, the optical spectrum looks like many molecular spectra but its origin is more subtle. Similarly the resonant Raman spectra fit the molecular model but their true origin is also likely to be explained in terms of the one-dimensional nature of the polymer backbone. Indeed the resonant Raman scattering in the diacetylenes gives us a unique opportunity to study the interaction of electrons and phonons in a one-dimensional situation. To this end we are continuing measurements on single crystals using a dye laser as the source of exciting radiation so that we can observe the wavelength dependence of the scattering in detail.

8.6 RESONANT RAMAN SCATTERING FROM AMORPHOUS DIACETYLENE POLYMERS

Though the resonant Raman spectrum provides information about the structure of crystalline diacetylene polymers more detailed information can always be obtained by X-ray structural determinations. However, some diacetylenes are produced in amorphous or quasi-crystalline forms. Direct X-ray methods cannot then easily provide precise information about the polymer backbone. We have, therefore, observed the resonant Raman scattering from such samples to assess its utility in providing structural information under these circumstances. We chose initially to work with samples of TSHD polymer obtained by solvent extraction from partially polymerized crystals as we have made a careful study of such samples by X-ray diffraction and scanning electron microscopy.[19]

Samples of TSHD polymer extracted at low conversions are amorphous and the crystallinity of the samples is found to increase with increasing conversion.[19] The absorption spectrum of an amorphous extract is shown in Figure 8.5. The absorption edge has been smoothed out but sharpens as the crystallinity of the samples increases. Thus, we expect the resonant enhancement of the Raman lines observed with the 647·1 nm laser line to increase as the crystallinity increases. Figure 8.11 shows the spectra recorded for samples obtained at 2%, 5% and 98% conversion. With increasing conversion the intensity of the Raman lines increases and the fluorescent background decreases. Despite the large change in morphology the frequencies of the prominent Raman lines do not change. This is in agreement with the X-ray observation that all extracted and single crystal TSHD polymers have the same lattice parameters. However, since the vibrational frequencies are less sensitive to long-range order than the X-ray measurements the resonant Raman spectra enable this constancy to be established even in the amorphous samples. For these samples additional peaks occur on the high-frequency side of the double- and triple-bond stretch peaks. Baughman et al.[18] have observed similar shifts for other diacetylenes modified by extraction or recrystallization which they attribute to electron delocalization. In this case we prefer an alternative explanation. Since the

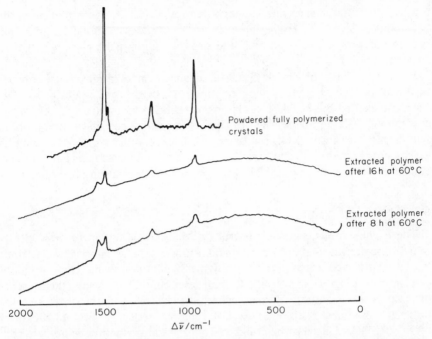

Figure 8.11 Raman spectrum of TSHD polymer extracted during polymerization with 647·1 nm excitation. Lower curve, sample extracted at 2% conversion; centre curve, sample extracted at 5% conversion; upper curve, sample extracted at 98% conversion

polymerization proceeds through a series of solid solutions[19–21] at low conversions the polymer chains occur in isolation and collapse on removal from the monomer lattice. Thus, though some of the backbone will retain the original *trans*-conformation other conformations will occur. The appearance of distinct

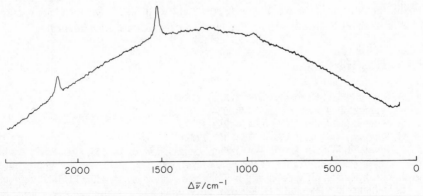

Figure 8.12 Raman spectrum of the amorphous polymer, obtained by thermal polymerization of 1-hydroxy-2,4-hexadiyne, with 647·1 nm excitation

peaks in the Raman spectrum suggests that one distorted conformation of the backbone is favoured. Higher frequencies imply shorter bond lengths and, because of the bulkiness of the side groups, a non-planar- rather than a *cis*-conformation.

Finally we have observed the Raman spectrum of amorphous polymer obtained by the thermal polymerization of 1-hydroxy-2,4-hexadiyne. The spectrum excited by the 647·1 nm laser line is shown in Figure 8.12. As for the amorphous TSHD polymer the principal Raman lines can be seen on a broad fluorescent background. On the basis of our observations of TSHD samples we can conclude that this polymer has the same bonding sequence in its backbone and, despite the low crystallinity, that one conformation is dominant.

8.7 CONCLUSIONS

The Raman spectra of single crystals of diacetylene polymers show clearly that the scattering is due to a resonant Raman process. Since the electronic transition involved is localized in the polymer backbone only backbone phonons can contribute significantly to the Raman scattering. Thus, despite the presence of complex side groups the Raman technique probes directly the structure of the polymer backbone. This enables the bonding sequence of the backbone ($=RC-C\equiv C-CR=$) to be confirmed in materials which have not yet been subjected to a full X-ray structural determination. It also enables the structure and conformation of the backbone to be studied in amorphous and poorly crystalline samples such as those obtained by the extraction of polymer from partially polymerized materials. Raman spectroscopy will be an important tool in structural investigations of diacetylene polymers and it will also allow the interaction of electronic excitations and phonons in a one-dimensional system to be studied in detail.

We wish to thank Mr. A. Hustings for recording the Raman spectra, Dr. R. H. Baughman for providing the PUDD polymer crystals and Dr. E. G. Wilson for useful discussions. This work was supported by an S.R.C. Major Centre grant. One of us (F.H.P.) thanks the S.R.C. for a postgraduate studentship.

8.8 REFERENCES

1. A. Baeyer and L. Landsberg, *Ber.*, **15,** 57, (1882).
2. G. Wegner, *Z. Naturforschung*, **24b,** 824, (1969).
3. G. Wegner, *Makromol. Chem.*, **134,** 219, (1970).
4. G. Wegner, *Makromol. Chem.*, **154,** 35, (1972).
5. J. Kaiser, G. Wegner and E. W. Fischer, *Israel J. Chem.*, **10,** 157, (1972).
6. K. Takeda and G. Wegner, *Makromol. Chem.*, **160,** 349, (1972).
7. J. Kiji, J. Kaiser, G. Wegner and R. C. Schultz, *Polymer*, **14,** 433, (1973).
8. E. Hadicke, E. C. Mez, C. H. Krauch, G. Wegner and J. Kaiser, *Angew. Chem. Int. Edit.*, **10,** 266, (1971).
9. D. Kobelt and E. F. Paulus, *Acta Cryst.*, **B30,** 232, (1974).

10. E. G. Wilson, *J. Phys.*, *C*, *Solid State Phys.*, **8**, 727, (1975).
11. D. Bloor, D. J. Ando, F. H. Preston and G. C. Stevens, *Chem. Phys. Letters*, **24**, 407, (1974).
12. J. Behringer, *Raman Spectroscopy—Theory and Practice*, Ed. H. A. Szymanski, Plenum Press, 1967, p. 168.
13. J. Behringer, loc. cit. p. 174 and 209.
14. G. Wegner, *Makromol. Chem.*, **145**, 85, (1971).
15. D. Bloor and G. C. Stevens, *J. Polymer Sci.*, *Polymer Phys.*, in press.
16. G. Wegner, private communication.
17. A. J. Melveger and R. H. Baughman, *J. Polymer Sci.*, *Polymer Phys.*, **11**, 603, (1973).
18. R. H. Baughman, J. D. Witt and K. C. Yee, *J. Chem. Phys.*, **60**, 4755, (1974).
19. L. Koski, G. C. Stevens, F. H. Preston, D. J. Ando and D. Bloor, *J. Materials Sci.*, in press.
20. R. H. Baughman, *J. Appl. Phys.*, **43**, 4362, (1972).
21. R. H. Baughman, *J. Polymer Sci.*, *Polymer Phys.*, **12**, 1511, (1974).

9

The Application of ESCA to Studies of Structure and Bonding in Polymers

D. T. Clark

University of Durham

9.1 INTRODUCTION

The past few years have witnessed an explosive growth in the application of the relatively newly developed technique of Electron Spectroscopy for Chemical Application (ESCA)* to studies of structure and bonding across a broad front in chemistry, physics and metallurgy.[1,2]

These studies have demonstrated that ESCA is an extremely powerful tool for investigations of structure and bonding with an information content per spectrum unsurpassed by any other spectroscopic technique. This distinctive attribute confers upon ESCA wide-ranging applicability and versatility in respect of studies on polymeric systems[3] and it is the purpose of this article to enlarge upon this theme.

It will become apparent that there are areas of study in which the required information can only at present be derived from ESCA studies, while in others the technique nicely complements the more established spectroscopic tools. In general, however, ESCA provides data at a much coarser level than most other spectroscopic tools and information pertaining for example to conformational effects may only be inferred rather indirectly. In many areas of application ESCA does not compare favourably in terms of resolution, sensitivity etc. with more established spectroscopic tools. The fact remains, however, that this is more than compensated by the great range of information available from a single ESCA experiment such that in the future one can envisage that ESCA will be the technique of choice for any initial investigation of a polymer sample. The ability to provide information straightforwardly on uncharacterized samples is unique to ESCA and gives the technique great potential (already exploited in some areas) for tackling not only academic problems but those of an applied 'trouble-shooting' nature.

The application of ESCA to studies of structure and bonding in polymers has largely been pioneered at Durham, however the field is already so large that in a short article such as this, it is not possible to present a complete picture.

---

* Originally designated '... for Chemical Analysis'.

Two primary objectives have therefore been kept to the fore in this review. Firstly, to delineate the current areas of applicability of the technique and hence to give the uninitiated some idea of the sort of problem to which ESCA might readily be applied. Secondly to indicate some of the likely 'growth areas' over the next few years. As a preliminary to this however, a brief introduction is given to ESCA as a spectroscopic technique together with examples of the sorts of information which may be derived from the ESCA experiments.

9.2 FUNDAMENTALS OF ESCA

9.2.1 Introduction

In common with most other spectroscopic methods ESCA is a technique originally developed by physicists and now gradually being taken over by chemists to be developed to its full potential as a tool for investigating structure and bonding. The technique has largely been developed by Professor Kai Siegbahn and his collaborators at the University of Uppsala over the past 20 years or so[1] and much of the early work has been extensively documented in the 'first ESCA book'. It is only within the last decade however, that the potential of the technique has been revealed with the development of spectrometers of sufficient resolution and sensitivity.[4] The field has been opened to chemists with the advent of commercially produced instruments (at the last count there were ten instrument manufacturers). In addition to the aesthetically pleasing designation as 'Electron Spectroscopy for Chemical Applications' originally coined by Siegbahn,[1a] the technique is also variously known as X-ray Photoelectron Spectroscopy (XPS), High Energy Photoelectron Spectroscopy (HEPS), Induced Electron Emission Spectroscopy (IEES), and Photoelectron Spectroscopy of the Inner Shell (PESIS).

The principal advantages of the technique may be summarized as follows:

(i) The sample may be solid, liquid or gas (it is as easy to study a high molecular weight polymer as it is to study a gas) and the technique is essentially non-destructive. One noticeable exception to this generalization is poly(thiocarbonyl fluoride) which depolymerizes rather rapidly under X-irradiation.

(ii) The sample requirement is modest, in favourable cases 1 mg of solid, $0.1\ \mu l$ of liquid or $0.5\ cm^3$ of gas (at s.t.p.). These sample requirements are based on the minimum readily handled with conventional techniques. As will become apparent the sensitivity of the technique is such that a fraction of a monolayer coverage may be detected.

(iii) The technique has high sensitivity and is independent of the spin properties of any nucleus and is applicable in principle to any element of the periodic table. H and He are exceptions, being the only elements for which the core levels are also the valence levels.

(iv) The information it gives is directly related to the electronic structure of a molecule and the theoretical interpretation is relatively straightforward.

(v) Information can be obtained on both the core and valence energy levels of molecules.

These particular advantages of ESCA as a technique make it eminently suitable for the study of polymers.

9.2.2 Properties of core orbitals

A clearer understanding of ESCA as a technique is obtained with some knowledge of the properties of core orbitals.

Traditionally chemists have discussed the electronic structure of molecules, dealing only with the valence electrons and neglecting inner shell or core electrons, the reasons for this being:

(i) core electrons are not explicitly involved in bonding (although most of the *total* energy of a molecule resides in the core electrons);
(ii) it is only in the past ten years that sufficient computing capability has become available to allow non-empirical quantum mechanical calculations on molecules in which the core electrons are explicitly considered.

It has become clear however, that although core electrons are not involved in bonding, the core energy levels of a molecule encode a considerable amount of information concerning structure and bonding.

We may summarize the important characteristics of core as opposed to valence orbitals as follows. Core orbitals are essentially localized on atoms, their energies are characteristic for a given element and are sensitive to the electronic environment of an atom. Thus for a given core level of a given element, whilst the absolute binding energy for that level is characteristic for the element (Table 8.1), differences in electronic environment of a given atom in a molecule give rise to a small range of binding energies (i.e. 'shifts' in binding energies are apparent) often characteristic of a particular structural feature.

Table 9.1 Approximate core binding energies for first and second row elements (in eV)

	Li	Be	B	C	N	O	F	Ne
1s	55	111	188	284	399	532	686	867

	Na	Mg	Al	Si	P	S	Cl	Ar
1s	1072	1305	1560	1839	2149	2472	2823	3203
2s	63	89	118	149	189	229	270	320
$2p_{\frac{1}{2}}$			74	100	136	165	202	247
	31	52						
$2p_{\frac{3}{2}}$			73	99	135	164	200	245

9.2.3 The ESCA experiment

ESCA involves the measurement of binding energies of electrons ejected by interactions of a molecule with a monoenergetic beam of soft X-rays. For a variety of reasons the most commonly employed X-ray sources are Al $K\alpha_{1,2}$

114

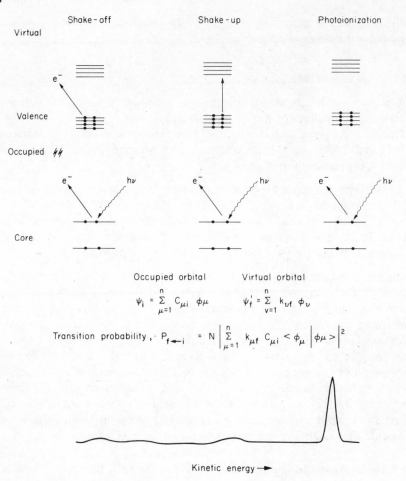

Figure 9.1 Shake-off, shake-up and photoionization processes (schematic). The three peaks in the kinetic energy curve correspond to the three processes

and Mg $K\alpha_{1,2}$ with corresponding photon energies of 1486·6 eV and 1253·7 eV respectively. In principle all electrons, from the core to the valence levels, can be studied and in this respect the technique differs from ultraviolet photo-electron spectroscopy (UPS) in which only the lower energy valence levels can be studied. The basic processes involved in ESCA are shown in Figure 9.1.

The removal of a core electron (which is almost completely shielding as far as the valence electrons are concerned) is accompanied by reorganization of the valence electrons in response to the effective increase in nuclear charge. This perturbation gives rise to a finite probability for photo-ionization to be accompanied by simultaneous excitation of a valence electron from an occupied to an unoccupied level (shake-up) or ionization of a valence electron (shake-off).[5] These processes giving rise to satellites to the low kinetic energy side of the

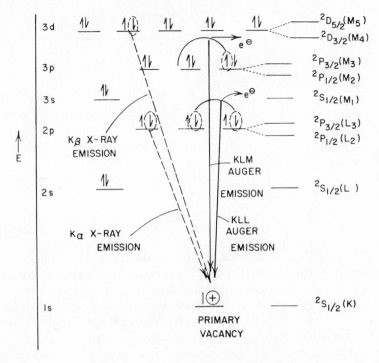

Figure 9.2 De-excitation of core hole states

main photoionization peak, follow monopole selection rules and may well be of considerable importance in the future in elucidating particular aspects of structure and bonding in polymer systems.*

De-excitation of the hole state can occur (Figure 9.2) via both fluorescence and Auger processes for elements of low atom number, the latter in fact being more probable.

Typically the lifetimes of the core hole states involved in ESCA are in the range $10^{-13} \sim 10^{-17}$ s emphasizing the extremely short time-scales involved in ESCA compared with molecular vibrations, and the process may fairly be called sudden with respect to the nuclear (but not electronic) motions. The basic experimental set up for ESCA is shown in Figure 9.3, and is largely self-explanatory. Most commercial instruments employ double-focussing electro-static analysers with a retarding lens system. Polymer samples are conveniently studies as films or powders mounted on a sample probe which may be taken into the spectrometer from atmosphere by means of insertion locks. Samples may thus be readily mounted and inserted into the spectrometer greatly facilitating

* There is a close analogy between shake-up processes for hole states and the more familiar electronic transitions for the neutral system as, for example, conventionally observed with an ultraviolet/visible spectrometer. However, whilst the former follow monopole selection rules (the computation of intensities therefore depending on overlap as indicated in Figure 9.1), the latter follow dipole selection rules.

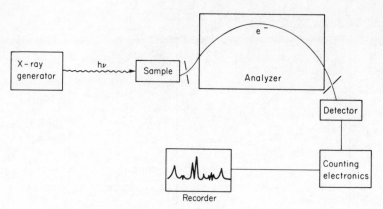

Figure 9.3 Basic experimental set up for ESCA (schematic)

routine analyses. Provision is usually made for the heating or cooling of samples *in situ* and an ancillary sample preparation chamber allows greater flexibility in terms of sample preparation or pretreatment (e.g. argon ion bombardment, electron bombardment, ultraviolet irradiation, chemical treatment such as oxidation etc.). Addition of a quadrupole mass spectrometer facilitates degradation studies and allows close control to be kept of the extraneous atmospheres in the sample region of the spectrometer. Typical operating pressures in the sample region are less than 10^{-7} Torr.

9.2.4 Photoelectric cross-sections

The probability for photoemission from a given core level depends on the photoelectric cross-section which represents the transition probability per unit time for exciting a photoelectron from a subshell with a photon flux of a given energy. Cross-sections are markedly dependent on photon energy and for conventional X-ray sources (e.g. Al $K\alpha_{1,2}$, Mg $K\alpha_{1,2}$) are such that for first row atoms for example, cross-sections are much larger for the core levels than for the valence levels. This is illustrated schematically in Figure 9.4.[6] It is clear also that cross-sections at a given photon energy depend on the symmetry of the orbitals involved and this is particularly useful in unravelling the complexities of the valence bands of molecular solids and crystals.

For example, it is clear from Figure 9.4 that the differential cross-section for photoionization of the 2s level of carbon relative to the 2p levels is larger at higher photon energies. This gives a convenient means for distinguishing between σ and π type valence orbitals in organic systems by monitoring the relative changes in peak intensities in going from a low energy (He I 21·2 eV to high energy Mg $K\alpha_{1,2}$ 1253·7 eV) photon source. In addition to being dependent on photon energy, cross-sections may also exhibit angular dependence. Two types of angular dependence studies may be envisaged and these are shown in Figure 9.5. In the first the analyzer and X-ray source are in a fixed

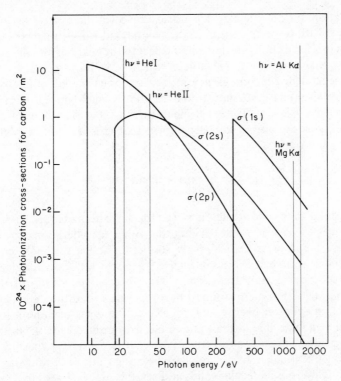

Figure 9.4 Calculated sub-shell photoionization cross-sections for carbon

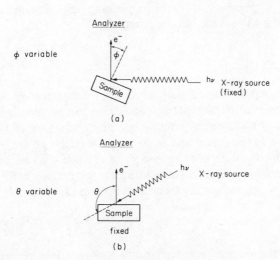

Figure 9.5 Angular distribution measurements (schematic)

configuration and the angular dependence involves the relative orientation of the sample with respect to these. In the second, more difficult arrangement the sample is in a fixed configuration with respect to the analyser and the position of the X-ray source is varied. The studies which have been reported to data on angular dependence of photoelectric cross-sections have predominantly referred to the gas phase and have demonstrated the utility of such studies for assignments of valence energy levels.[7] The experimental arrangement has then corresponded to (b) in Figure 9.5 and it has been shown that cross-sections conform to an equation of the form[5]

$$\frac{d\sigma}{d\Lambda} = A + B \sin^2 \theta$$

For molecular solids and crystals few attempts have been made to study angular dependences the reason for this being that other considerations considerably complicate the analysis.[8] For ESCA studies of solids the most convenient experimental arrangement is (a) in Figure 9.5 in which the angular dependence is introduced by rotating the sample. The observed intensities of the peaks corresponding to photoionization from core levels, in addition to being directly dependent on photon energy and angle (via the cross-section), also exhibit angular dependence due to photon and electron mean-free-paths in the solid; also if oriented single crystals are involved there may be specific electron channelling phenomena. In general, therefore, the study of the angular dependence of core levels in molecular solids and crystals may be stated to be still in its infancy. The most usual experimental arrangement is therefore merely to orient the sample to optimize the count rate on a given core level (e.g. C_{1s}, O_{1s}, F_{1s} etc.) and then operate at this fixed angle φ. In our particular experiments φ has typically between $\sim 30°$ or $45°$ depending upon whether a conventional or Henke type X-ray source has been employed.

9.2.5 Energy considerations

For photo-ejection of a core electron in a solid sample of an insulator (in electrical contact with the spectrometer) the energy considerations in the measurements of the electron binding energies are shown in Figure 9.6. The reference is the Fermi Level and if the work function of the spectrometer is known then the absolute binding energy of a given level may be calculated. This will differ from the absolute binding energy defined with respect to the vacuum level by approximately the work function of the sample.

Conservation of energy requires that

$$hv = E_b + \varphi_{sp} + T_{sp} + E_r \tag{1}$$

where hv is the photon energy, E_b the binding energy of the core level (Fermi Level as reference), φ_{sp} is the spectrometer work function, T_{sp} is the measured kinetic energy of the photoemitted electron and E_r is a recoil energy. Conservation of

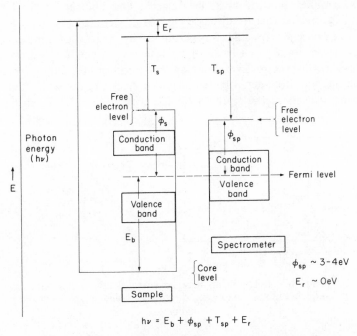

Figure 9.6 Energy considerations in measurement of electron binding energies

momentum dictates that for the commonly used X-ray sources E_r may effectively be ignored for atoms other than hydrogen.[1a]

More often than not the work function of the sample is not known and therefore energy levels referenced with respect to the Fermi Level of the sample are operationally convenient. When we come to consider the theoretical analysis of shifts in core binding energies, shake-up transitions etc., the more convenient energy reference is the vacuum level for an assembly of non-interacting molecules in the gas phase. Relating theory and experiment might therefore be expected to lead to considerable difficulties more particularly since for semi-conductors and insulators it is not easy to locate the Fermi Level which lies somewhere between the predominantly filled valence bands and predominantly empty conduction bands. A further problem is that in studying essentially insulating samples the assumption of thermodynamic equilibrium during photoemission may not be valid since insufficient charge carriers may be available to maintain overall electroneutrality at the sample surface. So called 'sample-charging effects' commonly occur and depending on spectrometer design can amount to tens of eV but more typically are of the order of 1–2 eV. These arise from net positive charging of the specimen, causing apparent shifts in the binding energy scale (to higher binding energy, due to net retardation of the electrons before entering the spectrometer). If absolute binding energies are required therefore, sample charging must be detected and allowed for and reliable methods of achieving this are discussed in a later section.

At this stage it is worthwhile considering briefly the relationship between the binding energies for a given level for an assembly of isolated (e.g. gas phase) molecules with the vacuum level as reference (as appropriate for direct theoretical computation) and those actually measured for the molecular solid. This is best considered in terms of the Born cycle shown in Figure 9.7. In this case we

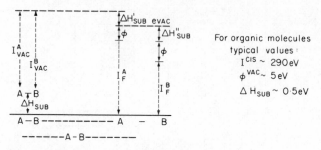

Figure 9.7 Relationship between energy levels of gaseous and solid samples (covalent, no strong intermolecular interactions)

consider a simple molecular solid AB with core levels of interest on fragments A and B. Considering firstly the right-hand side of Figure 9.7 the measured core binding energies for the solid with Fermi Level as reference are I_F^A and I_F^B. For the isolated molecules the measured binding energies (this may be hypothetical if AB represents a high molecular weight polymer) with respect to the vacuum level as energy reference, are I_{vac}^A and I_{vac}^B. In most cases the *shift* in binding energies are of prime interest rather than the absolute binding energies in which case it is clear from Figure 9.7 that the shifts for the two situations are related by differences of lattice sublimation terms involving the system AB with a core hole located on fragment A or B. Since typically for molecular solids, the absolute values of these terms would be expected to be small there should be a good correlation of shifts measured in the gas and solid phases. Studies of appropriate monomer systems in which intermolecular interactions are small shows that experimentally this is realized and computed *shifts* for the isolated molecules may be directly compared with experimentally determined shifts for the solids. In comparing one sample with another, however, further points arise, namely differences in work function, differences of lattice sublimation terms and any effects due to differences in sample charging (δ). For closely related materials the samples of which are either sufficiently thin for sufficient charge carriers to be available from say a metallic backing to maintain overall electroneutrality, or for which independent corrections for sample charging effects have been employed, a case can again be made for discussing shifts etc. in terms of an isolated molecule approach. As will become apparent the complexities of the model systems involved in quantitative discussions of core binding energies for polymers dictates that rigour be sacrificed to computational feasibility and the model employed is such that even if there are small differences in work function for polymers between which shifts are being compared it is still possible to use an isolated molecule approach to discuss the results, since such differences may be incorporated into the model parametrization. To put matters in perspective however, whereas the shift in core binding energy for say the C_{1s} levels of PTFE with respect to polyethylene is $\sim 7\cdot2$ eV, the work function difference is only $0\cdot85$ eV.

9.2.6 Linewidths

The measured linewidths for core levels (after taking into account spin orbit splittings if these are not resolved) may be expressed as:

$$(\Delta E_m)^2 = (\Delta E_x)^2 + (\Delta E_s)^2 + (\Delta E_{Cl})^2$$

where ΔE_m is the measured width at half height, the so-called full width at half maximum (FWHM), ΔE_x is the FWHM of the photon source, ΔE_s is the contribution to the FWHM due to the spectrometer (i.e. analyser), ΔE_{Cl} is the natural width of the core level under investigation. (For solids this includes solid state effects not directly associated with the lifetime of the hole state but rather with slightly differing binding energy due to differences in lattice environment.)

It has previously been pointed out that Mg $K\alpha_{1,2}$ and Al $K\alpha_{1,2}$ are the most useful photon sources from the standpoint of keeping the contribution of ΔE_x to the total linewidth small. With well-designed magnetic or electrostatic analysers the contribution ΔE_s can be reduced to negligible proportions so that the major limiting factors in terms of resolution are photon linewidths (which may be reduced by monochromatization) and the inherent width of the level itself. (For solids in which longer range interactions are important e.g. ionic lattices or hydrogen-bonded covalent solids, solid state effects can contribute to the overall linewidths.)

Table 9.2 Approximate natural widths of some core levels (in eV)[a]

Level		S	Ar	Ti	Mn	Cu	Mo	Ag	Au
					Atom				
	1s	0·35	0·5	0·8	1·05	1·5	5·0	7·5	54
	2p$_{\frac{3}{2}}$	0·10		0·25	0·35	0·5	1·7	2·2	4·4
Radiative widths	1s	0·04	0·07	0·2	0·33	0·65	3·6	6·0	50
Fluorescence yields	1s	0·1	0·14	0·22	0·31	0·43	0·72	0·8	0·93

[a] L. G. Parratt, *Rev. Mod. Phys.*, **31**, 616, (1959).

Some examples of natural linewidths (ΔE_{Cl}) derived from X-ray spectroscopic studies are given in Table 9.2. The uncertainty principle in the form

$$\Delta E\, \Delta t \geqslant \frac{h}{4\pi}$$

shows that for a hole state lifetime of $\sim 6.6 \times 10^{-16}$ s the linewidth, i.e. uncertainty in the energy of the state, is ~ 1 eV.

It is evident from Table 9.2 that there are large variations in natural linewidths both for different levels of the same element and for the same levels of different elements. These reflect differences in lifetimes of the hole state, the lifetime being a composite of radiative (fluorescence) and non-radiative (Auger) contributions, the importance of the former increasing with atomic number (approximately as Z^4). This is clearly shown in Table 9.2. This emphasizes the fact that there is no particular virtue in studying the innermost core level. For gold for example, the 1s level has a halfwidth of ~ 54 eV so that even if a monochromatic X-ray source with the requisite photon energy were available any subtle chemical shift effects we might wish to investigate would be swamped.

As has been previously indicated, lifetimes of core hole states are typically in the range 10^{-13}–10^{-17} s and the ESCA time scale can therefore be fairly described as sudden with respect to nuclear (but not electronic) motions. Very recent high resolution gas phase ESCA studies of simple molecules using a fine-focussed X-ray monochromation scheme has revealed that vibrational effects are indeed evident for core level photoionizations.[10] For molecular solids and crystals, however, it seems likely that these may well be masked and hence a detailed discussion of this fascinating area of the subject will not be given here.

For typical core levels of interest in the study of polymers, e.g. C_{1s}, N_{1s}, O_{1s} etc., and with conventional instrumentation employing unmonochromatized X-ray sources the single largest contribution to the overall linewidths arises from the inherent width of the X-ray source itself (Mg $K\alpha_{1,2} \sim 0.7$ eV, Al $K\alpha_{1,2} \sim 0.9$ eV).

9.2.7 Simple examples illustrating points discussed in 9.2.1–9.2.6

A typical wide-scan ESCA spectrum (i.e. plot of number of electrons of given kinetic energy arriving at the detector in unit time) is shown in Figure 9.8 for a PTFE sample.

The 'sharp' photoionization peaks due to F_{1s} and C_{1s} levels are readily identified by their characteristic binding energies. The group of three rather broader peaks whose kinetic energies are independent of the photon source are identified as Auger peaks arising from de-excitation of the F_{1s} hole states. As will become apparent one of the most important attributes of ESCA as a technique is the possibility of providing information relating to surface, sub-surface or essentially bulk properties, and this depends on differences in escape depths for photoemitted (or Auger) electrons corresponding to different

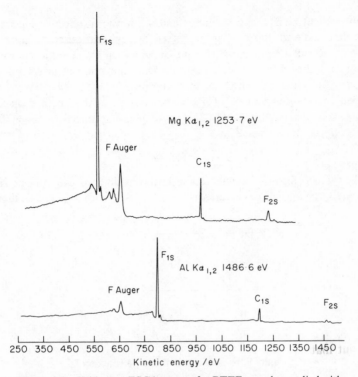

Figure 9.8   Wide scan ESCA spectra for PTFE samples studied with
Mg Kα$_{1,2}$ and Al Kα$_{1,2}$ radiation

kinetic energies. For the most commonly used soft X-ray sources the penetration
of the incident radiation for a solid sample is typically in excess of 1000 Å
(dependent on the angle of incidence etc.). However the number of photo
emitted electrons contributing to the elastic peak (i.e. corresponding to no
energy loss) is determined by the mean-free-path of the electron in the solid.
Thus in general the ESCA spectrum of a given core level consists of well-
resolved peaks corresponding to electrons escaping without undergoing energy
losses, superimposed on a background tailing to lower kinetic energy arising
from inelastically scattered electrons. (This is clearly evident in Figure 9.8.
With an unmonochromatized X-ray source as in this example, there is of course
a general contribution to the background arising from the bremsstrahlung.)

In the applications to be discussed below we may assume that the X-ray
beam is essentially unattenuated over the range of surface thickness from which
the photoelectrons emerge. The intensity of electrons of a given energy observed
in a homogeneous material may be expressed as

$$dI = F\alpha Nk \, e^{-x/\Lambda} \, dx$$

where F is the X-ray flux, α is the cross-section for photoionization in a given
shell of a given atom for a given X-ray energy, N is the number of atoms in

volume element, k is a spectrometer factor for the fraction of electrons that will be detected and depends on geometric factors and on counting efficiency, Λ is the electron mean-free-path and depends on the kinetic energy of the electron and the nature of the material that the photoelectron must travel through. The situation which is of common occurrence is that of a single homogeneous component or of a surface coating of thickness d on a homogeneous base. This is illustrated in Figure 9.9. The intensity for a film A of thickness d may be expressed as

$$I^A = I^A_\infty(1 - e^{-d/\Lambda_A})$$

whereas for the film B (considered infinitely thick) on the surface of which A is located, the intensity is given as $I^B = I^B_\infty e^{-d/\Lambda_B}$ (with $N_A = N_B$). Figure 9.10

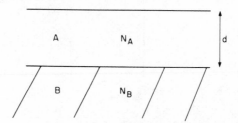

Figure 9.9   Peak intensities as a function of
film thickness

shows data pertaining to the escape-depth* dependence on kinetic energy for electrons, derived from Auger and ESCA studies on films of known thickness.[11] The striking feature clearly evident in Figure 9.10 is that the experimental data for solids, in which the contribution of various scattering processes might be expected to differ, fit closely onto a 'universal' curve of escape depth versus kinetic energy. The importance of this will become apparent in the ESCA studies relating to the differentiation of sub-surface and bulk phenomena, discussed in a later section. In general however, it is clear that for photon energy of 1253·7 eV or 1486·6 eV the mean-free-paths for photoemitted electrons should be <20 Å.

To put matters in perspective it is worth noting that for escape depths of 5 Å, 10 Å and 15 Å, 90% of the signal intensity from a homogeneous sample derives from the topmost 11·5 Å, 23·0 Å and 34·5 Å respectively.

The C_{1s} spectrum of benzotrifluoride (Figure 9.11) illustrates the large shifts in core levels which can occur. The electronegative fluorines withdraw electron density from the carbon of the trifluoromethyl group with concomitant large

---

* 'Mean-free-path' and 'escape depth' have been used interchangeably in this work; they both correspond to the depth at which 1/eth of the emitted electrons have not suffered any energy loss.

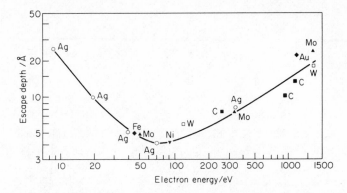

Figure 9.10 Escape-depth dependence on kinetic energy for electrons

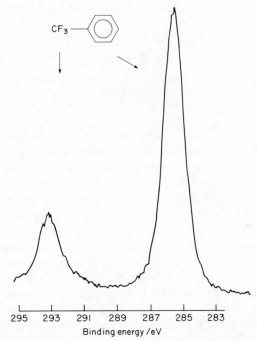

Figure 9.11 C_{1s} levels for benzotrifluoride

increase in C_{1s} core binding energy. As will become apparent a peak at ~ 294 eV binding energy may be used to 'fingerprint' CF_3 groups.

The C_{1s} and O_{1s} spectra of a film of poly(ethylene terephthalate) (Figure 9.12) illustrates that these effects may be as easily detected in polymeric as in monomeric samples, the O_{1s} spectrum clearly showing the two kinds of ester oxygen

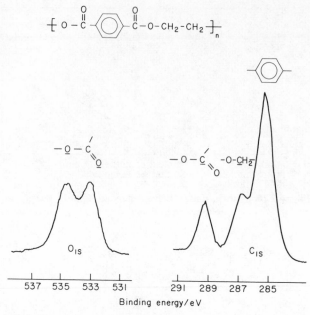

Figure 9.12 C_{1s} and O_{1s} levels for poly(ethylene terephthalate) films

and the C_{1s} spectra clearly distinguishing the ester, aliphatic and aromatic carbon levels. Again it can be seen that integrated intensities are directly proportional to the relative abundances of the various types of environment, namely $1:1$ for the oxygen levels and $1:1:3$ for the carbon levels.*

Compared with ^{13}C n.m.r. it should already be evident that ESCA is inferior with regard to shift range to linewidth ratio. In compensation however, it should be emphasized that ESCA has two distinctive advantages. Firstly, it is capable of differentiating surface from bulk effects and secondly, compared with n.m.r. the theoretical computation of shifts is infinitely more straightforward and the interpretation of data is therefore relatively simple. (The theory in fact shows that the diamagnetic contribution to the n.m.r. chemical shifts are related to the shifts in core binding energies.)[12]

An enlightening example of how the range of data levels available in an ESCA experiment compensates for the relative crudity of some of this data concerns simple hydrocarbon polymers. The C_{1s} binding energies for carbon in saturated (sp^3) and unsaturated (sp^2) environments are closely similar (typically within ± 0.2 eV). This is in striking contrast to ^{13}C n.m.r. where the shift range is of the order of 130 ppm.[13] In studying polyethylene, polypropylene

---

* This is not strictly correct since due allowance must be made for any shake-up or shake-off satellites. For most practical purposes however direct integration of the main photoionization peaks is an excellent approximation.

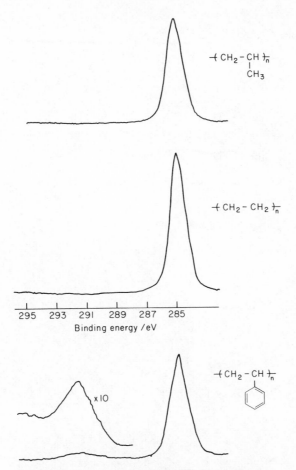

Figure 9.13 C_{1s} levels for polypropylene, polyethylene and polystyrene; showing shake-up peaks for polystyrene

and polystyrene for example, the C_{1s} core levels are almost identical, differing only slightly in the linewidths. They may however, be readily distinguished by studying the valence levels either directly or via shake-up satellites.

Figure 9.13 shows the C_{1s} levels for polyethylene, polypropylene and polystyrene. For polystyrene the C_{1s} core levels for the pendant aromatic groups are accompanied by $\pi^* \leftarrow \pi$ shake-up transitions readily distinguishing the polymer. As a further example, Figure 9.14 shows the C_{1s} levels for polystyrene, poly(1-vinylnaphthalene) and polyacenaphthylene. In each case a distinctive shake-up structure is apparent.

So far we have indicated the importance in ESCA measurements of, *inter alia*, the absolute binding energies and shifts in core levels, relative peak intensities, and their dependence on mean free path, kinetic energy, photoionization

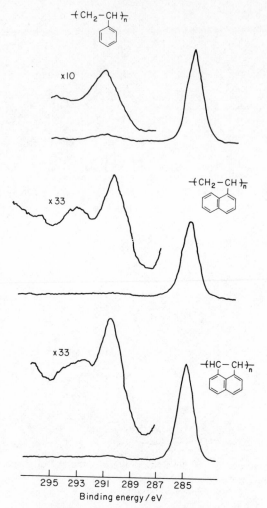

Figure 9.14 C_{1s} levels for polystyrene, poly-
(1-vinylnaphthalene) and polyacenaphthylene,
showing shake-up peaks

cross-sections etc. A further possible source of information arises from paramagnetic species due to multiplet splittings. One of the first demonstrations
of this, due to Siegbahn and coworkers, is illustrated in Figure 9.15 for the
paramagnetic NO and O_2 molecules.[1b] The theory in the cases of core ionization
from S levels is particularly simple as outlined in Figure 9.16. In the case of
NO the single unpaired electron is delocalized so that the magnitude of the
multiplet splitting of the O_{1s} and N_{1s} core levels will depend upon the unpaired
spin densities at the two atoms. The pronounced satellite for nitrogen indicates

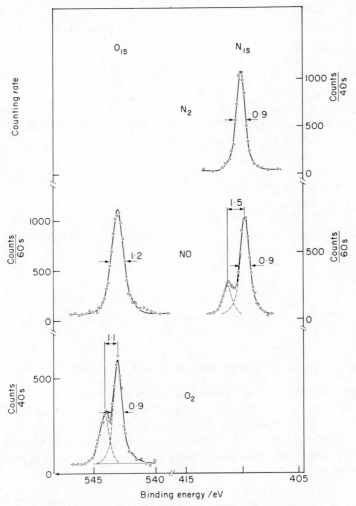

Figure 9.15 O_{1s} and N_{1s} levels showing multiplet splittings for the paramagnetic NO and O_2 molecules

that most of the unpaired spin density is on nitrogen and hence the magnitude of multiplet splittings can provide information comparable to that from e.s.r. studies. The best developed application of multiplet effects is in transition metal chemistry originally pioneered by Fadley and co-workers.[5] Possible applications in the polymer field might be the study of radicals localized at the surfaces of polymers or the incorporation of transition metal catalysts into the polymer chain.

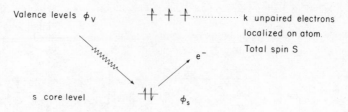

Valence levels ϕ_V — k unpaired electrons localized on atom. Total spin S

e^-

s core level — ϕ_s

If orbital contribution for valence levels is zero, energy difference between $(S + 1/2)$ and $(S - 1/2)$ states is

$$\Delta E = (2S + 1) K_{vs} \quad \text{where}$$

$$K_{vs} = \left\langle \phi_s (1) \, \phi_v (2) \left| \frac{1}{r_{12}} \right| \phi_s (2) \, \phi_v (1) \right\rangle$$

e.g. $k = 2$ $S = 1$ Two states

$$^4S_{3/2} \qquad ^2S_{1/2}$$

$$\Delta E = 3K_{vs}$$

Intensity ratio $2 : 1$

Figure 9.16 Multiplet splittings for photoionization from S hole states

9.2.8 Basic types of information available from ESCA studies

Although the emphasis in the discussion above has been in respect of core levels it should be pointed out that valence levels may be studied directly also. For molecular systems involving first row atoms however, the photoionization cross-sections are one or two orders of magnitude smaller than for core ionizations. The basic types of information available from ESCA studies are summarized in Table 9.3.

Table 9.3 Information from ESCA studies

Core Electrons	Valence Electrons
(1) Binding energy characteristic of a given level of a given element; therefore useful for analysis. Different KE dependences for different core levels provides means for analytical depth-profiles	(1) Can study valence energy levels of insulators. Densities of states for conduction bands of metals (of interest in study of metallized polymers)
(2) Absolute binding energy may be characteristic of particular structural features (e.g. CF_3, $C=O$, $-NH_2$ etc.). 'Shifts' can be related to electron distribution	(2) Studies of differential changes in cross-section with photon energy provides information on symmetries of orbitals (σ, π etc.)

Table 9.3—cont.

(3) *Multiplet Splittings*
For paramagnetic species observation
of multiplet splittings provides
information on spin states of atoms
or ions and distribution of unpaired
electrons

(4) Shake-up and shake-off satellites.
Information on excited states of hole
states

9.3 A PRELIMINARY APPRAISAL OF THE APPLICATION OF ESCA TO STUDIES OF STRUCTURE AND BONDING IN POLYMERS

9.3.1 Introduction

The brief introduction given above should serve to emphasize the fact that ESCA is in principle capable of providing information concerning the surface and immediate sub-surface of polymer samples as well as that pertaining to the bulk. As such, the technique has wide-ranging applicability much of which still remains to be exploited.

Since solids communicate with the rest of the universe by way of their surfaces, a non-destructive technique capable of delineating surface from sub-surface from bulk effects has important ramifications in many fields in which the physical, chemical, electrical and mechanical properties are of interest. Areas which immediately spring to mind are friction and wear phenomena, chemical or physical modification of polymer films initiated at the surface (encompassing oxidative degradation for example), and triboelectric and transport phenomena. ESCA has already given new insights into some of these areas and will undoubtedly contribute much more in the future.

For homogeneous polymeric systems information pertaining to the bulk is readily obtainable and in fact the study of simple homogeneous systems forms a necessary precursor to the development of ESCA to more complex areas. The applications outlined below therefore, fall logically into two sections. Firstly, the study of simple systems where the surface and immediate sub-surface are representative of the bulk; in this case ESCA is employed as a straightforward tool for studying structure and bonding for the polymer system as a whole. Secondly, the more complex situation of inhomogeneous samples where the unique (with regard to organic polymeric systems) depth-profiling capability of ESCA is used to delineate differences in structure and bonding as a function of depth into the sample.

The purpose of the ensuing discussion given in sections 9.4 and 9.5 is to highlight the development of the technique for detailed studies of structure and bonding in polymeric systems. It should be pointed out, however, that

ESCA also fulfils an important role in routine monitoring of samples prepared for other measurements. As a simple example, the spectra of PTFE recorded earlier, Figure 9.8, was obtained from a film pressed from PTFE powder between sheets of aluminium foil at the minimum temperature necessary for coalescence of the powder particles in a hand press ($\sim 200\,°C$). When the procedure is repeated using a higher temperature ($> 300\,°C$) films are produced which visually, chemically and by transmission infrared (TIR) and multiple attenuated total reflectance (MATR) measurements, appear to be the same as those produced at lower temperatures. The ESCA spectra of such films however are very revealing (Figure 9.17). It is clear from the appearance of peaks

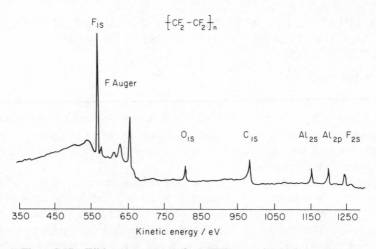

Figure 9.17   Wide scan spectrum for PTFE pressed at high temperature between Al foil

associated with the core levels of both oxygen and aluminium that in the high-temperature pressing process a contaminant surface layer of alumina Al_2O_3 is deposited on the PTFE. The thickness of the layer is almost certainly $< 10\,Å$. Although this is a trivial example it can readily be appreciated that if surfaces are mechanically prepared the possibility of contamination is always present and ESCA provides a convenient non-destructive means of monitoring this.

The application of ESCA to structure and bonding in polymer systems to date largely pertains to fluoropolymer systems.[3] The reasons for this are threefold. Firstly our research interests are centred around the halocarbon field (both monomeric and polymeric species) and hence samples and the expertise in preparing them are readily to hand. Even without an underlying research interest in these systems however, there are two further reasons why an initial research program into the application of ESCA to polymers, should concentrate initially on fluoropolymer systems. Thus the large shift in (e.g. C_{1s} core levels induced by fluorine gives the most favourable situation for delineating the likely areas of applicability of ESCA in this field. Finally, fluoropolymers are amongst the technologically most important systems of interest and are often difficult to study by other spectroscopic techniques.

Although the illustrative examples largely pertain to fluorocarbon polymers and are therefore limited in scope, the extension to other polymeric systems is under active investigation. With improvements in instrumentation, particularly that associated with the introduction of efficient X-ray monochromatization schemes, the applications to other systems will develop rapidly along the lines indicated here.

In developing ESCA as a spectroscopic tool for studying polymers the general guidelines adopted follow directly from our previous development of ESCA as a tool for studying structure and bonding in organic systems in general and halocarbon systems in particular;[14] namely, to start by studying simple, well-characterized systems to build up banks of data (on relative peak intensities, absolute binding energies and shifts) from which trends may be discerned and comparison drawn with simple monomers. The simplicity of these systems allows the development of rigorous theoretical models for the quantitative interpretation of the data. These models in turn point to the important electronic factors determining, for example, shifts in core binding energies, from which theoretically sound but less rigorous models may be developed to discuss larger systems for which more detailed computations are not feasible. The theoretical and experimental study of simple model systems is therefore an important adjunct to the full development of the application of ESCA to studies of complex polymeric systems. It is worthwhile pointing out at this stage that for much of the information obtained from ESCA measurements (e.g. absolute and relative binding energies) the theoretical interpretation is relatively straightforward and in this sense ESCA has a distinct advantage over many other spectroscopic techniques where inevitably the interpretation of data is more often than not in terms of analogy with model systems.

As a logical first step in the development of ESCA applied to polymers, we have studied simple, well-characterized homopolymer and copolymer systems. These 'homogeneous' samples then form a valuable yardstick for gauging the utility of ESCA for studies pertaining to information characteristic of the bulk.

In applying the technique to the investigation of simple polymer systems there are several distinct aspects about which one would hope to gain information. Firstly, the gross chemical composition of the polymers. This would include determination of elemental composition and in the case of copolymers the percentage incorporation of comonomers. Secondly, information concerning the gross structure, for example, for copolymers the block and/or alternating or random nature of the linkages. Thirdly, finer detail of structure such as structural isomerism, the nature of end groups, branching sites etc. Finally deductions about charge distributions and nature of the valence bands for polymers. It should be emphasized yet again that the data obtainable from ESCA is rather coarser in detail than that often obtainable by more conventional techniques (infrared, n.m.r., inelastic neutron scattering etc.) where information regarding for example, conformational aspects of polymer structure, may often be inferred and which are not amenable to direct study by ESCA. Although in favourable cases (see later) detailed interpretation of ESCA spectra can lead to conclusions concerning polymer chain conformations, ESCA is essentially a powerful technique providing information complementary to that from other branches of spectroscopy, but with unique advantages which mean that for many studies of polymer chemistry, such as dynamic studies of thermal or photochemical degradation, and in studies of polymeric films produced at surfaces by chemical reaction (e.g. fluorination, oxidation etc.) the information derived from ESCA studies is not obtainable by other techniques.

9.3.2 Sample preparation

Before discussing the results it is of some relevance to consider the ways in which polymer samples may be prepared for examination by ESCA. When the polymer is available as a powder it is often convenient to study it as such by applying the powder to double sided Scotch tape mounted on the sample probe. The pitfalls to beware of in this approach are that no extraneous signals are observed from the sample backing and also that no chemical reaction occurs between sample and substrate. The incomplete coverage and uneven topography of samples prepared in this way generally lead to lower signal/noise ratios than polymers studied directly as films. The most generally useful methods of preparing polymer films for ESCA studies may be classified as follows:

9.3.2.1 From solution

If the polymer is sufficiently soluble then thin polymer films may be deposited directly on a gold backing (ready for mounting on the sample probe), by conventional dip or bar coating, or spin casting. Since ESCA is such a surface-sensitive technique it is important to use clean apparatus and pure solvents containing no involatile residues (e.g. anti-oxidants etc.) which would segregate at the surface on evaporation of the solvent. With readily oxidized systems or with systems with sites capable of hydrogen bonding with extraneous water it is imperative to maintain a suitable inert atmosphere during the slow evaporation of solvent. Solvent entrainment can also be a problem, and indeed the technique lends itself well to studying diffusional problems in polymers.

9.3.2.2 From pressing or extrusion

Because of problems of contamination in solvent-cast films, it is convenient to study most polymers in the form of pressed or extruded films mounted on a suitable backing (e.g. gold). For stable thermopolastics of course it is often possible to 'melt' a small amount of the sample and allow it to spread in the form of a thin film on the tip of a sampling probe or to slice a thin film from a larger sample. In preparing samples from powders it is often convenient to press films between sheets of clean aluminium foil at an appropriate temperature and pressure. There are two precautions to be taken in doing this:

(i) the temperature and pressures used should be such that no decomposition or adhesion of surface contamination occurs;

(ii) since typically only the top ~ 50 Å of the sample is studied by ESCA it is important to avoid chemical reaction at the surface during preparation. Thus pressing polyethylene films in air at the minimum temperature and pressure necessary, results in considerable surface oxidation. This may be obviated by pressing in an inert atmosphere (e.g. N_2 or Ar). Surface contamination, e.g. hydrocarbon etc., arising for example by inadvertent handling during processing can most readily be removed by careful treatment with an appropriate solvent.

9.3.2.3 *Polymerization in situ*

ESCA is particularly suited to dynamic studies and would be useful for monitoring polymerizations carried out in situ at the sample probe.

9.3.3 Energy referencing

Of their very nature, most polymers of interest are extremely good insulators and therefore in studying samples as thin films in an ESCA experiment sufficient charge carriers will not normally be available to provide electrical contact between the sample and spectrometer. Sample charging will therefore occur, resulting in a shift of the energy scale, and some form of referencing back to the Fermi Level is therefore necessary. Referencing of the energy scale is most readily accomplished by depositing a thin coating of a suitable reference (e.g. C_{1s} signal from a hydrocarbon or $4f_{\frac{7}{2}}$ levels of gold with binding energies of 285 eV and 84 eV respectively) and monitoring the core levels of the sample and reference. To put matters in perspective however, it should be emphasized that with most commercial spectrometers (employing unmonochromatized X-ray sources and retarding lens systems, double-focussing hemispherical electrostatic analysers and associated slit systems), sample charging is not too serious a problem and at worst involves only a few eV correction to the energy scale. With slitless designs and a monochromatized X-ray source as convention-ally applied in dispersion compensation, sample charging can however reach serious proportions involving shifts in the energy scale of tens of eV for insulating samples. The problem can be alleviated to some extent by use of low-energy electron flood guns but this introduces other complications.

An interesting observation is that for homopolymers, even in the presence of overall sample charging, with suitable preparation of sample, linewidths for given core levels are comparable with those for monomers, Figure 9.18 (*cf.* poly(ethylene terephthalate), Figure 9.12). (Instrumental resolution set such that under comparable conditions (Mg K$\alpha_{1,2}$ photon source) the Au $4f_{\frac{7}{2}}$ level has FWHM 1·15 eV.) The range of applicability of ESCA in studying homo-polymers is thus essentially the same as in studying simple monomers. With thin films, prepared by slicing from a larger sample, or by pressing, or by

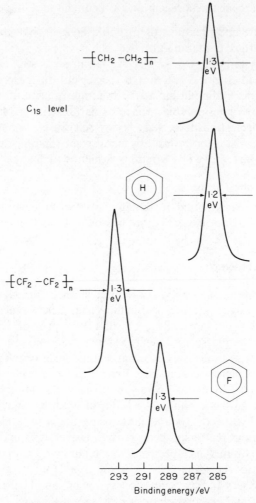

Figure 9.18 C_{1s} levels for homopolymers and typical monomers illustrating linewidth similarities

melting directly onto a probe tip, the less regular nature of the surface is often manifest in slight increase in linewidths arising from non-uniform distribution of charge on the sample. This is often of no consequence since it merely affects the overall resolution; however it may be obviated in many cases by modifying the way in which the film is produced. An example of this arising from our studies of copolymers of ethylene and tetrafluoroethylene will be discussed in more detail later on.

The first spectra recorded with these samples (Figure 9.19) seemed to indicate that the degree of fluorinated monomer incorporation was not as great as expected. Thus in the carbon-1s spectra the peaks at low binding energy due to $\underline{C}H_2$ groups were always bigger than the higher energy bands due to $\underline{C}F_2$ groups, even for samples where the degree of incorporation of tetrafluoro-ethylene was reliably known (from conventional carbon and fluorine elemental analysis) to be greater than 50 mol %. This observation is clearly indicative of surface contamination of the samples by hydrocarbon. When the surfaces of the films were cleaned with methylene chloride (by lightly rubbing the surface with a tissue dampened with solvent), the contaminants were removed and the elemental compositions of the cleaned surfaces could be shown (see later) to be the same as those measured by conventional combustion analysis of bulk samples. The effect of this surface cleaning on the carbon-1s spectrum of a copolymer sample is illustrated in Figure 9.19. Spectrum (a) shows the spectrum of the polymer prior to cleaning and spectrum (b) that of the same sample after it had been cleaned with methylene chloride. Two effects are apparent from inspection of the spectra, firstly the peaks due to hydrocarbon contamination of

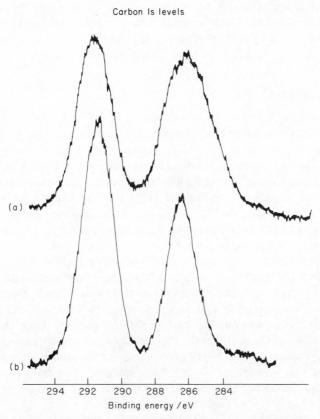

Carbon 1s levels

(a)

(b)

294   292   290   288   286   284

Binding energy / eV

Figure 9.19 C_{1s} levels for copolymer of ethylene-tetra-fluoroethylene illustrating the effect of removal of surface contamination: (a) before cleaning (b) after cleaning

the surfaces which occur at low binding energy have been removed by the cleaning process, and secondly both the peaks in the spectrum of the cleaned sample are narrower than those in the original untreated material. This second observation can be explained on the assumption that the surface contamination is unevenly distributed resulting in an uneven distribution of charges at the sample surface and a consequent line broadening, such an uneven distribution of contaminants could arise from handling of the sample.

9.4 APPLICATION TO STUDIES OF HOMOGENEOUS MATERIALS

In general an examination of an homogeneous material by ESCA is carried out in three stages. Firstly an initially wide-scan low-resolution spectrum may be used to establish which elements are present. Subsequent measurements of the relative intensities of the various signals, in conjunction with previously established sensitivity factors, allows the relative abundance of the elements (empirical formula) to be established. In the third step the absolute binding energies, taken with the relative intensities of resolved peaks within a given core level, allow deductions to be made concerning structure and bonding.

In this section we consider in detail this hierarchy of information levels for a series of well-characterized homopolymers and copolymers.

9.4.1 Homopolymers

The simplest systems to start with are the high molecular weight homo-polymers of the fluoroethylenes, for which complications due to branching, end groups etc., are minimal. The carbon-1s spectra for polyethylene (high density), poly(vinyl fluoride), poly(vinylidene fluoride), poly(vinylene fluoride), polytrifluoroethylene, and polytetrafluoroethylene are presented in Figures 9.20, 9.21 and 9.22.[15] (Prior inspection of the wide-scan spectra for the simple homopolymers qualitatively shows the increasing fluorine content along the series and also reveals in particular cases a small signal arising from O_{1s} levels associated with surface oxidation features.)

Typically measurements of all the core levels of a polymer take ~ 30 minutes whereas under the conditions employed (pressure in sample chamber typically $< 5 \times 10^{-7}$ Torr) hydrocarbon build-up on the surface only becomes appreciable after several hours. The technique employed for calibration of the energy scale therefore is to measure the core levels of a given polymer immediately on introduction of the sample into the spectrometer when hydrocarbon contamination is unimportant. After several hours in the sample chamber further spectra may then be recorded and the appearance of an extra peak, (clearly evident in the spectra in Figures 9.20–9.22) in the C_{1s} region at 285·0 eV binding energy may then be used to reference the energy scale.

The C_{1s} spectrum of poly(vinyl fluoride) shows two partially resolved peaks of equal area corresponding to $\underline{C}HF$ and $\underline{C}H_2$ carbons whilst for poly(vinylene

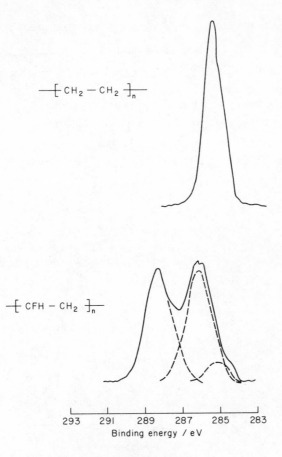

Figure 9.20 C_{1s} levels of polyethylene and poly-(vinyl fluoride)

fluoride) in addition to the main peak corresponding to $\underline{C}HF$ carbons and hydrocarbon calibration peak, there is a weak peak at 292.0 eV. This in fact corresponds to $\underline{C}F_2$ type carbon arising from contamination from the fluorocarbon soap ($H\overline{(CF_2)}_8COO^-NH_4^+$) used in the emulsion polymerization. For poly(vinylene fluoride) well resolved peaks of equal area corresponding to $\underline{C}F_2$ and $\underline{C}H_2$ carbons are evident and for polytrifluoroethylene partially resolved peaks of equal area corresponding to CF_2 and $\underline{C}FH$ carbons. The assignment of peaks in these spectra arising from $\underline{C}F_2$, $\underline{C}FH$ and $\underline{C}H_2$ structural units is straightforward (Table 9.4). By taking appropriate pairs of polymers it is possible to investigate both primary and secondary effects of replacing

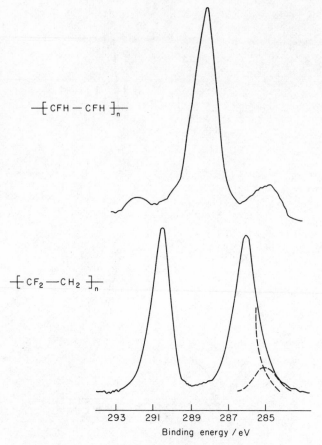

Figure 9.21 C_{1s} levels of poly(vinylene fluoride) and poly-(vinylidene fluoride)

Table 9.4 Binding energies of the homopolymers of ethylene and the fluoroethylenes (in eV)

		C_{1s}	$\Delta(C_{1s})$	F_{1s}	$\Delta(F_{1s})$
$+CH_2-CH_2+_n$		285·0	(0)	—	—
$+CFH-CH_2+_n$	$-\underline{C}FH-$	288·0	3·0	689·3	(0)
	$-\underline{C}H_2-$	285·9	0·9	—	—
$+CFH-CFH+_n$		288·4	3·4	689·3	0·0
$+CF_2-CH_2+_n$	$-\underline{C}F_2-$	290·8	5·8	689·6	0·3
	$-\underline{C}H_2-$	286·3	1·3	—	—
$+CF_2-CFH+_n$	$-\underline{C}F_2-$	291·6	6·6	690·1	0·8
	$-\underline{C}HF-$	289·3	4·3	690·1	0·8
$+CF_2-CF_2+_n$		292·2	7·2	690·2	0·9

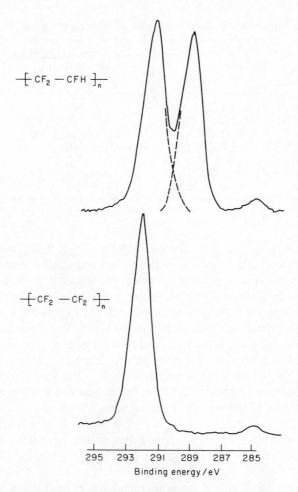

Figure 9.22 C_{1s} levels of polytrifluoroethylene and PTFE

Table 9.5 Primary substituent effects for C_{1s} levels

Polymer pairs	Shift in binding energy on replacing H by F/eV
$(\underline{C}HFCH_2)_n$, $(\underline{C}H_2CH_2)_n$	3·0
$(\underline{C}F_2CH_2)_n$, $(\underline{C}HFCH_2)_n$	2·8
$(CHF\underline{C}HF)_n$, $(CHF\underline{C}H_2)_n$	2·5
$(CF_1\underline{C}HF)_n$, $(CF_2\underline{C}H_2)_n$	3·0
$(\underline{C}F_2CF_2)_n$, $(CF_2\underline{C}HF)_n$	2·9
$(\underline{C}F_2CHF)_n$, $(\underline{C}HFCHF)_n$	3·2
Average	2·9

Table 9.6 Secondary substituent effects for C_{1s} levels

Polymer pairs	Shift in binding energy on replacing H by F (per substituent)/eV
$(CHF\underline{C}H_2)_n$, $(CH_2\underline{C}H_2)_n$	0·9
$(\underline{C}HFCHF)_n$, $(\underline{C}HFCH_2)_n$	0·4
$(CF_2\underline{C}H_2)_n$, $(CH_2\underline{C}H_2)_n$	0·7
$(\underline{C}F_2CHF)_n$, $(\underline{C}F_2CH_2)_n$	0·8
$(\underline{C}F_2CF_2)_n$, $(\underline{C}F_2CHF)_n$	0·6
$(CF_2\underline{C}HF)_n$, $(CHF\underline{C}HF)_n$	0·9
Average	0·7

hydrogen; Tables 9.5 and 9.6. The average primary (2·9 eV) and secondary (0·7 eV) substituent effects are in excellent agreement with those found for monomer systems and their relative constancy emphasizes the characteristic nature of substituent effects. It is interesting to note that these results are also in excellent agreement with detailed non-empirical LCAO-MO-SCF calculations on neutral model systems and the C_{1s} hole states. The rapid fall-off in effect of the fluorine substituent provides a crude but immediate manifestation of the σ inductive effect exerted by fluorine. As the degree of fluorine substitution increases in going through the series, the fluorine substituents to some extent compete for the sigma electron drift from the carbon and hydrogen atoms and this is clearly shown by the increase in binding energy for the F_{1s} levels (Table 9.4). It is gratifying to note that the shifts in core binding energies are qualitatively in agreement with what might be termed the organic chemist's intuitive ideas concerning charge distributions; this trend will be quantified in the next section.

In studying fluoropolymer systems in general, it may be noted that in addition to the F_{1s} and C_{1s} core levels and their associated Auger transitions, the F_{2s} levels at the bottom of the valence band are essentially core-like so that even with a single photon source the escape-depth dependences on kinetic energy span a large range. For Mg $K\alpha_{1,2}$ kinetic energies for F_{1s}, C_{1s} and F_{2s} are ~ 560 eV, ~ 960 eV and ~ 1220 eV respectively whilst the Auger processes for de-excitation of the F_{1s} and C_{1s} hole states are in the kinetic energy range 610–660 and 240–270 eV respectively. This being the case it is a relatively easy matter to use ESCA to 'depth profile' samples to investigate their homogeneity. For the homopolymers of the fluorinated ethylenes the C_{1s} spectra consist of two peaks of equal area. Since the kinetic energies for the photoemitted electrons are virtually identical the sampling depths must be the same and the results therefore suggests a homogeneous sample. This may be confirmed by comparison with data from the F_{1s} level where the sampling depth will be different. Table 9.7 shows the relative carbon-1s to fluorine-1s peak areas for the four samples (with our particular instrumental arrangement).[16]

Table 9.7 Relative C_{1s} to F_{1s} peak areas for homopolymers

Polymer	Carbon-1s peak area : fluorine-1s peak area	Number of carbon atoms : number of fluorine atoms	Relative carbon-1s peak area : fluorine-1s peak area[a]
$(CH_2CHF)_n$	1:1	2:1	1:2·0
$(CH_2CF_2)_n$	1:2·08	1:1	1:2·1
$(CHFCF_2)_n$	1:3·15	2:3	1:2·1
$(CF_2CF_2)_n$	1:3·97	1:2	1:2·0

[a] With probable error limits of ± 0.05.

The constancy of the ratio demonstrates unambiguously the uniformity of the samples since the escape-depth dependences are significantly different (mean-free-paths $F_{1s} \sim 7\,\text{Å}$ (kinetic energy $\sim 564\,\text{eV}$), $C_{1s} \sim 10\,\text{Å}$ (kinetic energy $\sim 965\,\text{eV}$)). This determination of the intensity ratios of core levels for model homogeneous systems is an important precursor in discussing more complex systems. The intensity ratio of 2.05 ± 0.05 for these homopolymers is in complete agreement with that previously established from studies of simple monomers [17] Studies of simple monomers has also established apparent* relative intensity ratios for other core levels of major interest (viz. N, O, S, P, Cl).

Although the emphasis in the work to data has been on fluoropolymers the extension to other polymers is straightforward. As an example the C_{1s} and O_{1s} spectra shown in Figure 9.12 for poly(ethylene terephthalate) readily establish the gross features of the composition and structure. Comparison, with appropriate model compounds studied under the same conditions, of the relative peak areas for the two core levels establishes the composition, whilst identification of the various structural features may be accomplished by comparison of absolute binding energies also from appropriate model systems.

$$O_{1s} \qquad\qquad C_{1s}$$

$$\underline{O}-C{=}O \sim 534{\cdot}6 \qquad -\underline{C}\!\!\begin{array}{l}{}^{\displaystyle O}\\[-2pt]{}_{\displaystyle O}\end{array} \sim 289{\cdot}2$$

$$-O\underline{C}H_2 \sim 286{\cdot}7$$

$$O{-}C{=}\underline{O} \sim 533{\cdot}0 \qquad \langle\bigcirc\rangle \quad 285\,\text{eV}$$

As simple examples of non-fluorinated systems which may readily be studied, Figure 9.23 shows data pertaining to simple vinyl homopolymers (poly(vinyl

---

* It should be emphasized here and indeed be apparent from previous discussions that the measured intensity ratios are partly determined by instrumental factors.

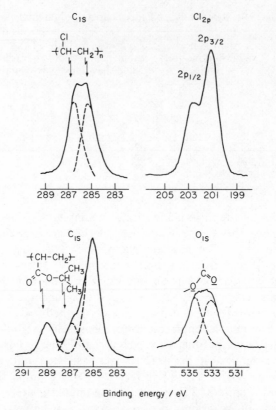

Figure 9.23 C_{1s}, Cl_{2p} and O_{1s} levels for poly-(vinyl chloride) (upper curves) and poly(isopropyl acrylate) (lower curves)

chloride) and poly(isopropyl acrylate). In each case the relative intensities of different core levels and shifts within the latter establish gross features of structure and bonding.

For hydrocarbon polymers as we have previously indicated the system is at first sight informationally underdetermined (viz. only one core level and shifts between carbons formally in sp^3, sp^2 and sp bonding situations being extremely small). However, if the polymer backbone and/or pendant groups contain unsaturated systems then the possibility of obtaining information via the observation of shake-up satellites presents itself and we will return to this point in a later section.

Although we have emphasized in the discussion the various levels of information available from ESCA, the observation (or lack thereof) of a core level corresponding in binding energy to a particular structural feature may be sufficient in itself to distinguish between alternative structures for a polymer system. A simple example is provided by poly(ω-pentachlorocyclopentadienyl alk-1-enes).

Diels–Alder polymers can be prepared by heating derivatives of pentachloro-cyclopentadiene of the form **A**. A report in the patent literature[18] describes the

$$\text{A} \qquad\qquad \text{B} \qquad\qquad \text{C}$$

A: pentachlorocyclopentadiene with $(CH_2)_nCH=CH_2$ and Cl_5 substituents

B: Cl_2; Cl, Cl, $(CH_2)_nCH=CH_2$ substituents

C: Cl, Cl, $(CH_2)_nCH=CH_2$, Cl, Cl substituents

preparation of the polymer from allyl pentachlorocyclopentadiene and assigns the structure as implying either that the monomer has the structure **B** (n = 1)

[structure: ring with CH_2, CCl_2, and Cl, Cl, Cl substituents]

or that if the monomer has the expected structure **C** (n = 1) then a molecular rearrangement involving a chlorine migration must have occurred during or following the polymerization. Attempts to establish the structure of the monomer by conventional chemical and spectroscopic methods were unsuccessful[19]; however, by means of ESCA the monomer structure was unambiguously assigned as **C** (n = 1).[20] The presence of a $\underline{C}Cl_2$ unit in related systems is indicated by a C_{1s} peak at *ca*. 289·1 eV, whereas $\underline{C}-Cl$ units give C_{1s} peaks at *ca*. 287·7 eV. Shifts of this magnitude are usually fairly easily distinguished, and the absence of a $\underline{C}Cl_2$ unit establishes the structure as **C**. When polymers from monomers **C** (n = 1 and n = 9) were examined by ESCA no peak at 289·1 eV could be detected, and consequently no $\underline{C}Cl_2$ units are present. The reported structure of the polymer must therefore be incorrect and the correct structure must include one or more of the isomeric units represented by

[structure: bicyclic ring with Cl, Cl, Cl, Cl, Cl and (CH_2) substituents, subscript n]

this reassignment of structure being consistent with other chemical and spectro-scopic (i.r. and n.m.r.) data for the polymers and related model compounds.[21]

9.4.2 Theoretical models for a quantitative discussion of results

At this stage we consider models for the interpretation of absolute and relative binding energies for core levels. Although the spectra of the homopolymers discussed above can be understood in terms of the qualitative description

146

advanced, it is essential to establish a reliable quantitative interpretation of observed shifts if headway is to be made with more complicated systems. The details of the theory have been discussed elsewhere[22] and only a brief summary is presented here. The photoionization process itself, at least for systems containing first and second row atoms, can in general be quantitatively described within the Hartree–Fock formalism although this is not true for the accompanying shake-up and shake-off processes for which detailed considerations of electron correlation are necessary.

The energy consideration for discussion of relative or absolute binding energies within the Hartree–Fock formalism are shown in Figure 9.24. For the

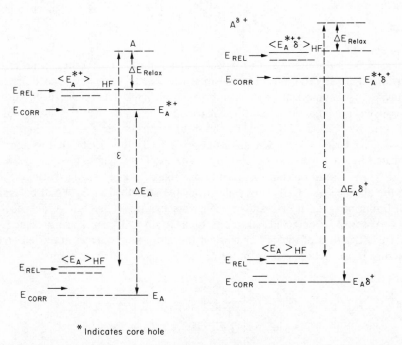

*Indicates core hole

Figure 9.24 Energy considerations in the theoretical interpretation of relative and absolute binding energies

C_{1s} level of methane for example, the exact agreement between theory (calculated as an energy difference between neutral molecule and hole state with a basis set approaching the Hartree–Fock limit) and experiment indicates that both relativistic and correlation energy considerations are small.[23] By contrast comparison with Koopmans' Theorem shows that electronic reorganization accompanying core ionization is substantial (e.g. $\sim 14\,\mathrm{eV}$ for the C_{1s} level.)

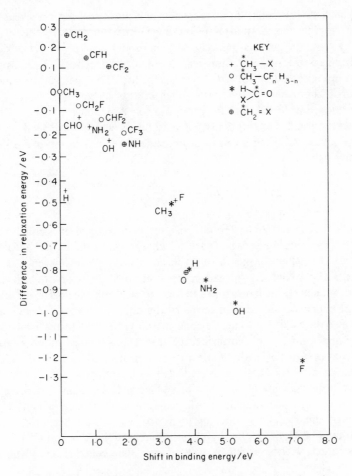

Figure 9.25 Shifts in binding energy *vs.* difference in relaxation
energy for C_{1s} levels

A further point of interest recently illuminated by detailed theoretical studies is
that the magnitude of such relaxation energy is strongly dependent on the
electronic environment of a given atom.[24] Figure 9.25 illustrates this rather
nicely for C_{1s} levels and moreover a relationship between difference in relaxation
energy and shift in core binding energy is clearly evident.[24]

Detailed non-empirical LCAO MO SCF calculations can therefore provide
a quantitative discussion of experimental ESCA data* (other than for shake-up

* It should be emphasized that in many senses the quantitative discussion of both absolute
and relative binding energies is somewhat more straightforward for core than for valence
levels. This arises from the fact that the core orbitals being so localized, correlation energy
corrections tend to remain constant for a given core level whereas for valence levels of
widely differing localization characteristics this is not the case.

148

and shake-off phenomena), however, such computations are only feasible on relatively simple systems. In attempting to quantify theoretically the results for polymer systems the great complexity of even model systems renders a rigorous approach impossible. A less sophisticated but still theoretically valid approach based on an expansion for the expression for the Fock eigenvalues (i.e. Koopmans' Theorem) in the zero differential overlap approximation has therefore been developed.[1b,2b] By grouping terms which are essentially independent of the local electron density it may be shown that the binding energy of a given core level of an atom in a molecule is related to the overall charge distribution by:

$$E_i = E_i^0 + kq_i + \sum_{j \neq i} \frac{q_j}{r_{ij}}$$

The so-called 'charge potential model' originally developed by Siegbahn et al.[1b] relates the core binding energy E_i to the charge on atom i and the potential provided by the other charges within the molecule. E_i^0 is a reference level and k represents approximately the one-centre coulomb repulsion integral between a core and valence electron on atom i. In dealing with complex organic molecules, analysis of experimental data in terms of the charge potential model and all-valence-electron SCF MO calculations in the CNDO/2 formalism (formally an approximation to a non-empirical treatment) has proved highly rewarding. From studies of series of related molecules values of k and E_i^0 may be established for a given core level of a given atom.[14] It is interesting to note that although the charge potential model may be derived from Koopmans' Theorem which does not take into account electronic relaxation, the latter in fact follows a relationship remarkably similar in form to the charge potential equation (cf. the data in Figure 9.25). The charge potential model with k and E^0 determined from experiment works extremely well overall. The central role of the charge

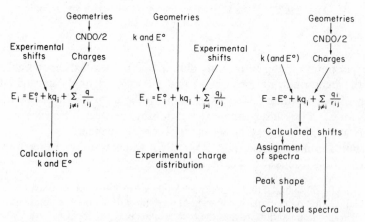

Figure 9.26 Central role of the charge potential model in quantifying experimental data

potential model in quantifying experimental data on complex molecules is shown in Figure 9.26. Starting on the left-hand side of the figure, with given geometries and appropriate charge distributions, (e.g. from CNDO/2 SCF MO calculations) experimental shifts on series of model compounds may be used to obtain values of k and E^0 for a given level of a given element. On the right-hand side, if k and E^0 values are available, then theoretical charge distributions may be used for the assignment of spectra and if peak shapes and widths have been established then theoretically calculated spectra may be simulated. Last but not least, if appropriate values of k and E^0 and geometries are available it is possible to invert the charge potential model to obtain 'experimental' charge distributions; this application to polymer systems is of particular importance and will be discussed more fully later on.

The short-range nature of substituent effects on core binding energies in saturated systems suggests that it may be feasible quantitatively to discuss the results for polymers in terms of calculations on simplified model systems which contain the essential structural features and accommodate all short-range interactions. The success of the charge potential model, coupled with CNDO SCF MO calculations of charge distributions, in quantitative discussions of data for simple monomers, suggests that this is a feasible approach since this model accounts well for the short-range nature of substituent effects.

For polyethylene and poly(vinylene fluoride) the representative model units were taken as being the monomer linked to other monomer units and then appropriate end groups.[15] In this way substituent effects over three carbon atoms were taken into account. (The calculations showed that longer range effects were negligible so that the model incorporated all of the important short-range interactions.) All-valence-electron SCF MO calculations were then carried out and the electron distributions obtained were used, with appropriate values for k's and E^0's, to calculate absolute binding energies for the core levels of the representative structural unit of polymer. The adequacy of the theoretical treatment for polyethylene and poly(vinylene fluoride) is apparent from Table 9.8 which also includes the results of calculations on a model for PTFE. In the latter case computer limitations at that time dictated that a smaller model system than optimum be studied, and this accounts for the small discrepancy between calculated and observed binding energies; in fact, subsequent improve-

Table 9.8 Absolute C_{1s} binding energies calculated for model systems of polyethylene, poly (vinylene fluoride) and polytetra-fluoroethylene

| | Binding energy/eV | |
Model compound	Calc.	Obs.[a]
$CH_3CH_2CH_2CH_2CH_2CH_2CH_2CH_3$	284·9	285·0
$CH_2FCHFCHFCHFCHFCHFCH_2F$	288·6	288·4
$CF_3CF_2CF_2CF_2CF_2CF_3$	292·6	292·0

[a] For polymer.

ments in computing facilities allowed calculations on a larger model compound and there is now quantitative agreement between theory and experiment in this case as well.

For the unsymmetrical monomers poly(vinyl fluoride), poly(vinylidene fluoride) and polytrifluoroethylene the possibility arises of structural isomerism by way of head-to-tail and/or head-to-head addition. In fact ^{19}F n.m.r. studies[25] show that structural isomerism does occur in both of the first two and indeed provides information on tacticity which is not available by ESCA studies. Since C_{1s} spectra of the three polymers are relatively well resolved it is evident that information regarding structural isomerism, if available, must be encoded in the lineshapes and/or linewidths.

Theoretical calculations on suitable models incorporating the relevant structural features viz. head-to-tail and head-to-head linkages gives the results shown in Table 9.9. Considering firstly poly(vinyl fluoride), the calculated

Table 9.9 Absolute C_{1s} binding energies calculated for model systems of poly(vinyl fluoride) having head-tail and head-head structures respectively

Model compound	Calculated binding energy/eV	
	CFH	CH$_2$
CFH$_2$CH$_2$CFHCH$_2$CFHCH$_2$CFHCH$_2$CFHCH$_3$	288·1	285·4
CFH$_2$CH$_2$CH$_2$CFHCFHCH$_2$CH$_2$CFHCFHCH$_3$	288·0	285·6
Observed for polymer	288·0	285·9

binding energies for both types of structural arrangements are the same within experimental error and are in fact in excellent agreement with the observed values. It seems clear therefore that for this particular polymer ESCA is unable to provide information on structural isomerism along the chain. (This contrasts with the situation to be described later concerning nitroso rubbers.) The same considerations apply to both poly(vinylidene fluoride) and polytrifluoro-ethylene, the theoretical models again being in excellent agreement with experiment.

With a background knowledge of intensity ratios for different core levels, of absolute binding energies as a function of electronic environment, and of quantitative theoretical models for describing the latter, ESCA can be applied to problems which have proved intractable by more conventional spectroscopic techniques. A particularly interesting example is provided by the study of the product of the fluoride-ion-initiated polymerization of hexafluorobut-2-yne.[26] The rapid polymerization of perfluorobut-2-yne in the presence of fluoride ion has been investigated and described by several groups of workers; the product is an off-white, insoluble solid which cannot be pressed into films. The material is not oxidized by $KMnO_4$ in acetone which is known to oxidize C=C bonds in fluorinated systems, neither does the compound display any absorption in the double-bond region of the infrared spectrum. At first sight these observations would appear to preclude the linear polyene structure,

$$\text{---}\text{C(CF}_3)\text{=}\text{C(CF}_3)\text{---}_n$$

which could reasonably be expected as the reaction product, in favour of an extensively cross-linked material or a saturated cyclobutane ladder polymer.

The carbon-1s spectrum of the polymer is shown in Figure 9.29. The hydrocarbon peak at ~ 285 eV arises from solvent and/or the Scotch tape backing.

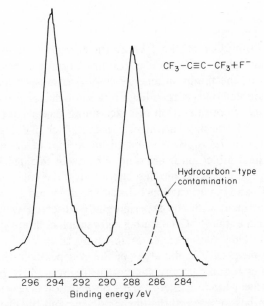

Figure 9.27 C_{1s} levels for the polymer produced in the fluoride-ion-initiated polymerization of hexafluorobut-2-yne

(It is difficult to remove the last traces of solvent used in the reaction.) The spectrum for the polymer is therefore extremely simple with just two peaks of approximately equal areas at 287·7 eV and 294·1 eV. From our studies of substituent effects (and since the polymer contains only C and F) the peak at high binding energy may be unambiguously assigned to $-CF_3$ groups. The assignment of the other peak is not as straightforward (but is entirely consistent with a vinylic carbon with attached perfluoroalkyl group) and a complete assignment of the polymer structure can only be made on the basis of a careful examination of the spectrum, illuminated by calculations on model compounds and studies of shake-up phenomena in fluorinated systems. Since approximately half the C_{1s} peak area due to the polymer is accounted for by CF_3 groups and the remaining peak area consists of a single narrow line the structure must be regular and the most probable candidates were the expected linear polyene or the inherently unlikely cyclobutane ladder structure. The question was decided on the basis of two lines of investigation. Firstly calculations on the model compound

gave rise to two useful conclusions. When the energy of the model system was computed as a function of the angle of rotation about the single bonds linking the ethylene units, the minimum energy conformation was shown to be that with adjacent double bonds perpendicular, this conclusion is eminently reasonable on the basis of consideration of space-filling models and invalidates the objection to the polyene structure arising from its lack of colour. The chemical inertness of the material is also accounted for in that in the minimum energy conformation the double bonds are quite effectively shielded by the trifluoromethyl groups. The predicted binding energies (on the basis of model calculations) for the $\underline{C}F_3$ and vinylic carbon-1s levels are 293·9 and 287·8 eV respectively in excellent agreement with the experimentally observed values of 294·1 and 287·7 eV. Experimental (ESCA) studies of model compounds also confirm this assignment. The second factor establishing the linear polyene structure arises from the inequality in the areas of the two peaks in the C_{1s} spectrum, the area of higher binding energy component being slightly greater than that of the peak at lower binding energy. This observation is crucial to the unambiguous differentiation between the linear polyene and the cyclobutane ladder structure. The observed area difference is attributed to a shake-up process associated with the unsaturated units of the linear polyene. Investigations of a large number of unsaturated fluorocarbon systems shows, in all cases, low-intensity satellite peaks to the low-kinetic-energy side of the main photoionization peak for vinylic-type carbons.[26] As a typical example the energy separation in the unconjugated perfluorocyclohexa-1,4-diene is ~ 7 eV. A satellite at approximately the same energy separation but much reduced in intensity is also apparent in the F_{1s} spectrum and this, coupled with theoretical investigations of shake-up transition probabilities within the sudden approximation, identifies the shake-up transition as arising from a $\pi^* \leftarrow \pi$ transition. The energy separation between the principal components of the C_{1s} spectrum of the polymer are 6·4 eV and for a polyene structure the shake-up transition arising from $\pi^* \leftarrow \pi$ excitation accompanying core ionization for the vinylic carbons would fortuitously overlap that arising from the $\underline{C}F_3$ levels. For a ladder structure of course the analogous shake-up process would be expected to have a very low intensity and much higher excitation energy. The unequal areas of the two peaks in the C_{1s} region are therefore readily understandable only in terms of a polyene structure for the polymer.

9.4.3 Copolymers

It is evident from the previous section that substituent effects on core binding energies in polymers can be understood both qualitatively and quantitatively on the same basis as those for simple monomeric systems. With this knowledge it is possible to proceed to more complex systems such as copolymers. The first feature of interest is the determination of compositions, viz. percentage comonomer incorporations. The applicability of ESCA in this field is illustrated by reference to our work on Viton, Kel F, ethylene, tetrafluoroethylene, and styrene/pentane copolymers. The use of ESCA for studying finer details of structure such as structural isomerism and sequence distribution is exemplified by studies of nitroso rubbers and ethylene/tetrafluoroethylene copolymers.

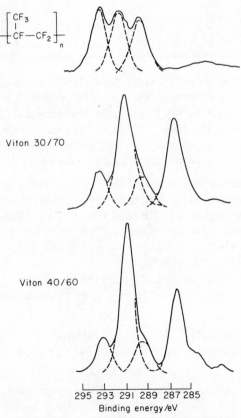

Figure 9.28 C_{1s} levels for polyhexafluoropropene and Viton polymers

9.4.3.1 Copolymer compositions

For *Viton polymers* the C_{1s} levels for the parent polyhexafluoropropene and the 30/70 and 40/60 copolymers with vinylidene fluoride are shown in Figure 9.28. The binding energies are tabulated in Table 9.10 and may again be understood in terms of simple substituent effects.[16] In obtaining the binding energies of the $\underline{C}F_2$ and $\underline{C}F$ carbon-1s levels a simple deconvolution is necessary.

Table 9.10 Binding energies of polyhexafluoropropene and the Vitons (in eV)

		C_{1s}	$\Delta(C_{1s})$	F_{1s}	$\Delta(F_{1s})$
40/60 Viton	CF_3	293·3	8·3	690·2	0·9
	CF_2	291·1	6·1	690·2	0·9
	CF	289·4	4·4	690·2	0·9
	CH_2	286·6	1·6	—	—
30/70 Viton	CF_3	293·4	8·4	689·9	0·6
	CF_2	290·9	5·9	689·9	0·6
	CF	289·3	4·3	689·9	0·6
	CH_2	286·4	1·4	—	—
$\left[CF{-}CF_2\right]_n$ $\quad\vert$ $\quad CF_3$	CF_3	293·7	8·7	690·2	0·9
	CF_2	291·8	6·8	690·2	0·9
	CF	289·8	4·8	690·2	0·9

In any estimation of copolymer compositions, however, it is obviously desirable to avoid even such a minor complication. The procedure adopted, therefore, was to measure the area of the $\underline{C}F_3$ peak, the total area of the $(\underline{C}F_2 + \underline{C}F)$ peak and the area of the $\underline{C}H_2$ peak. The degree of incorporation of hexafluoropropene (HFP) in the copolymer was then calculated from the percentage of the total area due to C_{1s} levels represented by each peak:

(I) The mole percent incorporation of HFP is three times the area of the peak due to $\underline{C}F_3$, on the basis of the stoichiometry of the HFP unit.

(II) The area of the peak due to $\underline{C}F_2$ and $\underline{C}F$ (A) is equivalent to half the total C_{1s} peak area due to vinylidene fluoride ($\frac{1}{2}VF_2$) and two thirds the total C_{1s} peak area due to HFP, i.e. $A = \frac{1}{2}VF_2 + \frac{2}{3}HFP$ and $100 = VF_2 + HFP$, by definition, hence % HFP = $6(A\text{-}50)$.

(III) The area of the peak due to $\underline{C}H_2$ is half the total area due to VF_2 and hence the mole percent incorporation of HFP is given by the expression $100(2 \times \%$ Area due to $\underline{C}H_2)$.

Using these three methods to determine the degrees of incorporation of HFP gives an internal check on the reliability of the method. The results are as follows:

Method of calculation	% HFP incorporation		
	(I)	(II)	(III)
Sample 40/60	39	42	40
Sample 30/70	33	30	32

The internal consistency is good to within 3 % and the values obtained are in good agreement with those obtained by combustion analysis.

For *Kel F polymers* the C_{1s} spectrum of the parent polychlorotrifluoroethylene consists of a single, broad peak with a flattened top corresponding to overlapping lines from $\underline{C}F_2$ and $\underline{C}FCl$ carbons; see Figure 9.29. The absolute binding energies and shifts, shown in Table 9.11, can again be rationalized in terms of simple substituent effects.[15] The C_{1s} spectra for the two copolymers differ considerably, the most noticeable feature being the high proportion of $-\underline{C}H_2$ units in the 30/70 copolymer coupled with the drastically reduced linewidth for the composite line at high binding energy.

In estimating the compositions of these copolymers it is again desirable to avoid reliance on deconvoluted peak areas. Two methods are available: measurements of the total $(-\underline{C}F_2-) + (\underline{C}FCl)$ peak area, and measurement of

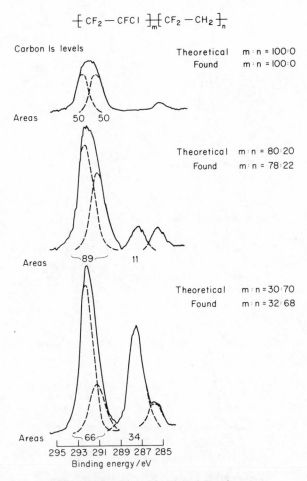

Figure 9.29 C_{1s} levels for Kel-F polymers

Table 9.11 Binding energies of the Kel-F polymers (in eV)

		C_{1s}	$\Delta(C_{1s})$	F_{1s}	$\Delta(F_{1s})$	$Cl_{2p_{3/2}}$	Cl_{2s}
80/20 Kel-F	CF_2	291·7	6·7	690·3	1·0	—	—
	CFCl	290·5	5·5	690·3	1·0	201·4	272·8
	CH_2	286·8	1·8	—	—	—	—
30/70 Kel-F	CF_2	291·5	6·5	690·5	1·2	—	—
	CFCl	290·6	5·6	690·5	1·2	201·6	272·5
	CH_2	286·9	1·9	—	—	—	—
$+CF_2-CFCl+_n$	CF_2	291·9	6·9	690·8	1·5	—	—
	CFCl	290·8	5·8	690·8	1·5	201·1	272·2

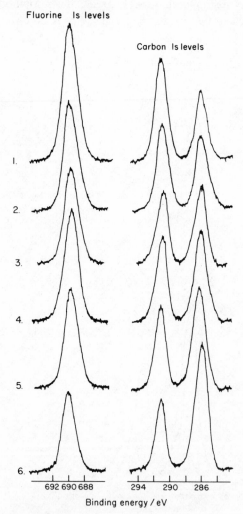

Figure 9.30 C_{1s} and F_{1s} spectra of ethylene/ tetrafluoroethylene copolymers; for composition, see Table 9.12

the ($\underline{C}H_2$) peak area. Since the total $\underline{C}F_2$ content of each polymer is 50 mole %, the difference between the percentage of the total C_{1s} peak area attributable to ($\underline{C}F_2 + \underline{C}FCl$) and 50 gives the percentage of the total C_{1s} area due to $\underline{C}FCl$; twice this figure gives the amount of chlorotrifluoroethylene units in the polymer. For the 80/20 and 30/70 copolymers, this gives 78 % and 32 % respectively, both being within 2 % of the values of 80 % and 30 % based on elemental analysis. If the areas of the $\underline{C}H_2$ peaks are used, the proportions of chlorotrifluoroethylene so obtained are again 78 % and 32 %. Thus both methods give exactly the same composition within 2 % of the quoted values.

The C_{1s} and F_{1s} spectra of a series of *ethylene/tetrafluoroethylene copolymers* are recorded in Figure 9.30. From the ESCA data the copolymer compositions may be calculated in two independent ways[16]: firstly, from the relative ratios of the high to low binding energy peaks in the C_{1s} spectrum attributable to CF_2 and $\underline{C}H_2$ type environments respectively; secondly, from the overall C_{1s}/F_{1s} intensity ratios taken in conjunction with the relative intensity ratio established previously for the homopolymers (Table 9.7). The results of these two analyses from ESCA data and those derived from conventional analysis for carbon and fluorine, and predicted from monomer reactivity ratios are compared in Table 9.12.

Table 9.12   Analysis of ethylene/tetrafluoroethylene comonomer incorporations

Sample	Composn. monomer mixture mol % C_2F_4	Copolymer composition (mol % C_2F_4)				
		Predicted from monomer reactivity ratios	Calc. from C analysis	Calc. from F analysis	Calc. from area ratio C_{1s} peak: F_{1s} peak	Calc. from $C_{1s}(CH_2$ peak): $C_{1s}(CF_2$ peak)
1	94	63	61	61	63	62
2	80	53	52	54	52	52
3	65·5	50	49	48	47	46
4	64	50	47	45	44	45
5	35	45	41	40	42	40
6	15	36	—	—	32	31

The agreement between the two sets of ESCA data is striking and demonstrates the uniformity (within the outermost ~ 50 Å) of the copolymer since the escape-depth dependences for the F_{1s} and C_{1s} levels are significantly different. This is indicative of a largely alternating structure for the copolymers since the PTFE domains of a block copolymer would be expected (on the basis of their lower free energy) to predominate at the surface. In this hypothetical case, estimates of TFE incorporation based on either ESCA method would be unlikely to agree and both procedures would give values higher than those computed from elemental analysis of bulk samples. Comparison of the ESCA data with classical bulk analyses reveals overall good agreement and also demonstrates that for these systems ESCA is competitive with microanalytical techniques in terms of accuracy and reproducibility, with the added advantage of being non-destructive and much faster.

Finally in this section we show how copolymer compositions may be deduced from shake-up satellites for some styrene hydrocarbon copolymers. ESCA would not at first sight seem particularly well suited for studies of composition of such copolymers since the shifts within the single core level observed (C_{1s}) are so small. In particular cases however the well-developed shake-up structure arising from $\pi^* \leftarrow \pi$ transitions within the pendant groups can provide sufficient additional information to allow compositions to be determined. As simple examples we have studied copolymers of general composition.[27,28]

$$\{CH-CH_2-CH_2-CH-(CH_2)_x\}_n \qquad x = 1, 2, 3, 4, \ldots$$
$$\quad | \qquad \qquad \qquad |$$
$$\quad Ph \qquad \qquad \quad Ph$$

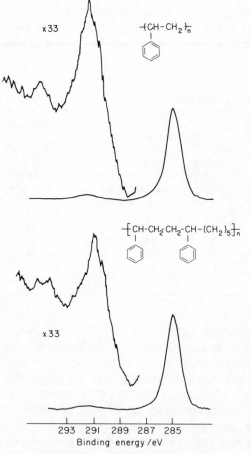

Figure 9.31 C_{1s} spectra, showing shake up peaks for polystyrene and styrene/pentane co-polymer

The C_{1s} levels for the styrene/pentane copolymer (x = 5) are shown in Figure 9.31 and the spectrum of polystyrene has been included for comparison. Since the shake-up structure shown in the spectra arises from the phenyl rings and since these are effectively insulated from one another by the saturated backbone it is not unreasonable to start with the working hypothesis that the shake-up probabilities for the phenyl rings should be additive. For polystyrene the ratio of intensities of the main photoionization peak and shake-up satellites for cast films is found to be 17·3. Since only the carbon atoms of the unsaturated system contribute to this it may readily be shown that for a copolymer of composition corresponding to the styrene/pentane copolymer the calculated ratio of intensities should be 22·7. This is in excellent agreement with the measured value of 22·7 indicating that ESCA may be used for quantitative discussions of copolymer compositions in such systems.

9.4.3.2 Structural details

The examples presented in the previous section illustrate the utility of ESCA in the determination of copolymer compositions and in providing information on gross details of structure and bonding. The latter information may be most straightforwardly derived from shifts in core binding energies; however, we have previously noted that for saturated systems the factors governing shifts are relatively short range in nature. This being so it is not infrequently the case that shifts in core binding energies for structurally isomeric systems are within experimental error the same and therefore indistinguishable as far as the core levels are concerned. This is apparent from the previous discussions of simple homopolymers involving unsymmetrical vinyl monomers. For poly(vinyl fluoride), poly(vinylidene fluoride) and polytrifluoroethylene, for example, both theoretical calculations and experimental data show that shifts in binding energies for the C_{1s} levels are the same for head-to-tail, or tail-to-tail sequences.[15]

In other cases the shifts between possible structurally isomeric units are sufficiently large to be measured and in these cases ESCA may be used to provide finer details of molecular structure. As an example we outline pertaining to nitroso rubbers.[29]

The C_{1s} spectra for copolymers of trifluoronitrosomethane with tetrafluoroethylene, chlorotrifluoroethylene and trifluoroethylene are shown in Figure 9.32. The assignment of peaks is straightforward and the shifts are understandable in terms of simple substituent effects. The 1:2 area ratio for $\underline{C}F_3$ with respect to $\underline{C}F_2$ carbons for the copolymer involving tetrafluoroethylene together with the relevant binding energies demonstrates the 1:1 alternating nature of this copolymer. Similar arguments apply to the other polymers. The possibility of structural isomerism exists for the copolymers with trifluoroethylene and chlorotrifluoroethylene and its presence can be demonstrated by a careful analysis of the ESCA spectra. For example, for the C_{1s} levels of the polymer involving trifluoroethylene although the three peaks are of equal area there are substantial differences in linewidth that are not apparent in the spectrum for the polymer formed from tetrafluoroethylene (Figure 9.32). To investigate the possibility that the lineshapes may contain information concerning structural

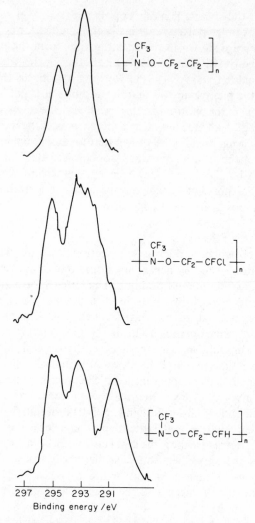

Figure 9.32 C_{1s} spectra for some nitroso rubbers

isomerism, calculations were performed on model systems; these calculations leading to predictions of binding energy shifts for the possible isomeric structures. The broadened peaks in the C_{1s} spectra were then deconvoluted using the lineshape observed for the trifluoromethyl carbon level as a standard, and in the case of the trifluoroethylene copolymer a satisfactory deconvolution is only possible if the two isomeric structures occur with equal frequency; see Figure 9.33. This deconvolution gives experimental shift differences between $\underline{C}F_3$ and $\underline{C}F_2$ levels for the two isomeric structures of 1·6 and 2·2 eV, to be compared with the theoretically predicted differences of 1·4 and 2·1 eV (Table 9.13), the exper-

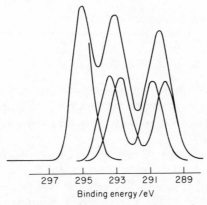

Figure 9.33 Deconvolution of C_{1s} spectra for nitroso rubber containing both $\text{+N(CF}_3)\text{OCF}_3\text{CFH+}$ and $\text{+N(CF}_3)\text{OCFHCF}_2\text{+}$ units

Table 9.13 Calculated and experimental shifts for models of the nitroso rubbers

Copolymer	Model compound		Shifts in relative binding energy/eV	
			Calc.	Obs.
CF_3NO and $CF_2{=}CF_2$	$CF_3{+}CF_2{-}\overset{\overset{\displaystyle CF_3}{\vert}}{N}{-}O{-}CF_2{+}CF_3$	$\Delta\underline{C}F_3{-}\underline{C}F_2(N)$	1.7	1.8
		$\Delta\underline{C}F_3{-}\underline{C}F_2(O)$	1·5	1·8
CF_3NO and $CF_2{=}CFCl$	$CF_3{+}\overset{\overset{\displaystyle CF_3}{\vert}}{N}{-}O{-}CFCl{+}CF_3$	$\Delta\underline{C}F_3{-}\underline{C}FCl$	3·4	3·1
	$CF_3{+}CFCl{-}\overset{\overset{\displaystyle CF_3}{\vert}}{N}{-}O{+}CF_3$	$\Delta\underline{C}F_3{-}\underline{C}FCl$	3·6	3·1
CF_3NO and $CF_2{=}CFH$	$CF_3{+}CFH{-}\overset{\overset{\displaystyle CF_3}{\vert}}{N}{-}O{-}CF_2{+}CF_2H$	$\Delta\underline{C}F_3{-}\underline{C}FH$	4·0	4·4
		$\Delta\underline{C}F_3{-}\underline{C}F_2$	1·4	1·9
	$CF_2H{+}CF_2{-}\overset{\overset{\displaystyle CF_3}{\vert}}{N}{-}O{-}CFH{+}CF_3$	$\Delta\underline{C}F_3{-}\underline{C}FH$	4·3	4·4
		$\Delta\underline{C}F_3{-}\underline{C}F_2$	2·1	1·9

imental $\underline{C}F_3$ to $\underline{C}FH$ shifts being 4·1 and 4·9 eV (theory 4·0 and 4·3 eV). The deconvolution for the chlorotrifluoroethylene copolymer is less straightforward and leads to the conclusion that

$$\overset{\overset{\displaystyle CF_3}{\vert}}{{+}N}{-}O{-}CF_2{-}CFCl{+}$$

is the major contributing structure and that

$$
\begin{array}{c}
\text{CF}_3 \\
|
\end{array}
$$
$$
+\text{N}-\text{O}-\text{CFCl}-\text{CF}_2+
$$

is the minor structural feature and present to an extent of between 20 and 33%. These conclusions are in reasonable agreement with earlier investigations of the structure of these copolymers which were based on an analysis of the products of pyrolysis of the polymers.[30]

The series of copolymers of ethylene with tetrafluoroethylene discussed previously provide a good example of how the details of molecular structure can be established by ESCA. In this case the information to be established, once the composition in terms of comonomer incorporation is known, is the sequence distribution. For these particular copolymers the problem has also been studied by the classical kinetic approach and so it is possible to compare the results of the two different approaches.[16] Preliminary observations (section 9.4.3.1) indicated that an essentially block structure was unlikely and careful deconvolution of the C_{1s} spectra (Figure 9.30, Table 9.14) and assignment of

Table 9.14 Analysis of component peaks, a–g, for C_{1s} levels for ethylene/tetrafluoroethylene copolymer. The table shows the binding energy/eV for each peak, with the % total area shown in brackets

Sample No.	a	b	c	d	e	f	g
1	292·1(25)	291·2(31)	290·4(5)	287·8(3)	286·3(30)	285·7(5)	285·0(1)
2	292·1(12)	291·0(33)	290·2(7)	287·6(4)	286·3(32)	285·7(6)	285·0(6)
3	292·3(4)	291·0(37)	290·1(5)	287·7(5)	286·4(36)	285·7(9)	285·0(4)
4	292·4(6)	291·1(35)	290·4(4)	287·7(4)	286·3(35)	285·7(9)	285·0(7)
5	292·2(2)	291·0(35)	289·4(3)	287·8(4)	286·4(34)	285·6(17)	285·0(5)
6	292·1(3)	291·1(27)	289·5(1)	287·3(3)	286·4(26)	285·7(30)	285·0(10)

the deconvoluted peaks to particular structural features with the aid of calculations for model compounds and the data banks established from earlier studies of the fluorethylene homopolymers allows the distribution of the various structural possibilities to be quantified. This procedure was complicated by the discovery that in addition to the expected methylene and difluoromethylene units there were small but significant concentrations of carbonyls in the surface layers of these copolymers; however, this could be satisfactorily taken into account and the analysis of sequence distribution in terms of pentads is in good agreement with that derived from a consideration of copolymerization statistics in the light of established monomer reactivity ratios. The core levels corresponding to the central triads of each pentad sequence were assigned to the six distinct peaks ($a \rightarrow g$) obtained from the deconvolution of the C_{1s} spectra (after excluding the surface carbonyl contributions); this assignment was relatively

Table 9.15 Assignment of peaks $a–g$ of the C_{1s} core levels to the most probable pentad sequences in ethylene (B)/tetrafluoroethylene (A) copolymer

Sequence	Assignment
ABAAA	$-CF_2CF_2(CH_2CH_2CF_2CF_2CF_2CF_2)CF_2CF_2-$ e e b a a a
ABAAB	$-CF_2CF_2(CH_2CH_2CF_2CF_2CF_2CF_2)CH_2CH_2-$ e e b a a b
ABABA	$-CF_2CF_2(CH_2CH_2CF_2CF_2CH_2CH_2)CF_2CF_2-$ e e b b e e
BBABA	$-CH_2CH_2(CH_2CH_2CF_2CF_2CH_2CH_2)CF_2CF_2-$ f e c b e e
BABAA	$-CH_2CH_2(CF_2CF_2CH_2CH_2CF_2CF_2)CF_2CF_2-$ b b e e b a
BABAB	$-CH_2CH_2(CF_2CF_2CH_2CH_2CF_2CF_2)CH_2CH_2-$ b b e e b b
ABBAB	$-CF_2CF_2(CH_2CH_2CH_2CH_2CF_2CF_2)CH_2CH_2-$ e f f e c b
BBBAB	$-CH_2CH_2(CH_2CH_2CH_2CH_2CF_2CF_2)CH_2CH_2-$ g g f ϵ c c

straightforward and representative examples are given in Table 9.15. Using these assignments the probabilities for particular sequences (derived from reactivity ratios) were compared directly with the deconvoluted experimental spectra, and the results of this comparison are recorded in Table 9.16. The measure of agreement between these completely different approaches to sequence distribution is remarkably good and well within the bounds of the likely experimental error of either method.

Table 9.16 Comparison of observed % areas of peaks $a–g$ with those calculated from pentad assignments with the aid of known reactivity ratios (cf. Tables 9.14 and 9.15)

Sample	Peak:	a	b	c	e	f	g
1	Experimental	25	31	5	30	5	1
	Theory	26	37	—	37	—	—
2	Experimental	12	33	7	32	6	6
	Theory	8	43	1	45	1	—
3, 4	Experimental	5	37	5	37	9	4
	Theory	4	46	2	46	3	1
5	Experimental	2	35	3	34	17	5
	Theory	1	39	6	42	10	3
6	Experimental	3	27	1	26	30	10
	Theory	—	24	10	31	23	12

9.5 APPLICATION TO SURFACE STUDIES

In previous sections the application of ESCA to homogeneous materials has been outlined. It should be evident from this discussion that ESCA nicely complements other more conventional techniques for establishing composition and details of structure and bonding. However in the introductory section the unique attribute of ESCA in being capable of differentiating surface from sub-surface from bulk structures for inhomogeneous materials was outlined. In this section representative examples of the application of ESCA to surface studies of polymers are given.

Some of the earliest applications of ESCA in polymer chemistry in fact were aimed specifically at investigation of the surface. This is not surprising in view of the particular suitability of ESCA for examination of surfaces and the considerable technological importance of surface modifications of polymers.

9.5.1 Surface structure of block copolymers

AB block copolymers frequently exhibit phase separation which typically gives rise to a dispersed phase consisting of one block type in a continuous matrix of the second block type. The detailed morphology of the domain structure of block copolymers depends upon such factors as the relative proportions of the two components, the molecular weight, the thermal and physical history of the polymer and for solvent-cast films, upon the solvent and temperature.[31-33] Earlier in this article (section 9.4), in relation to an examination of ethylene/tetrafluoroethylene copolymers, it was briefly indicated how ESCA could be used to demonstrate the absence of any surface segregation of polytetrafluoroethylene block sequences; and consequently the unlikelihood of such sequences being present in the copolymers. In the example discussed below, the surface structure of AB block copolymers of polydimethylsiloxane (PDMS) and polystyrene (PS) is examined by ESCA and the nature and thickness of the outer layer of the copolymer established.

Solvent-cast films of PDMS/PS AB block copolymers have been extensively studied by electron micrography in recent years.[34] The extreme incompatibility of the two molecular segments has been shown to result in the formation of discrete and continuous regions and a variety of domain structures have been observed for the polymer bulk. The large differences in surface free energy of the two components makes this copolymer system particularly interesting from the point of view of an investigation of surface structures by both contact-angle measurements and by ESCA, more particularly for the latter since a variety of core levels are available for study. Films of both homopolymers and two AB block copolymers containing respectively 23 and 59 wt. % PS were prepared by pressing and solvent casting. The contact angles of water and glycerine on all the copolymer films were identical to those determined on pure PDMS and differed significantly from the corresponding contact angles on polystyrene. This suggests that the immediate outer boundaries of the copolymer films are independent of the composition or physical treatment of the polymers, the PDMS component always predominating in at least an outer monolayer of the solid.

Accurate values of the intensity ratios $I_{\infty C_{1s}}:I_{\infty O_{1s}}:I_{\infty Si_{2p}}$ (see p. 124 for meaning of symbols) determined from spectra of polydimethylsiloxane are given in Table 9.17, the subscripts referring to the appropriate core electron levels.

Table 9.17 Experimental intensity ratios in the ESCA spectra of films of polydimethylsiloxane (PDMS) and polystyrene (PS)-polydimethylsiloxane AB block copolymers. The values for polydimethylsiloxane were taken as the I_∞ value (see text)

Polymer film	$I_{C_{1s}}/I_{Si_{2p}}$	$I_{C_{1s}}/I_{O_{1s}}$	$I_{O_{1s}}/I_{Si_{2p}}$
PDMS (I_∞ values)	1.66 ± 0.05^a	1.20 ± 0.05	1.38 ± 0.05
PS-PDMS copolymers:			
copolymer 1	1.7 ± 0.1	1.3 ± 0.1	1.35 ± 0.05
copolymer 2 cast from			
toluene	2.59 ± 0.05	1.71 ± 0.04	1.51 ± 0.01
benzene	2.78 ± 0.03	1.84 ± 0.10	1.51 ± 0.10
cyclohexane	2.27 ± 0.10	1.57 ± 0.04	1.44 ± 0.03
bromobenzene	1.68 ± 0.01	1.24 ± 0.01	1.36 ± 0.03
styrene	1.77 ± 0.11	1.28 ± 0.08	1.38 ± 0.01

[a] Errors are standard deviations.

The corresponding intensity ratios for the block copolymer films studied are also listed in this table. It is evident that copolymer 1 and the films of copolymer 2 cast from bromobenzene and styrene (selective solvents for PS) are indistinguishable from pure PDMS by the ESCA technique, indicating that in these samples the PDMS component tends to form the surface. With a knowledge of the appropriate electron escape depths, it is possible to set an approximate lower limit to the thickness of this surface coverage; taking $\Lambda_{C_{1s}} = 10$ Å, $\Lambda_{O_{1s}} = 8$ Å, and $\Lambda_{Si_{2p}} = 11.5$ Å, the PDMS layers must be at least ~ 40 Å deep.

The intensity ratios for the films of copolymer 2 cast from toluene, benzene (solvents for both components), and cyclohexane (preferential solvent for PDMS) indicate that a small amount of PS is being detected along with the PDMS component; assuming that the PS is distributed uniformly throughout the surface layer, the C_{1s}/Si_{2p} and C_{1s}/O_{1s} intensity ratios can be used to calculate the mole fractions of each component (styrene or dimethylsiloxane) present in the surface layer observed by ESCA. In terms of the mol fraction of the PS segment, x, such a treatment gives x (toluene) $= 0.12 \pm 0.02$, x (benzene) $= 0.13 \pm 0.03$ and x (cyclohexane) $= 0.08 \pm 0.03$, where the casting solvent is indicated.

However, the assumption of a uniform copolymer composition throughout the surface region is questionable. The contact-angle measurements suggest that in all the copolymer films the immediate outer boundary is predominantly PDMS. Also, the O_{1s}/Si_{2p} intensity ratios for the films cast from toluene, benzene and cyclohexane are significantly larger than would be observed for a uniform surface region. Since the escape depth of electrons originating from the Si_{2p} atomic level is greater than that of O_{1s} electrons, the experimental intensity ratios indicate that the concentration of the PDMS component decreases with depth into the polymer.

Assuming then that the copolymer films consist of an outer layer of pure PDMS, the intensity ratios of the various core levels determined by ESCA can be used to estimate the thickness of this layer. As a simple model, suppose each film consists of a layer of PDMS of thickness d covering the polymer bulk, which is composed of 60 mol % PS, 40 mol % PDMS. Using the experimentally determined intensity ratios listed in Table 9.17, with the values for the electron escape depths given above, three independent estimates of d are available. The results of such a treatment are given in Table 9.18, and the consistency in the values obtained for a given copolymer film can be taken as an indication of the reliability of the treatment and the constants used.

Table 9.18 Calculated thickness of the surface PDMS layer in PS-PDMS AB block copolymers

Polymer film	Thickness (in Å) of PDMS surface calculated from		
	$I_{O_{1s}}/I_{Si_{2p}}$	$I_{C_{1s}}/I_{Si_{2p}}$	$I_{C_{1s}}/I_{O_{1s}}$
copolymer 1	$\gtrsim 40$		
copolymer 2 cast from			
toluene	13	17	17
benzene	13	15	15
cyclohexane	25	20	23
bromobenzene	$\gtrsim 40$		
styrene	$\gtrsim 40$		

Of course this model probably does not correspond exactly to the morphology of the actual system. There is undoubtedly a transition region of composition intermediate between that assumed for the surface layer and the bulk, just as the bulk domain structure of an AB block copolymer undoubtedly consists of regions of pure A, regions of pure B, and transition regions of mixed A and B composition. But in the PS-PDMS system regions of mixed compositions are likely to be of minor importance since the high incompatibility of the two components is expected to lead to sharp phase separation. At any rate, calculations done on model systems incorporating such regions have shown that these considerations have only a small effect on the calculated thickness of the outer PDMS layer.

It is important to note that although the compositions of the surfaces of these films are all very similar, the bulk morphologies undoubtedly differ. The predominance of the PDMS component in the surface undoubtedly reflects the substantially lower surface free energy of PDMS compared with PS rather than a change in the domain morphology. The origin of the domain structures of block copolymers cast from solutions has been attributed to the formation of micelles of the copolymer in solution.[32,35] The formation of particular types of micelles depends principally upon interactions between the block components and the differential solvation of these sub-units in a given solvent. Such considerations can account for the various bulk morphologies, but on formation of the polymer-air interface in the casting process, it is necessary also to consider the surface free energy in order to rationalize the surface morphology. This study showed that the domain structure in the surface of an AB block copolymer may differ considerably from that in the bulk.[36]

9.5.2 CASING

The formation of good adhesive bonds to plastics is a problem of some technological importance. Several surface treatments of polymers have been

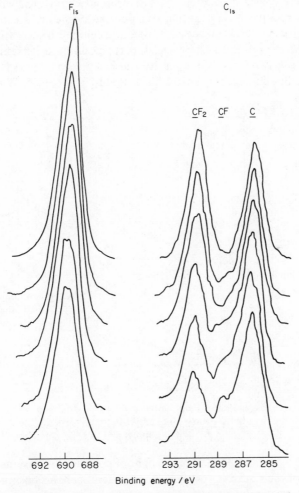

Figure 9.34 F_{1s} and C_{1s} levels for argon-ion-bombarded samples of ethylene/tetrafluoroethylene copolymer

168

developed in attempts to overcome this problem. One such treatment, in which the material is exposed to activated species of an inert gas, is particularly effective and has been designated CASING (Cross-linking by Activated Species of Inert Gases).[37] Studies of polyethylene submitted to this treatment suggested that the modified surface properties arose from extensive cross-linking in the outer layers of the sample.

ESCA has been used to study this phenomenon in copolymers of ethylene with tetrafluoroethylene. Samples were irradiated with a low energy (2 kV) beam of argon ions with a beam current of 5 μA for successive periods of 5 s and the C_{1s} and F_{1s} core levels monitored. The results, shown in Figure 9.34, are quite striking and it is clear that argon ion bombardment causes extensive rearrangement in the outermost 50 Å or so of the sample. The main features are as follows. The F_{1s} signal decreases whilst the total C_{1s} signal remains approximately constant in intensity. The original copolymer, with its largely alternating structure of ethylene and tetrafluoroethylene units, exhibits initially the characteristic doublet structure in the C_{1s} region. On successive argon-ion treatment the high-binding-energy peak corresponding to tetrafluoroethylene units progressively decreases in intensity and is replaced by peaks in a region of intermediate binding energy. This is illustrated in Figure 9.35. With the results

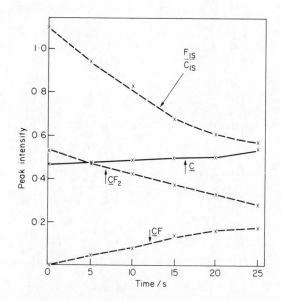

Figure 9.35 Peak intensities as a function of irradiation time for ethylene/tetrafluoroethylene copolymer

from Figures 9.34 and 9.35, and in the light of previous discussions of the likely mechanism of the CASING process, some general conclusions may be drawn. The ESCA data clearly show the decrease in F_{1s} signal corresponding to loss of fluorine; and in the process $\underline{C}F_2$ sites are converted to $\underline{C}F$ type. It had previously been established that polytetrafluoroethylene requires prolonged treatment for any detectable CASING effect, so it is reasonable to assume that the reduction of F_{1s} signal intensity results from inter- and/or intra-chain elimination of HF either synchronously or in a stepwise radical process in which the initial step involves the formation of a hydrogen atom and a —CH· radical. More detailed studies of these processes are in progress.

9.5.3 Alkali metal etching of PTFE

The remarkable physical properties and the difficulties of fabricating PTFE are well known. For many applications of PTFE it is essential to form a good adhesive bond between PTFE and some other material and the low surface energy of the fluorocarbon presents very great difficulties. Several methods of modifying the surface so that good adhesion can be obtained have been developed and one of these consists of a treatment with sodium in liquid ammonia, or sodium naphthalene in tetrahydrofuran (THF). PTFE surfaces modified in this way can be bonded to other materials with an adhesive, or can have another polymer surface grafted on them. Earlier work indicated that the modified surfaces contained carbon-carbon double bonds and a C:F stoichiometry of 1:1.[38,39] The nature of these surfaces has been examined by ESCA by two groups, with qualitatively similar results. The most striking feature immediately evident from the ESCA spectra is the total disappearance of all fluorine signals as a result of the surface treatment, from which it can be concluded that in the outermost 50–100 Å all the fluorines are removed; this layer of course is critical in determining the surface properties. The details of the residual surface are dependent on the exact treatment of the surface after etching with alkali metal and before presentation to the spectrometer. Neither of the groups who have reported investigations were successful in examining the freshly etched surface and it is clear that the surface reacts readily with atmospheric oxygen and oxygen-containing solvents. There can be several sorts of oxygen environment (depending on the surface treatment), and carbonyl, carboxyl and oxonium structures have been postulated. The C_{1s} spectra reveal several carbon environments in the range 285 to 288 eV. The freshly etched surface appears to be carbon and must be extensively unsaturated and contain a number of highly reactive sites. Ultraviolet irradiation or heating in the presence of air results in the removal of the surface layer and regeneration of the PTFE surface.

9.5.4 Oxidation of polymer samples

ESCA can be used to show the extent of oxidation at a polymer surface at a stage where other techniques fail to detect any change occurring. This is illustrated by the examination of various polyethylene samples.[40]

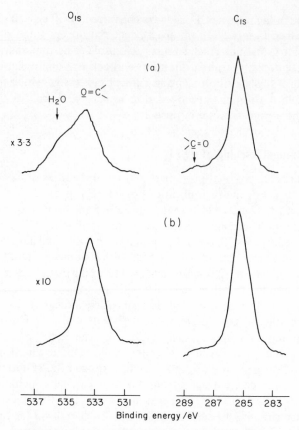

Figure 9.36 C_{1s} and O_{1s} levels for low density polyethylene (a) before desiccation, (b) after desiccation

Figure 9.36 shows the C_{1s} and O_{1s} spectra for commercially low-density polyethylene films both before (a) and after (b) desiccation. Two features are clearly evident. Firstly, the substantial signals corresponding to the O_{1s} levels exhibit an unresolved structure, one component of which disappears on storing the samples over P_2O_5. Comparison with previously measured binding energies indicates that the higher binding energy component of the O_{1s} levels corresponds to H_2O whilst the lower binding energy peak corresponds to $>C=\underline{O}$, although the peak width suggests that there are probably several differing carbonyl environments. Secondly, the C_{1s} levels show a small shoulder to the high-binding-energy side of the main peak, the shift of ~ 3 eV again being consistent with carbon in $>C=O$ environments. Comparison with MATR measurements employing single crystal germanium is instructive in this respect since strong absorptions are present in the carbonyl region (but not in the OH region). Careful double beam TIR experiments on the $200\,\mu$ films also shows the presence of carbonyl groups. The three techniques taken together therefore emphasize that whilst the carbonyl groups are distributed throughout the bulk of the sample, hydrogen bonding involving extraneous water is localized at the surface. The carbonyl groups may be attributable in part to various additives such as anti-oxidants, anti-blocking agents etc. By comparison, when a sample of unstabilized high-density polyethylene was examined as made (that is as a fine powder) no trace of O_{1s} signal could be detected. Preparation of films of this same polymer give different surface features according to the method used. Three films were pressed with a hand press (~ 200 °C) between sheets of clean aluminium foil at the minimum temperature necessary for plastic flow. The conditions of pressing were as follows: (a) pressed in air; (b) the samples were pressed in a nitrogen atomosphere; (c) the polyethylene

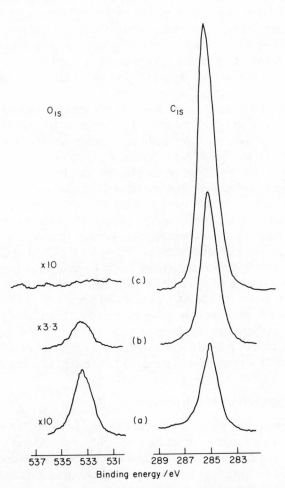

Figure 9.37 C_{1s} and O_{1s} levels for pressed films of high density polyethylene. For conditions of pressing see text

powder was pumped down (10^{-4} Torr) and let up to a pure nitrogen or argon atmosphere for several cycles before being transferred in an inert atmosphere into the press.

The O_{1s} and C_{1s} spectra corresponding to samples from these three modes of preparation are shown in Figure 9.37 and are quite striking. ATR and TIR experiments did not reveal the presence of any oxygen function ($-OH$, $>C=O$, $C-O-C$ etc.) and in fact the spectra were virtually identical. This in itself demonstrates the great power of ESCA in distinguishing minute differences between samples when such differences are localized at or near the surface. Comparison with the data for low-density polyethylene and with model monomer systems shows that the O_{1s} signal arises from $>\underline{C}=O$ type environments. The three methods of preparation clearly indicate that 'unoxidized' surfaces may most readily be prepared by excluding all traces of oxygen during the pressing stage. On leaving samples exposed to the atmosphere for some time hydrogen bonding to the surface $>C=O$ groups from extraneous water in the atmosphere occurs and the O_{1s} peak then acquires the characteristic doublet nature. Again MATR and TIR do not reveal any changes since the hydrogen bonding is localized at the surface.

It is of interest to obtain some information on the relative frequency of $>C=O$ features in the oxidized films and this may be achieved as follows. From extensive studies of compounds containing both carbon and oxygen, infinity values (I_α) may be obtained for the C_{1s} and O_{1s} core levels in homogeneous solids. If we make the reasonable assumption that because of temperature gradients, diffusional phenomena, and the relatively low concentration of oxygen (the only available oxygen being that entrained in the powdered polyethylene sample), oxidation in the sample prepared by method (II) is likely to be limited to the first monolayer, we may then with a knowledge of the likely escape-depth dependences (Λ_1 and Λ_2 for the O_{1s} and C_{1s} core levels) calculate the surface concentration of $>C=O$ structural features. Taking Λ_1 and Λ_2 as 8 Å and 10 Å (corresponding to Mg K$\alpha_{1,2}$ exciting radiation) it is calculated that if all the O_{1s} signals correspond to $>C=\underline{O}$ located in the first monolayer then there is $\sim 9\%$ monolayer coverage of this structural feature. Films prepared in this manner were used in an investigation of the surface fluorination of polyethylene and detailed analysis of the results of that work confirmed this preliminary estimate of the extent of surface oxidation.[40]

9.5.5 Surface fluorination of polyethylene

Direct fluorination of polyethylene has been the subject of several investigations,[41-45] the main incentive being the possibility of producing a fluorocarbon layer, with its associated desirable chemical and physical properties, on the surfaces of articles fabricated from a relatively cheap and malleable material. Despite extensive investigation of this process little information has been acquired concerning the early stages of the reaction of fluorine with polyethylene films which must, of course, be initiated at the surface. A priori, therefore, one might expect a dependence of stoichiometry (and hence structure) on depth as a function of the fluorinating conditions (e.g. concentration, temperature and duration). Indeed there is evidence for this from studies of extensive fluorination of films of poly(vinyl fluoride) of different thicknesses. Although, in general, previous investigations have established overall stoichiometries for the extensively fluorinated films, because of the depth dependence of composition it is not possible to derive unambiguously either the stoichiometry at the surface and immediate sub-surface or more particularly the actual molecular structure in this region which is obviously crucial not only to an understanding of the kinetics and mechanism of the fluorination process but also to an understanding of the physico-chemical and mechanical properties of the surface. Application of ESCA to the investigation of this process allows attention to be directed to the following important aspects:

(i) establishing the depth of fluorination as a function of reaction conditions,
(ii) stoichiometries as a function of depth,
(iii) molecular structure of the immediate sub-surface,
(iv) kinetics and mechanism of the process.

Films of polyethylene, prepared and characterized as described in the previous section, were exposed to an atmosphere of 10 % fluorine in nitrogen. Extensive modification of the surface could readily be demonstrated by ESCA

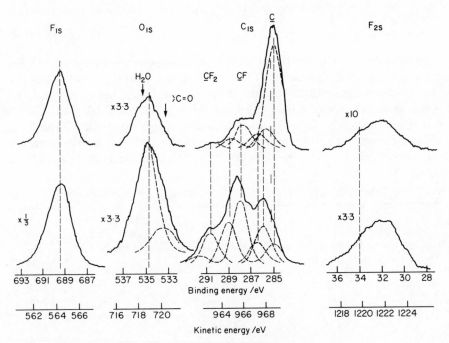

Figure 9.38 C_{1s}, O_{1s}, F_{1s} and F_{2s} levels for high density polyethylene films fluorinated for 0·5 (upper curves) and 30 s (lower curves)

at a stage where other techniques (transmission and MATR infrared spectroscopy) failed to distinguish any differences between the original and the treated films; see Figure 9.38.

The detailed analysis of these results is lengthy and involved and is to be published elsewhere.[40] However, the conclusions emerging from the analysis may be summarized as follows:

(i) The thickness of the fluorinated layer can be established either from the intensities of the F_{1s} and F_{2s} peaks, or from the deconvoluted C_{1s} spectra. In the 0–50 Å region for which these analyses are applicable, agreement between the two approaches is very good; see Table 9.19.

Table 9.19 Calculated film thickness d for surface-fluorinated polyethylene films

Time of exposure to fluorine/s	$\dfrac{A_{C_{1s}}(F)^a}{A_{C_{1s}}(E)}$	$d/Å$		
		from C_{1s} spectra	from F_{1s} and F_{2s} spectra	Av.
0·5	0·71	5·5	3·5	4·5
2	1·50	9·0	6·0	7·5
15	3·76	15·0	16·0	15·5
30	10·10	24·0	30·0	27·0
300	15·67	28·0 (46)[b]	36·0	32 (41)[b]

[a] $A_{C_{1s}}(F)$ denotes area of peak from fluorinated polymer; $A_{C_{1s}}(E)$ is corresponding area for polyethylene.
[b] Revised estimate, see text.

(ii) The overall stoichiometries of the fluorinated layers can be established from the C_{1s}/F_{1s} peak intensities and from the deconvoluted C_{1s} spectra; for the latter method the contribution due to surface oxidation can be allowed for. Again agreement between the two approaches is good; see Table 9.20.

(iii) When the depth of fluorination (using the average values from Table 9.19) is plotted versus log t, it is evident that the points for the first four experiments in the range 0·5–30 s fit on a smooth curve whilst that for the longest run of 300 s

Table 9.20 C/F stoichiometries for fluorinated polyethylene films

Time of exposure to fluorine/s	Method I: C/F from C_{1s}/F_{1s} spectra	Method II: C/F from C_{1s} spectra	Method III: C/F from C_{1s} spectra taking surface oxidation into account	Av. I and III
0·5	2·01	1·60	2·05	2·03
2	1·43	1·40	1·54	1·48
15	1·37	1·27	1·40	1·38
30	1·16	1·12	1·23	1·20
300	1·17	1·04	1·12 (1·16)[a]	1·16

[a] Revised estimate, see text.

would appear to indicate too low a fluorinated film thickness. The most likely cause of this would be a sub-monolayer contaminant hydrocarbon film arising after the fluorination experiment. Such hydrocarbon contaminant would have the same binding energy (285 eV) as polyethylene and unless allowed for would lead to a low estimate for the fluorinated film thickness. Clearly the longer the reaction and flushing time the greater the chances of contaminating the surface in this way, and further this effect will cause increasing errors with increasing fluorinated film thickness. Blank experiments showed that hydrocarbon surface contaminant does accumulate on fluorocarbon surfaces under our reaction conditions and a detailed analysis of the results leads to the conclusion that for the 300 s experiment this contamination amounts to *ca.* 10% monolayer coverage of the surface with hydrocarbon. Allowing for this in a recalculation of the depth of fluorination gives a value of 46 Å (footnote Table 9.19) which brings the point for the 300 s experiment onto the smooth curve for *d* versus log *t*; further surface stoichiometry for this point can also be recalculated giving the revised value shown in Table 9.20, column III.

(iv) It is to be expected that as the fluorination proceeds into the sample the initially lightly fluorinated surface monolayer will become progressively more fluorinated. Information concerning this aspect of the process can also be derived from a detailed analysis of the data. Figure 9.39 shows the derived first monolayer stoichiometry as a function of time. This analysis indicates that although for the 300 s experiment the overall stoichiometry for the fluorinated

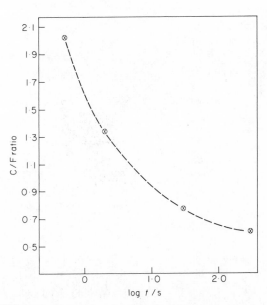

Figure 9.39 First monolayer stoichiometry as a function of time of fluorination of polyethylene

176

layer is 1·16 (Table 9.20), corresponding to a composition intermediate between $(C_2FH_3)_n$ and $(C_2F_2H_2)_n$, at this stage the surface monolayer has reached a stoichiometry intermediate between that of $(C_2F_3H)_n$ and $(C_2F_4)_n$, which is an entirely reasonable conclusion.

(v) Finally, some comment on the dynamics of the fluorination process can be made. Although the abstraction reaction which initiates the fluorination process has a low activation energy, successive replacement of hydrogen by fluorine will be expected to increase the activation energy both because of the concomitant strengthening of C—H bonds and because of adverse polar effects in the transition state. Of some relevance in this connection are the relative rates of fluorination at different sites reported for gas phase studies of 1-fluorobutane, $CH_2FCH_2CH_2CH_3$, which are 0·3:0·8:1:1 for the four positions α, β, γ and δ to the fluorine atom. It is clear that introduction of fluorine deactivates both α and β sites. This data may be used as a basis for discussing the fluorination of polyethylene.

For the initial surface fluorination, corresponding to the 0·5 s experiment, an overall C/F stoichiometry of the surface film of ~ 2 is observed. Figure 9.40 shows the calculated relative rates of fluorination for a four-carbon fragment of the polymer chain based on the data for 1-fluorobutane. For a stoichiometry C/F of exactly 2, the possibilities for this simple model would be one CF_2

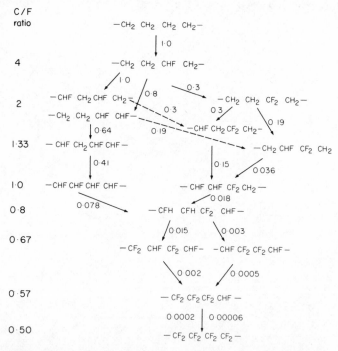

Figure 9.40 Scheme to illustrate the progress and relative rates of fluorination in a four-carbon fragment of a polyethylene chain

Table 9.21 C_{1s} levels for fluorinated films of polyethylene (normalized contributions)

Time of fluorination/s	1		2		3		4		5		6	
	BE[a]	%	BE	%	BE	%	BE	%	BE	%	BE	%
0·5	—	—	290·6	9·6	288·9	14·9	287·9	25·5	286·5	21·2	285·8	28·7
2	—	—	290·9	12·2	289·0	21·1	287·9	24·4	286·5	14·4	285·8	27·8
30	291·6	3·3	290·7	17·4	289·0	21·7	287·9	25·0	286·4	12·0	285·9	20·7
300	291·9	6·5	290·8	17·7	289·1	27·4	288·0	19·4	286·5	18·3	285·9	10·8

[a] BE denotes binding energy/eV.

group or two $\underline{C}$HF groups in the four carbon chain. It is clear that the preference for the latter should be about three times that for the former. The data in Table 9.21 provides striking confirmation of this.

The two major peaks of approximately equal intensity at 287·9 eV and 285·8 eV correspond extremely well in binding energy with those previously reported for poly(vinyl fluoride) (288·0 eV, 285·9 eV). By contrast the peak at 290·6 eV, corresponding in binding energy to $\underline{C}F_2$ groups with adjacent $-CH_2-$ groups (cf. 290·8 eV in poly(vinylidene fluoride)), represents approximately 0·3 times the intensity of these peaks. It might also be inferred from Figure 9.40 that progressive fluorination would tend to proceed relatively uniformly so that for a stoichiometry C/F of 1·0, a $(CHF-CHF)_n$ backbone would predominate. The relatively large increase in intensity of the peaks at $\sim 289·0$ eV (Table 9.21) would tend to support this. Clearly however, a great deal more experimental work is required before a more detailed discussion is possible.

My sincere thanks go to Dr. W. J. Feast who has made major contributions to the successful development of the ESCA polymer program in Durham. I would like to thank also a succession of extremely willing and able research fellows and graduate students who have participated at various stages in establishing the polymer program in the ESCA lab and to the Science Research Council and Institute of Petroleum for provision of equipment and operating funds. Thanks are also due to Professor W. K. R. Musgrave for his support and encouragement throughout the development of the research program. The greatest debt however, must be to Professor Kai Siegbahn and his co-workers at Uppsala for providing the scientific community with such a versatile and exciting technique.

9.6 REFERENCES

1. (a) K. Siegbahn *et al.*, *Nova Scta R. Soc. Sci.*, Uppsala Ser. IV, 20 (1967). (b) K. Siegbahn *et al.*, *ESCA Applied to Free Molecules*, North Holland Publishing Co., Amsterdam, 1969.
2. (a) D. M. Hercules, *Analytical Chemistry*, **44**, 106R, (1972). (b) D. T. Clark, 'Chemical Aspects of ESCA' in *Electron Emission Spectroscopy*, ed. W. Dekeyser, D. Reidel Publishing Co., Dordrecht, Holland, pp. 373–507, 1973, and references therein.

3. (a) D. T. Clark, in *Advances in Polymer Friction and Wear*, ed. L. K. Lee, **5, A,** 241, (1974), Plenum Press, New York. (b) D. T. Clark and W. J. Feast, *J. Macromol. Sci., Reviews in Macromol. Chem.*, **C, 12,** 191, (1975). (c) D. T. Clark, 'Structure and Bonding in Polymers as Revealed by ESCA', in Proceedings of the NATO Advanced Study Institute, NAMUR, Belgium, 1974, ed. J. Ladik and J. M. Andre, Plenum Press, New York, 1975.

4. (a) B. Wannberg, U. Gelius and K. Siegbahn, *J. Phys.* **E, 7,** 149, (1974). (b) U. Gelius, H. Fellner-Feldegg, B. Wannberg, A. G. Nilsson, E. Basilier and K. Siegbahn, University of Uppsala, Institute of Physics Report UUIP No. 855 (1974).

5. C. S. Fadley, 'Theoretical Aspects of X-ray Photoelectron Spectroscopy', in *Electron Emission Spectroscopy*, ed. W. Dekeyser, D. Reidel Publishing Co., Dordrecht, Holland, pp. 151–224, 1973.

6. Cf. H. Fellner-Feldegg, U. Gelius, K. Siegbahn, C. Nordling and K. Thimm, UUIP Report No. 856, (1974).

7. Cf. R. M. White, T. A. Carlson and D. P. Spears, *J. Electron Spectroscopy*, **3,** 59, (1974) and references therein.

8. C. S. Fadley, *J. Electron Spectroscopy*, **5,** 725, (1974).

9. L. G. Parrat, *Rev. Mod. Phys.*, **31,** 616, (1959).

10. (a) U. Gelius, S. Svensson, H. Siegbahn, E. Basilier, A. Faxalv and K. Siegbahn, UUIP Report No. 860, (1974). (b) H. Siegbahn, L. Asplund, P. Kelfire, K. Hamrin, L. Karlsson and K. Siegbahn, *J. Electron Spectroscopy*, **5,** 1059, (1974).

11. J. C. Tracey, 'Auger Electron Spectroscopy for Surface Analysis', in *Electron Emission Spectroscopy*, ed. W. Dekeyser, D. Reidel Publishing Co., Dordrecht, Holland, pp. 295–372, 1973.

12. B. J. Lindberg, *J. Electron Spectroscopy*, **5,** 149, (1974).

13. G. C. Levy and G. L. Nelson, *Carbon*-13 *Nuclear Magnetic Resonance for Organic Chemists*, Wiley–Interscience, 1972.

14. Cf. Ref. 2(b) and D. T. Clark, D. B. Adams and D. Kilcast, *J. Chem. Soc. Disc. Faraday Soc.*, **54,** 182, (1972).

15. D. T. Clark, D. Kilcast, W. J. Feast and W. K. R. Musgrave, *J. Polymer Sci., Polymer Chemistry Edn.*, **11,** 389, (1973).

16. D. T. Clark, W. J. Feast, I. Ritchie, W. K. R. Musgrave, M. Modena and M. Ragazzini, *J. Polymer Sci., Polymer Chemistry Edn.*, **12,** 1049, (1974).

17. D. T. Clark, D. Kilcast, D. B. Adams and W. K. R. Musgrave, *J. Electron Spectroscopy*, **1,** 227, (1972).

18. E. T. McBee, U.S. Patent 3,317,497, May 2 (1967).

19. W. L. Dilling, Thesis, Purdue University (1962); *Diss. Abstr.* **B27(4),** 1083, (1966).

20. D. T. Clark, W. J. Feast, M. Foster and D. Kilcast, *Nature Physical Science*, **236,** 107, (1972).

21. W. J. Feast and M. Foster, unpublished; report in part at the 3rd International Symposium on Polyhalogen Compounds, Barcelona, 1973.

22. Cf. Ref. 5, and U. Gelius, *Physica Scripta*, **9,** 133, (1974).

23. E. Clementi and H. Popkie, *J. Amer. Chem. Soc.*, **94,** 4057, (1972).

24. D. T. Clark, I. W. Scanlan and J. Muller, *Theoretica Chim. Acta*, **35,** 341, (1974).

25. C. W. Wilson and E. R. Santee, in Analysis and Fractionation of Polymers (*J. Polymer Sci.* **C., 8**), J. Mitchel and F. W. Billmeyer, eds., Interscience, New York, p. 97, 1965.

26. R. D. Chambers, D. T. Clark, D. Kilcast and S. Partington, *J. Polymer Sci., Polymer Chem. Ed.*, **12,** 1647, (1974).

27. F. J. Burgess, A. V. Cunliffe, D. H. Richards and P. Shadbolt, *European Polymer J.*, **10,** 193, (1974).

28. D. T. Clark and A. Dilks, *J. Polymer Sci.*, in press.

29. D. T. Clark, D. Kilcast, W. J. Feast and W. K. R. Musgrave, *J. Polymer Sci.*, **10,** 1637, (1972).

30. D. A. Barr, R. N. Haszeldine and C. J. Willis, *J. Chem. Soc.*, **1351**, (1961) and R. E. Banks, R. N. Haszeldine, H. Sutcliffe and C. J. Willis, *J. Chem. Soc.*, 2506, (1965).
31. J. V. Dawkins, in *Block Copolymers*, ed. D. C. Allport and W. M. Jones, Applied Science Publishers, London, p. 363, 1973.
32. H. Kawai, T. Soen, T. Inoue, T. Ono and T. Uchida, Memoirs of Faculty of Engineering, Kyoto University, **33**, 383, (1971).
33. S. L. Aggarwal, R. A. Livigni, L. F. Marker and T. J. Dudek, in *Block and Graft Copolymers*, Syracuse University Press, Syracuse, New York, 1970.
34. (a) J. C. Saam, D. J. Gordon and S. Lindsey, *Macromolecules*, **3**, 1, (1970). (b) J. C. Saam and F. W. S. Featon, *Ind. Eng. Chem. Prod. Res. Develop.*, **10**, 10, (1971).
35. T. Inoue, T. Soen, T. Hashimoto and M. Kawai, *J. Polymer Sci.*, **A2, 7**, 1283, (1969).
36. D. T. Clark, J. Peeling and J. M. O'Malley, in press.
37. R. H. Hansen and H. Schonhorn, *Polymer Letters*, **4**, 203, (1966).
38. H. Brecht, F. Mayer and H. Binder, *Angew. Makromol. Chem.*, **33**, 89, (1973).
39. W. M. Riggs and R. P. Fedchenko, *American Laboratory*, **65**, (1972).
40. D. T. Clark, W. J. Feast, W. K. R. Musgrave and I. Ritchie, *J. Polymer Sci., Polymer Chemistry Edn.*, **13**, 857, (1975).
41. A. J. Rudge, British Patent, 710,523 (June 1959).
42. J. L. Margrave and R. J. Lagow, *Chem. Eng. News*, **48**, 40, (1970).
43. K. Tanner, *Chimica*, **22**, 176, (1970).
44. H. Schonhorn, P. K. Gallagher, J. P. Luongo and F. J. Padden, *Macromolecules*, **3**, 800, (1970).
45. H. Shinohara, M. Twasaki, S. Tsujimwa, K. Watanabe and S. Okazaki, *J. Polymer Sci.*, **A1, 10**, 2129, (1972).

10
High-Resolution Carbon-13 Studies of Polymer Structure

F. A. Bovey

Bell Laboratories, New Jersey

10.1 INTRODUCTION

In this contribution, a few selected topics illustrating the use of carbon-13 nuclear magnetic resonance spectroscopy in the study of polymer structure are discussed. We will consider first the chemical shifts of paraffinic hydrocarbon chains in a general way, and discuss briefly some of the rules and regularities deduced from them which can be usefully extended to polymers. We shall then discuss certain aspects of the carbon-13 spectra of polyolefins, particularly the effects of *stereochemical configuration*, and will show to what extent it appears possible to rationalize these observations in terms of conformation. We shall then consider the measurement of *branching* in polyethylene and polyvinyl chloride, and, finally, some applications to polypeptides, in particular, poly-L-proline.

10.2 C-13 CHEMICAL SHIFTS OF PARAFFINIC HYDROCARBONS

Carbon-13 n.m.r. is especially powerful in the study of paraffinic hydrocarbons because of its great sensitivity to branching and other structural features. In contrast to paraffinic protons, which embrace a range of only about 2 ppm, carbon chemical shifts are spread over 45 ppm. This is essentially what gives the method its power. Here we shall briefly review the major chemical shift effects and correlations that have been observed in open-chain paraffins. This will help to clarify the interpretation of the polymer spectra which we shall discuss a little later. These regularities were first noticed and systematized over a decade ago by Spiesecke and Schneider[1] and by Grant and Paul[2] and later refined by Lindeman and Adams[3] and by Dorman *et al.*[4] The particular way in which they will be presented here is a somewhat simplified version of that presented by the last two groups of authors.

The empirical ordering of chemical shift effects is done in terms of α, β and γ effects. (δ and ε effects are also recognized, but as these are small we shall omit them here for brevity.) Table 10.1 shows a series of data for simple hydrocarbons which illustrate the α-*effect*. Here, we are to keep our eye on the °C

Table 10.1 α-effect

	ppm from TMS	α-effect ppm
(a) °CH₃—H	−2·1	—
(b) °CH₃—ᵅCH₃	5·9	8·0
(c) °CH₂(—ᵅCH₃)(—ᵅCH₃)	16·1	10·2
(d) °CH(—ᵅCH₃)(—ᵅCH₃)(—ᵅCH₃)	25·2	9·1
(e) °C(—ᵅCH₃)(—ᵅCH₃)(—ᵅCH₃)(—ᵅCH₃)	27·9	2·7

carbon and ask what happens on adding carbons α to this one. (Chemical shifts are expressed in ppm vs. tetramethylsilane (TMS) as zero; positive values represent *decreased* shielding; on some of the C-13 spectra in this paper, two shift scales

Table 10.2 β-effect I

	ppm from TMS	β-effect ppm
(a) °CH₃—ᵅCH₃	5·9	—
(b) °CH₃—ᵅCH₂—ᵝCH₃	15·6	9·7
(c) °CH₃—ᵅCH(—ᵝCH₃)(—ᵝCH₃)	24·3	8·7
(d) °CH₃—ᵅC(—ᵝCH₃)(—ᵝCH₃)(—ᵝCH₃)	31·5	7·2

are provided: with TMS as zero and with CS_2 as zero (TMS $= 192.8$); on others, only the CS_2 scale is given.) Here we see a regular de-shielding of about 9 ± 1 ppm for each added carbon, except in neopentane, where apparently crowding reduces the effect.

In Table 10.2 we see examples of the *β-effect*, i.e. the effect of added carbons β to the one observed, $^\circ C$. The effect is of similar magnitude to that for α-carbons.

In Table 10.3 it is shown that the β-effect is the same even when $^\circ C$ is *non-terminal*. However, if $^\circ C$ is a *branch point*, it is reduced in magnitude.

Table 10.3 *β-effect II*

	ppm from TMS	β-effect ppm
(e) $^\alpha CH_3$—$^\circ CH_2$—$^\alpha CH_3$	16.1	—
(f) $^\alpha CH_3$—$^\circ CH_2$—$^\alpha CH_2$—$^\beta CH_3$	25.0	8.9
(g) $^\alpha CH_3$—$^\circ CH_2$—$^\alpha CH{\big<}^{^\beta CH_3}_{^\beta CH_3}$	31.8	6.8
(h) $^\alpha CH_3$—$^\circ CH_2$—$^\alpha C{\Big<}^{^\beta CH_3}_{^\beta CH_3}$ (—$^\beta CH_3$)	36.7	4.9
(i) $^\alpha CH_3{\big>}^{}_{^\alpha CH_3}{\!}^\circ CH$—$^\alpha CH_3$	25.2	—
(j) $^\alpha CH_3{\big>}^{}_{^\alpha CH_3}{\!}^\circ CH$—$^\alpha CH_2$—$^\beta CH_3$	29.9	4.7
(k) $^\alpha CH_3{\big>}^{}_{^\alpha CH_3}{\!}^\circ CH$—$^\alpha CH{\big<}^{^\beta CH_3}_{^\beta CH_3}$	34.1	4.2
(l) $^\alpha CH_3{\big>}^{}_{^\alpha CH_3}{\!}^\circ CH$—$^\alpha CH{\Big<}^{^\beta CH_3}_{^\beta CH_3}$ (—$^\beta CH_3$)	38.1	4.0

Table 10.4 γ-effect

		ppm from TMS	γ-effect ppm
(a)	$^{\circ}CH_3 \!-\! ^{\alpha}CH_2 \!-\! ^{\beta}CH_3$	15·6	—
(b)	$^{\circ}CH_3 \!-\! ^{\alpha}CH_2 \!-\! ^{\beta}CH_2 \!-\! ^{\gamma}CH_3$	13·2	−2·4
(c)	$^{\circ}CH_3 \!-\! ^{\alpha}CH_2 \!-\! ^{\beta}CH \big<^{^{\gamma}CH_3}_{_{\gamma}CH_3}$	11·3	−1·9
(d)	$^{\circ}CH_3 \!-\! ^{\alpha}CH_2 \!-\! ^{\beta}C \big<^{^{\gamma}CH_3}_{_{\gamma}CH_3} {-}\,^{\gamma}CH_3$	8·8	−2·5
(e)	$^{\gamma}CH_3 \!-\! ^{\circ}CH_2 \!-\! ^{\alpha}CH_2 \!-\! ^{\beta}CH_3$	25·0	
(f)	$^{\gamma}CH_3 \!-\! ^{\circ}CH_2 \!-\! ^{\alpha}CH_2 \!-\! ^{\beta}CH_2 \!-\! ^{\gamma}CH_3$	22·6	−2·4
(g)	$^{\gamma}CH_3 \!-\! ^{\circ}CH_2 \!-\! ^{\alpha}CH_2 \!-\! ^{\beta}CH \big<^{^{\gamma}CH_3}_{_{\gamma}CH_3}$	20·7	−1·9
(h)	$^{\gamma}CH_3 \!-\! ^{\circ}CH_2 \!-\! ^{\alpha}CH_2 \!-\! ^{\beta}C \big<^{^{\gamma}CH_3}_{_{\gamma}CH_3} {-}\,^{\gamma}CH_3$	18·8	−1·9

We finally must consider the γ-*effect*, which, although smaller than the α- and β-effects, is in some respects the most significant. It is a *shielding* rather than a de-shielding effect (Table 10.4). We observe that the γ-effect is the same whether $^{\circ}C$ is primary or secondary. Further data, which are not presented here, indicate that it is also operative when $^{\circ}C$ is *tertiary*, i.e. has one remaining proton, but that when it is quarternary, the effect vanishes. This suggests the conclusion that this is a conformational phenomenon resulting from non-bonded steric interaction of the carbons mediated through their attached protons. This hypothesis was suggested by Grant and Cheney[5] and now seems to be more or less accepted. One should note that the γ-effect is a much broader phenomenon and can be observed for a variety of chains, not just paraffinic hydrocarbons. However, we shall adopt the Grant–Cheney hypothesis for the present and see where it leads.

For a clearer view of the γ-effect, consider the staggered conformations of three compounds selected from the previous data.[6] Viewed as a conformational

Butane:

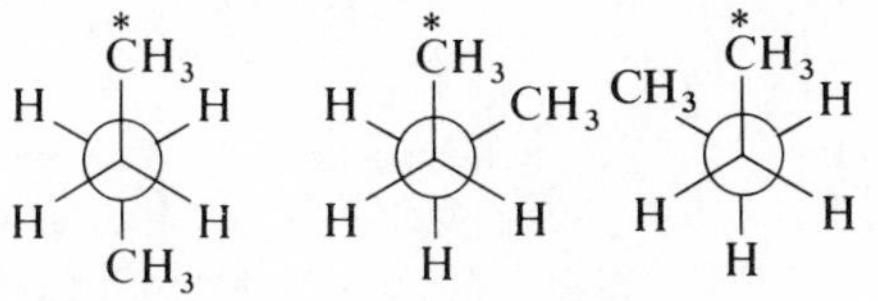

trans   gauche (~ 0.35)

C* shielding
vs. propane

-2.4 ppm
or
~ 6.9 ppm
per contact

2-Methylbutane:

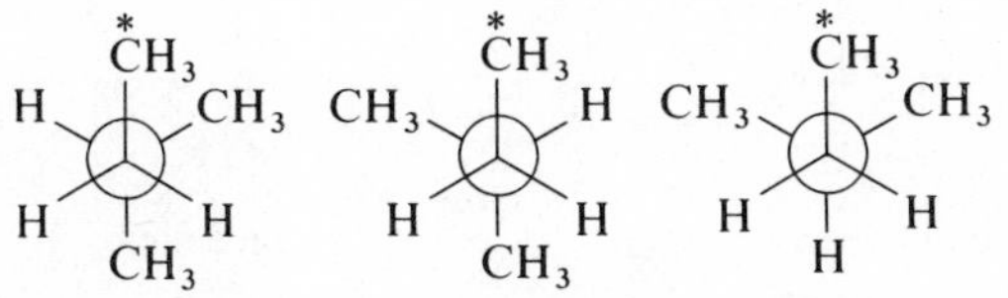

~ 0.5 ~ 0.5 ~ 0.0

-4.3 ppm
or
~ 4.3 ppm
per contact

2,2-Dimethylbutane:

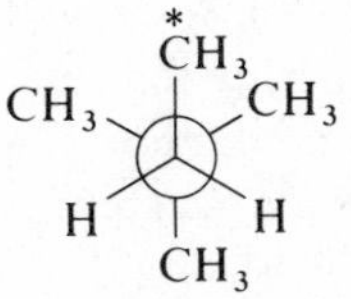

-6.8 ppm
or
~ 3.4 ppm
per contact

or *gauche* interaction, the γ-effect, despite its apparent constancy in Table 10.4, actually decreases in going from butane to 2-methylbutane to 2,2-dimethyl-butane. The effect per CH$\cdots$CH contact decreases from 6.9 to 4.3 to 3.4 ppm when adjusted for conformational behaviour. For the interpretation of the C-13 spectrum of polypropylene, we may adopt -4.3 ppm ($+4.3$ ppm on the CS$_2$ scale) per gauche contact as a reasonable value. Below are shown the corresponding 4-carbon interactions of a polypropylene chain and of a more general vinyl polymer chain in which R is assumed *not* to be CH and therefore to have a negligible γ-effect.

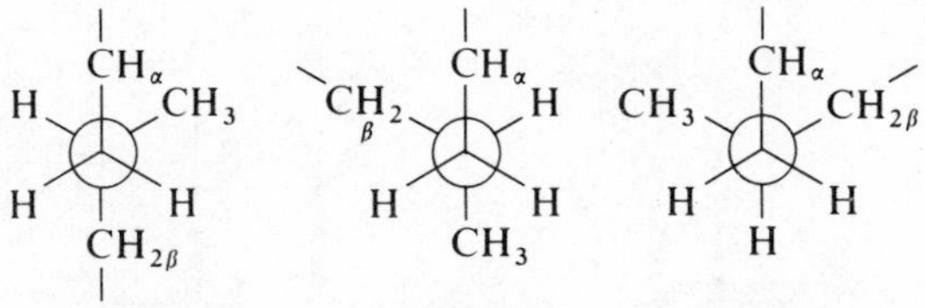

trans
$C_\alpha - C_\beta$: T
$C_\alpha - CH_3$: G

gauche +
$C_\alpha - C_\beta$: G
$C_\alpha - CH_3$: T

gauche −
$C_\alpha - C_\beta$: G
$C_\alpha - CH_3$: G

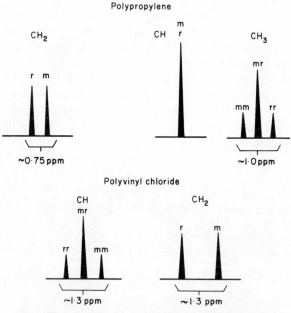

The top diagrams show three Newman projections labelled:

trans
$C_\alpha - C_\beta : T$

gauche +
$C_\alpha - C_\beta : G$

gauche −
$C_\alpha - C_\beta : G$

The proportions of *trans*, *gauche* + and *gauche* − conformations are influenced by the relative configurations about neighbouring C_α atoms:

meso (*m*) dyad

racemic (*r*) dyad

The *gauche/trans* ratio is higher for *m* dyads than for *r* dyads. A quantitative treatment of this situation,[6] assuming that only the *gauche* γ-interactions affect the chemical shift, leads to the predictions shown in Figure 10.1 for (a) polypropylene and (b) polyvinyl chloride, where R = Cl. The 68 MHz C-13 spectrum

Figure 10.1 Predicted C-13 spectra for polypropylene and poly(vinyl chloride), both assumed to be atactic

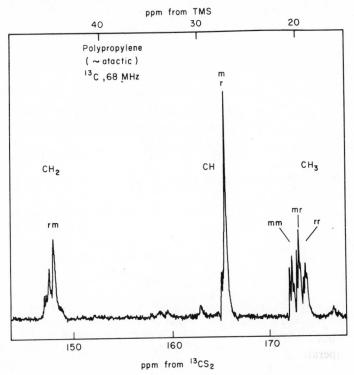

Figure 10.2 68 MHz C-13 spectrum of nearly atactic polypropylene (observed using a 20% solution in 1,2,4-trichlorobenzene at 135 °C; 2124 scans)

of a nearly atactic polypropylene is shown in Figure 10.2. There is fine structure not accounted for in this highly simplified treatment,[8] but the general features of the experimental spectrum are at least approximately as predicted.

The experimental poly(vinyl chloride) spectrum (Figure 10.3) clearly raises some problems. The interpretation of the spectra of the model compounds *meso*- and *dl*-2,4-dichloropentane[9] and of the polymer[10,11] in terms of *gauche* carbon–carbon interactions has already been considered by Carman and co-workers.[9,10,11] In the present calculation, it has been assumed that *gauche* CH···Cl interactions have no effect. (If this interaction were assumed equivalent to the corresponding CH···CH interaction, the configurational chemical shift effects for PVC would be similar to those for polypropylene, which is clearly not the case.) The predicted spectrum is then expected to exhibit α-carbon triads in the sequence *rr*, *mr*, *mm* in order of increasing shielding. This is the order observed in Figure 10.3, in which the assignments shown are those of Carman.[11] The β-carbon spectrum should show *m* more shielded than *r*. This is broadly true, but there is a very marked splitting indicating sensitivity to tetrad sequences. This seems to suggest that the chirality of next-nearest

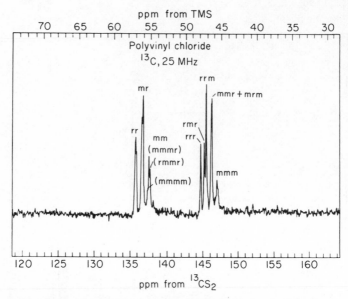

Figure 10.3 25 MHz C-13 spectrum of poly(vinyl chloride) (commercial product, $P_m \simeq 0.43$; observed using a 20% solution in 1,2,4-trichlorobenzene at 132 °C; 1024 scans). The assignments are those of Carman[11]

asymmetric carbons has a very marked effect on the *gauche/trans* ratio of the bonds of the central dyad, a question which we are currently exploring by conformational energy calculations.

10.3 C-13 OBSERVATIONS OF CHAIN STRUCTURE AND BRANCHING IN POLYETHYLENE AND POLY(VINYL CHLORIDE)

It is well known that the morphology and solid state properties of *polyethylene* are critically dependent on the frequency of short-chain branches, which reduce the extent of crystallinity. The same may also be true, but in a considerably weaker degree, for poly(vinyl chloride). PVC is normally only marginally crystalline but the crystalline fraction may be quite high for polymer prepared at low temperatures. The greater regularity of the PVC chain no doubt is the reason, and this greater regularity may include not only a higher degree of syndiotacticity (which is now very well documented[12]), but also a lower frequency of branching. It has been long known that methyl groups, presumed to be branch ends, can be detected by infrared in PVC after reduction to polyethylene with LiAlH$_4$, and that the branch frequency decreases at lower polymerization temperatures.[13-15] Rigo *et al.*[16] reported that the number of branches increases with polymerization temperature up to 3% at 50 °C, but is

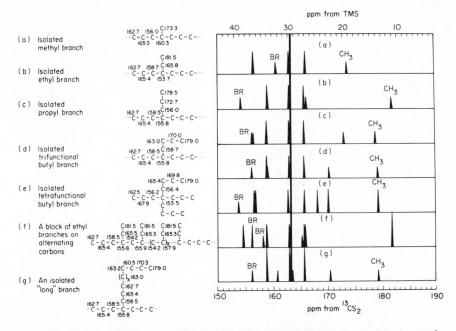

Figure 10.4 Predicted C-13 chemical shifts and schematic spectra for various types of possible branch structures in a polyethylene chain

independent of monomer conversion, strongly suggesting that they are generated by an intra-molecular mechanism.

Our understanding of this problem is severely hampered by not knowing what these branches are, in particular their length, whether they are tri- or tetrafunctional, and whether they occur in groups or at random. We have recently demonstrated,[17] using C-13 n.m.r., that the branches in commercial low-density polyethylene are predominantly *n*-butyl, a result independently confirmed by Randall.[18] This was possible by application of the chemical shift rules described at the beginning of this paper, which lead to a prediction of the spectral patterns shown in Figure 10.4 for various types of branches. In these spectra, the very large peak near the centre corresponds to the strongly predominant fraction of CH_2 groups which are four carbons or more removed from a branch point, branch end, or chain end. Of the branch carbons, the methyl is always the most shielded and the branch-point carbon usually the least shielded; the others may be identified by reference to the scheme at the left of the spectra in Figure 10.4. It is noteworthy that there is a chemical shift spread of nearly 30 ppm and that each structure has a very distinctive pattern. In Figure 10.5 is shown the C-13 spectrum of a typical low-density polyethylene (DYNK). The principal pattern of minor peaks corresponds to that calculated for isolated trifunctional *n*-butyl branches ((d) in Figure 10.4). This result is

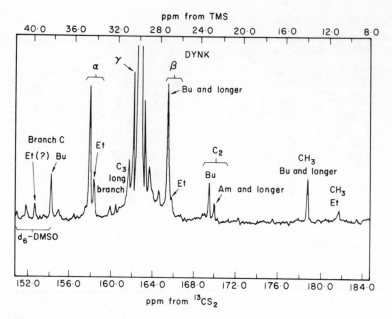

Figure 10.5 Experimental 25 MHz C-13 spectrum of a low density polyethylene, DYNK; (observed using a 25% solution in 1,2,4-trichlorobenzene at 135 °C; 21,081 scans)

consistent with the Roedel 'back-biting' mechanism, proposed over 20 years ago[19]:

There are also indications of minor proportions of ethyl branches and of amyl, hexyl, and possibly longer branches. Very long branches, presumed to be produced by intermolecular chain transfer to polymer, cannot be clearly distinguished (at least not at 25 MHz) from hexyl, and are probably generally too few in number to detect under the conditions employed here. (But see the discussion of poly(vinyl chloride) below.)

One might suppose that branches in poly(vinyl chloride) could be directly determined by C-13 n.m.r., but inspection of Figure 10.3 shows that the spectrum is made so complex by configurational irregularity that there is little chance of detecting minor structural anomalies unless they differ markedly in chemical shift. A better procedure is to eliminate all configurational complexity by reduction to polyethylene with LiAlH$_4$, as in earlier infrared studies, taking care to remove chlorine as completely as possible. The spectrum of a reduced PVC ($\overline{DP}_n \simeq 450$) is shown in Figure 10.6.[20] The original polymer was

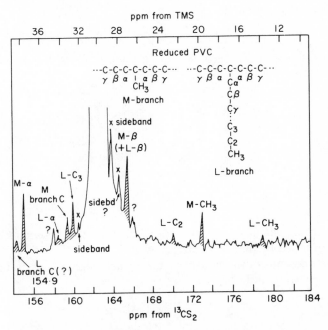

Figure 10.6 25 MHz C-13 spectrum of a reduced commercial poly(vinyl chloride); (observed using a 20% solution in 1,2,4-trichlorobenzene at 135 °C; 26,108 scans)

prepared in the laboratory at 75 °C; reduction does not cleave the chains. The pattern of minor peaks corresponds closely to that of *methyl* ('M') branches ((a) in Figure 10.4), rather than *n*-butyl. These cannot be generated by back-biting, but nevertheless must, as we have seen, be generated by some intra-molecular mechanism. Rigo *et al.*[16] suggested the plausible mechanism shown below. This begins with a radical formed by a head-to-head monomer addition. The detection of such units in the finished polymer chain has been claimed,[21] but from the data on model pentanes[10] it is clear that the methine carbons of such structures should appear at 131–133 ppm. There are no detectable resonances in this region in Figure 10.3. Rigo *et al.*, however, do not propose its stable existence but rather that the initial radical is unstable and rapidly isomerizes by 1,2 chlorine migration to the more stable radical; further propagation then yields a chloromethyl branch, which is reduced to a methyl group by $LiAlH_4$:

$$-CH_2CHCH\dot{C}H_2 \quad (Cl, Cl)$$

$$\xrightarrow{CH_2=CHCl} \quad -CH_2CHCHCH_2CH_2CH\cdot \quad (Cl\ Cl \quad Cl)$$
(not observed by C-13)

$$\xleftrightarrow[\text{migration}]{1.2\ Cl} \quad -CH_2CH\dot{C}HCH_2Cl \quad (Cl)$$

$$\xrightarrow{CH_2=CHCl} \quad -CH_2CH-CH-CH_2-CH\cdot \quad (Cl \quad CH_2Cl \quad Cl)$$

$$\xrightarrow{LiAlH_4\ \text{reduction}} \quad -CH_2CH_2CHCH_2CH_2- \quad (CH_3)$$

This mechanism, which was in part speculative when proposed, seems to be strongly supported by the C-13 n.m.r. results.

A simpler mechanism would involve merely an occasional rearrangement of the normal growing radical:

$$-CH_2CHCl\cdot \ \rightarrow \ -\dot{C}HCH_2Cl$$

From direct spectral measurements of the relative areas of the M-CH_3, M-α, M-β, and unperturbed CH_2 peaks of Figure 10.6, the methyl branch frequency is found to be 3·2 CH_3/1000 CH_2. It will be noted that in addition to the methyl branch resonances, weak peaks appear at all the expected positions[17,18] (although not with all the correct relative intensities) for 'long' branches, designated L. 'Long' branches may be defined for this purpose as *n*-amyl, *n*-hexyl, or longer, although the presence of a minor proportion of *n*-butyl is not excluded. The frequency of such 'branches' is about 1 per 1000 CH_2. However, we must remember that the number average degree of polymerization of this polymer is *ca* 450 or 900 CH_2 groups, so that this 'long branch' may actually correspond to one paraffinic chain-end per molecule. The accepted mechanism of vinyl chloride polymerization postulates one $-CH=CHCl$ chain-end per molecule (such groups are not reducible by

LiAlH$_4$). The other end would be expected to be paraffinic in the reduced polymer. At present, we cannot clearly distinguish between this end and a long branch; the presence of a resonance at 158·3 ppm, in the correct position for a long branch C$_\alpha$, is the best evidence for the latter. De Vries et al.[22] concluded from extensive solution measurements that poly(vinyl chloride) probably contains only about 0·5 branch per 1000 CH$_2$.

The total methyl group content (M + L) found by C-13 measurements, ca 4 CH$_3$/1000 CH$_2$, is about half the value found by infrared for this sample. It is in approximate agreement with the findings of Nakajima et al.[23] for a similar material, but is much lower than the infrared estimates of Rigo et al.,[16] Braun and Schureck,[24] Burnett et al.,[25] and Baijal et al.[26] We are at present unable to resolve this discrepancy.

10.4 C-13 STUDY OF BIOPOLYMERS

Carbon-13 spectroscopy has also been very widely applied to the study of biopolymers, particularly polypeptides and proteins. In such studies one normally knows the covalent structure of the chain and the object of the n.m.r. studies is to obtain conformational information. Proton n.m.r. has been successfully applied for the last 5 or 6 years to such investigations, but carbon-13 studies have so far yielded only rather limited conformational conclusions.

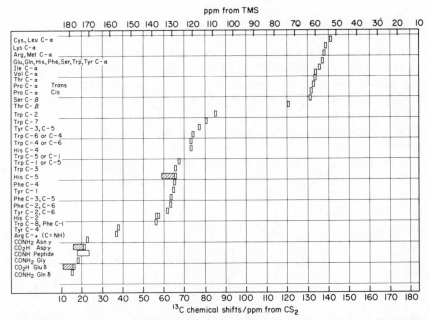

Figure 10.7 Carbon-13 chemical shifts of amino-acid residues in an open-chain environment in aqueous solution, ca 10–140 ppm (CS$_2$ as zero). Chemical shift titration ranges are shown as hatched areas, the low pH value being at the right

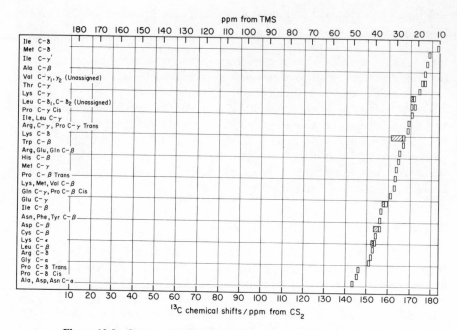

Figure 10.8 Same as Figure 10.7, but for the *ca* 140–180 ppm range

In Figures 10.7 and 10.8 are shown the C-13 shifts of the individual amino-acids. All identifiable chemical shifts are summarized in these figures; at present there are very few remaining ambiguities. It may be noted that the carbon-13 chemical shifts are observed over a range (expressed in ppm) approximately 20-fold greater than that of protons. The shielding behaviour of the various groups is very similar for carbon nuclei and for protons, with aromatic and electronegative groups showing the least shielding and aliphatic hydrocarbon side chains the greatest. One also has the advantage of being able to observe carbonyls.

The problem of determining the conformations of polypeptide chains is that of determining the three rotation angles, ϕ, ψ and ω, shown below:

The measurement of angle ϕ can be accomplished by measurement of the J coupling of the α-CH and NH protons. Carbon-13 is often very useful in the determination of ω. The determination of this angle provides the answer to the question of whether the peptide bond is *cis* or *trans*:

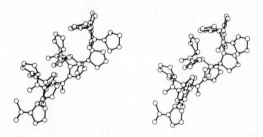

trans   *cis*

It is well known that amino-acid peptide bonds are normally *trans*, but that imino-acid peptide bonds, in particular the peptide bond between a proline unit and a preceding residue, may be *cis* or *trans* and that the energy difference is small and strongly solvent dependent although the barrier between them is *ca* 8 kJ mol^{-1}

The well known and much studied conformational transition of polyproline form I to polyproline form II is a transition from an all-*cis* to an all-*trans* chain, a process that can be readily followed by proton spectroscopy.[27]

Normally, in their equilibrium state, polyproline chains are either all-*cis* or all-*trans*. These conformations are shown in Figures 10.9 and 10.10. In water,

Figure 10.9 Stereo views of the conformation of form I of poly-L-proline

only *trans* residues are stable. In both form I (*cis*) and form II (*trans*), the

$C_\alpha - C$ rotational angle, ψ, is approximately 300° (1966 convention[28]; 120° in the 1970 convention[29]), i.e. the conformation is as shown at the left-hand

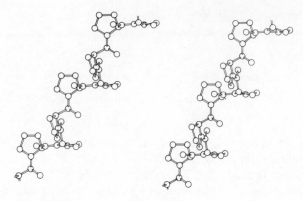

Figure 10.10 Stereo views of the conformation of form
II of poly-L-proline

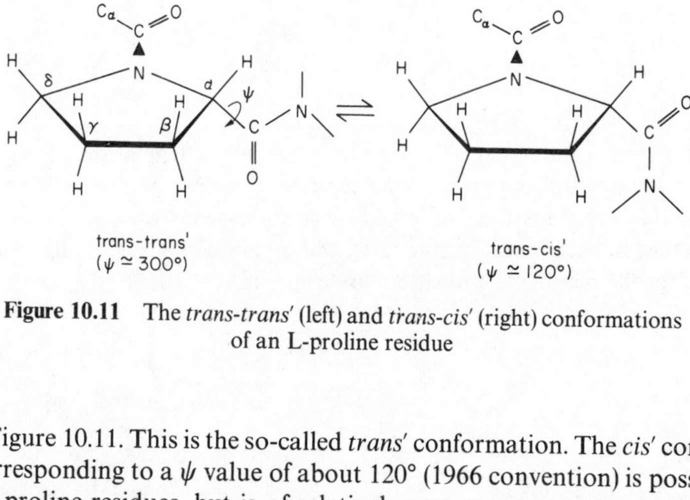

trans-trans'
($\psi \simeq 300°$)

trans-cis'
($\psi \simeq 120°$)

Figure 10.11 The *trans-trans'* (left) and *trans-cis'* (right) conformations
of an L-proline residue

side of Figure 10.11. This is the so-called *trans'* conformation. The *cis'* conformation, corresponding to a ψ value of about 120° (1966 convention) is possible for isolated proline residues, but is of relatively rare occurrence compared to the *trans'* conformation.

In contrast to what is observed in water, one finds that in strong aqueous salt solutions the polyproline helix appears to collapse, as judged from viscosity measurements accompanying this process. One can see by n.m.r. both *cis* and *trans* residues in stable coexistence, corresponding to a disordered structure with respect to the peptide bonds. Proton spectra are rather unsatisfactory for such measurements because of the great breadth of the peaks, which tends to frustrate quantitative and even qualitative interpretation. We therefore turned to carbon-13 n.m.r. Model compounds show that *cis* and *trans* conformations can be very clearly distinguished. It is known for simple *n*-alkyl amides that the alkyl carbon *syn* to the carbonyl oxygen is more shielded than the *anti* carbon.

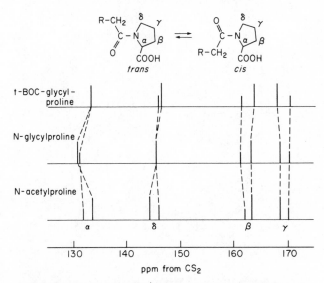

The same is true for proline derivatives. Figure 10.12 shows schematic C-13 spectra of three simple proline derivatives.[30] The relative shieldings and intensities of the β and γ carbons provide a reliable measure of the *cis-trans* conforma-

Figure 10.12 Carbon-13 spectra of the rings of three L-proline derivatives. The peak heights denote the approximately relative proportions of *cis* and *trans* conformers, *trans* being preferred. "t-BOC" is t-butyloxycarbonyl,

$$\text{(CH}_3)_3\text{C}-\text{O}-\overset{\overset{\displaystyle O}{\|}}{\text{C}}-$$

tional preference, *trans* being generally preferred. In Figure 10.13 are shown 25 MHz carbon-13 spectra of a sample of poly-L-proline of molecular weight 2500.[31] The *cis* and *trans* β and γ carbon resonances are narrow and well-resolved. It is evident that the chemical shifts reflect primarily the conformation of the local peptide bond and not those of more distant peptide bonds, which is presumably the source of the broadening in the proton spectra.[27] Spectrum (a) is the spectrum of poly-L-proline from I, taken immediately after dissolution in cold water (9 °C), under which conditions earlier kinetic studies indicate that the isomerization should be strongly retarded. As expected, the spectrum shows only one major isomer. The chemical shift of the γ-carbon, which *must* correspond to the *cis* isomer, is in agreement with the conclusions of the model studies. Spectrum (c) is that of a solution in which the isomerization has been

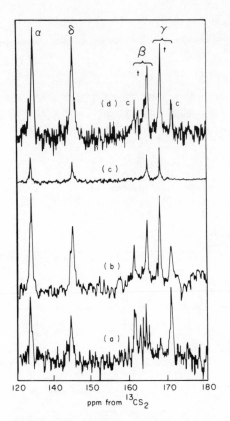

Figure 10.13 25 MHz carbon-13 spectra of form I poly (L-proline) of *ca* 2500 molecular weight; (a) immediately after dissolution in H_2O at 9 °C; the multiplet centered near 163 ppm is that of acetone-d_6; (b) same as (a) but allowed to isomerize partially (*ca* 2/3) to form II; (c) after complete isomerization to form II; (d) same polymer dissolved as form II in 5 M aqueous KI, observed at 25 °C

allowed to proceed completely to form II. The γ-carbon shift is consistent with that of the *trans* isomers of the models. Spectrum (b) corresponds to the initial form I having been allowed to go about two-thirds of the way toward form II. Both forms are clearly evident. Spectrum (d) is the spectrum of poly-L-proline, initially in form II, dissolved in 5 M aqueous KI and observed at about 25 °C. It is obvious that spectra (b) and (d) are virtually identical, and that the disordering of the polymer results from the insertion, probably nearly at random, of *cis* units among the *trans*. One can easily show, using molecular models, that mixed *cis* and *trans* sequences can form suitable 'baskets' of carbonyl groups for complexing cations. It is well known that cyclic polypeptides and depsipeptides of appropriate structure can complex cations quite strongly in this way.

10.5 A FINAL WORD

In this contribution I have chosen some particular examples out of what is now quite a large literature of C-13 studies of polymers. These have been selected from work carried out at Bell Laboratories, and it is hoped they will serve to show some of the ways in which this powerful technique can provide important structural and conformational information concerning macromolecules.

10.6 REFERENCES

1. H. Spiesecke and W. G. Schneider, *J. Chem. Phys.*, **35**, 722, (1961).
2. D. M. Grant and E. G. Paul, *J. Amer. Chem. Soc.*, **86**, 2984, (1964).
3. L. P. Lindeman and J. Q. Adams, *Anal. Chem.*, **43**, 1245, (1971).
4. D. E. Dorman, R. E. Carhart, and J. D. Roberts, private communication.
5. D. M. Grant and B. V. Cheney, *J. Amer. Chem. Soc.*, **89**, 5315, 5319, (1967).
6. See P. J. Flory, *Statistical Mechanics of Chain Molecules*, Interscience, New York, 1969, Chapters V and VI for discussion and references.
7. A. E. Tonelli, private communication; see also J. E. Mark, *J. Chem. Phys.*, **56**, 451, (1972) and references therein.
8. For example, the CH resonance is not strictly the expected singlet, and the CH_3 resonance is clearly resolved into pentad sequences; see A. Zambelli, D. E. Dorman, A. I. R. Brewster and F. A. Bovey, *Macromolecules*, **6**, 925, (1973); J. C. Randall, *J. Polymer Sci., Polymer Physics Ed.*, **12**, 703, (1974).
9. C. J. Carman, A. R. Tarpley and J. H. Goldstein, *J. Amer. Chem. Soc.*, **93**, 2864, (1971).
10. C. J. Carman, A. R. Tarpley and J. H. Goldstein, *Macromolecules*, **4**, 445, (1971).
11. C. J. Carman, *Macromolecules*, **6**, 725, (1973).
12. See F. A. Bovey, *High Resolution N.M.R. of Macromolecules*, Academic Press, New York, 1972, pp. 148–149.
13. J. D. Cotman, *Ann. N.Y. Acad. Sci.*, **57**, 417, (1953).
14. M. H. George, R. J. Grisenthwaite and R. F. Hunter, *Chem. Ind. (London)*, 1114 (1958).
15. G. Boccato, A. Rigo, G. Talamini and F. Zilio-Grandi, *Makromol. Chem.*, **108**, 218, (1967).
16. A. Rigo, G. Palma and G. Talamini, *Makromol. Chem.*, **153**, 219, (1972).
17. D. E. Dorman, E. P. Otocka and F. A. Bovey, *Macromolecules*, **5**, 574, (1972).
18. J. C. Randall, *J. Polymer Sci., Polymer Physics Ed.*, **11**, 275, (1973).
19. M. J. Roedel, *J. Amer. Chem. Soc.*, **75**, 6110, (1953).
20. K. B. Abbås, F. A. Bovey and F. C. Schilling, *Macromol. Chem., Suppl.* **1**, 227, (1975).
21. M. Shimizu and T. Otsu, *J. Chem. Soc. Japan, Ind. Chem. Sec.*, **67**, 966, (1964).
22. A. J. de Vries, C. Bonnebat and M. Carrega, *Pure and Applied Chem.*, **25**, 209, (1971).
23. A. Nakajima, H. Hamada and S. Hayashi, *Makromol. Chem.*, **95**, 40, (1966).
24. A. Braun and W. Schureck, *Angew. Makromol. Chem.*, **7**, 121, (1969).
25. M. Baijal, T. Wang and R. Diller, *J. Macromol. Sci.*, **A, 4**, 965, (1970).
26. G. M. Burnett, F. L. Ross and J. N. Hay, *J. Polymer Sci.*, **A-1, 5**, 1467, (1967).
27. D. A. Torchia and F. A. Bovey, *Macromolecules*, **4**, 246, (1971).

28. J. T. Edsall, P. J. Flory, J. C. Kendrew, A. M. Liquori, G. Nemethy and G. N. Ramachandran, *Biopolymers*, **4,** 121, (1966); *J. Biol. Chem.*, **241,** 1004, (1966); *J. Mol. Biol.*, **15,** 399, (1966).
29. J. C. Kendrew, W. Klyne, S. Lifson, T. Miyazawa, G. Nemethy, D. C. Phillips, G. N. Ramachandran and H. A. Scheraga, *Biochemistry*, **9,** 3471, (1970); *J. Biol. Chem.*, **245,** 489, (1970); *J. Mol. Biol.*, **52,** 1, (1970).
30. D. E. Dorman and F. A. Bovey, *J. Org. Chem.*, **38,** 2379, (1973).
31. D. E. Dorman, D. A. Torchia and F. A. Bovey, *Macromolecules*, **6,** 80, (1973).

11

The Analysis of ^{13}C N.M.R. Relaxation Experiments in Polymers

J. Schaefer

Monsanto Company, St. Louis, Missouri

11.1 INTRODUCTION

Carbon-13 spins are isolated from dipolar interactions with one another by their low natural abundance. In addition, they are, in a certain sense, isolated even from those protons to which they are dipolar-coupled because of the difference in the ^{13}C and ^{1}H resonance frequencies. Because of this isolation, individual carbons have individual spin-lattice relaxation times, which are not averaged to a single value by spin-spin interactions, as often are the proton spin-lattice relaxation times of the same systems. Thus, the various ^{13}C spin-lattice relaxation times of polymer solutions and solid elastomers are determined by the dynamics of the individual carbon environments, and so provide information of the details of polymeric segmental motion. We will illustrate the kind of chemically useful information which can be obtained from ^{13}C spin-lattice relaxation experiments with results from studies of solutions of polystyrene, of some linear alkanes, and from studies of the solid elastomers, poly(propylene oxide) and polybutadiene.

Complementary information about segmental motion is often obtained from another relaxation parameter, the ^{13}C Overhauser enhancement, which, since it determines observed integrated line intensities, is also of importance in high-resolution experiments. However, theoretical descriptions in simple terms of both the spin-lattice relaxation time and the Overhauser enhancement of the same non-rigid polymer system are generally not successful. This failure is the result òf the fact that most polymer chains engage in a wide variety of involved cooperative motions. We will show that taking some of these complications into account, even in a modest approximate way, leads to a theoretical description of segmental motion which not only is in agreement with observation, but also provides insight into the extent of cooperativity in *non-rigid* solid polymers.

Relaxation experiments, performed with strong ^{1}H decoupling and magic-angle spinning, exploit the isolation of the ^{13}C spin, and are, in addition, well suited for the study of *rigid* solid polymers at temperatures well below the glass transition. High-resolution ^{13}C n.m.r. spectra of rigid amorphous and crystalline polymers are obtained because the line broadening due to near-static hetero-nuclear dipolar interactions is removed by the strong decoupling, and the inherent chemical shift anisotropy is removed by the magic-angle spinning; just as in more conventional experiments, individual spin-lattice relaxation times are determined by the dynamics of the individual carbon environments, and this information can be combined with resolved line-shape information in order to describe details of polymeric segmental motion in the solid state. We will illustrate the unique information available from such ^{13}C experiments with results from studies, at room temperature, of solid poly(methyl methacrylate) and solid polystyrene.

11.2 RELAXATION EXPERIMENTS ON SOLUTIONS OF POLYMERS AND ON SOLID POLYMERS ABOVE T_g

11.2.1 Spin-lattice relaxation times

The spin-lattice relaxation times of ^{13}C nuclei in polymers are predominantly determined by ^{13}C—^{1}H dipolar interactions.[1] The ^{13}C nuclei couple to fluctuating ^{1}H magnetic dipoles and thereby return to their steady-state magnetization, following an n.m.r. sampling perturbation, with a characteristic time constant called the spin-lattice relaxation time, T_1. The fluctuating ^{1}H dipolar field is generated by the microscopic segmental *rotational* motion of the polymer. These rotational motions are associated either with genuine changes in chain conformation, or with the torsional motions occurring within a conformation. Apparently, enough different kinds of motion of these types are possible, that an approximate description of segmental rotational motion as an isotropic random walk characterized by a single effective correlation time, τ, is often found to be valid.[1] This is especially true in describing relaxation experiments on solutions of non-elastomeric polymers at temperatures above 50 °C. Under these conditions, a well-known expression[2] for a ^{13}C T_1 is

$$\frac{1}{T_1} = \tfrac{1}{10}\hbar^2\gamma_C^2\gamma_H^2 \sum_N r_N^{-6}[f(\omega_H - \omega_C) + 3f(\omega_C) + 6f(\omega_H + \omega_C)] \qquad (1)$$

where γ_C and γ_H are the carbon and proton gyromagnetic ratios, respectively, r_N is the internuclear C—H distance to the Nth proton, ω_C and ω_H are the respective carbon and proton Larmor frequencies, and

$$f(\omega_i) = \frac{\tau}{1 + \omega_i^2\tau^2} \qquad (2)$$

An expression analogous to equation (1) can be written for the relaxation parameter used to describe the line-width, the spin-spin relaxation time T_2.

From this analysis we can see that if all the carbons in the repeating unit of the polymer chain have the same effective correlation time, τ, then the number of directly bonded or nearby protons determines the T_1's. Thus, a methylene-carbon T_1 will be just half as long as a methine-carbon T_1 in a vinyl-type polymer, since it has twice as many directly bonded protons, and hence its spin-lattice relaxation is just twice as efficient.

However, such a simple description of ^{13}C spin-lattice relaxation is not always successful. In general, the failures result from one of three causes: (1) a direct long-range contribution to T_1, through dipolar interactions with ostensibly distant protons; (2) an indirect long-range contribution to T_1, through an influence on the local effective isotropic segmental correlation time by distant units in the chain; and (3) the domination of T_1 by the effects of multiple correlation times, resulting from distinctly non-isotropic segmental rotational motion. While the presence of any or all of these complications means we cannot use a simple analysis of relaxation, this is actually a desirable situation. Clearly, if we can adequately describe the cause of the failure of the simple analysis, we will, in fact, have gained some genuine insight into the details of polymeric segmental motion. We will now consider three examples of situations in which relatively detailed descriptions of the ^{13}C spin-lattice relaxation can lead to physically interesting conclusions.

11.2.1.1 Polystyrene

The anomalously small T_1 observed for the quaternary carbon of polystyrene is an example of direct long-range contributions to spin-lattice relaxation through dipolar interactions with distant protons.[1] The 550-ms quaternary carbon relaxation time is about eight times longer than that of the methine carbon. Nevertheless, this relaxation time is still shorter, by a factor of 3, than is predicted. The prediction is based on equation (1), the effective correlation time obtained from the observed methine- and methylene-carbon T_1's and the various known proton-carbon distances within the monomer unit (the smallest of which, the distance to the two ortho protons, is 2·15 Å, or about twice an average C—H bond distance). This discrepancy exists despite the fact that, by using the same procedure, calculated values for the relaxation times of the side-chain aromatic meta and para carbons are in good agreement with experiment,[3] results which make unlikely unusual effects resulting from pronounced anisotropies in the motion of the phenyl side group relative to the main chain. Non-dipolar contributions to the quaternary carbon spin-lattice relaxation time are known not to be important. Spin diffusion is not effective in these experiments in equalizing ^{13}C spin-lattice relaxation times (and so lowering the quaternary-carbon T_1) because the ^{13}C nuclei are isolated from one another by their low natural abundance.

The apparent discrepancy arises because the spin-lattice relaxation time of the quaternary carbon results from dipolar coupling with many protons, including those of nearest-neighbour monomer units in the chain. For example, an estimate of the distance of the quaternary carbon to the closest methine proton in a nearest-neighbour monomer unit in crystalline polystyrene is comparable to estimates of the shortest quaternary carbon-proton distances within the monomer. In addition, space-filling molecular models of both iso-tactic and syndiotactic polystyrene point to this proton as being particularly close to the aromatic quaternary carbon. Considering protons of the same monomer unit, as well as protons of nearest- and next-nearest-neighbour monomer units, there are some 6–10 protons which, in solution, are probably within 2–3 Å of a given quaternary carbon. All of these protons are capable of making a significant dipolar contribution to the quaternary carbon T_1.

Because of the likely importance of dipolar interactions with protons of nearest-neighbour monomer units, the similarities of relaxation times for the quaternary carbons of both isotactic and atactic polystyrene in solution require that the average distance between a quaternary carbon and its nearest-neighbour protons be about the same for the two polymers. Furthermore, these protons and carbons in both polymers must be involved in about the same kind of segmental motion. This is consistent with calculations of conformational populations in polystyrene.[4] These calculations show that the very few highly preferred conformations of isotactic and syndiotactic polystyrene dominate all other contributors (some of which have more strained geometries and less normal internuclear distances) and that, as a result, the average distance between a quaternary carbon and its nearest-neighbour protons is about the same for both isotactic and syndiotactic placements. Furthermore, in each of the preferred conformations, the phenyl group is constrained by about the same steric interactions of its ortho protons and carbons with various main-chain protons. The physics of the steric interactions of the phenyl group is of primary importance for polystyrene. Thus, while different chemical environments can give rise to differential ^{13}C chemical shifts, the dominant physical interactions controlling the segmental motions for the different preferred conformations are similar, and this leads to similar relaxation times.

11.2.1.2 Short-chain alkanes

The variation in T_1 along the interior of a linear alkane chain is an example of an indirect contribution to T_1, through an influence on the local effective isotropic segmental correlation time by distant units near the end of the chain.[5] The observed T_1 for interior carbons (five carbons removed from the chain end), along with the calculated effective isotropic correlation time (obtained using equation (1)), for some linear alkanes are shown in Table 11.1. At first, it is tempting to attribute the observed decrease in T_1 and the increase in the effective correlation time of the interior carbons to some kind of growing inflexibility of the centre of the chain with increasing chain length. However, this intuitive interpretation is complicated by the fact that for medium- and low-molecular weight chains, the observed T_1 depends upon the overall tumbling rotation of the entire chain. Thus,

$$[\tau_{eff}(i)]^{-1} = \tau_0^{-1} + [\tau_{int}(i)]^{-1} \tag{3}$$

where $\tau_{eff}(i)$ is the effective correlation time of the ith carbon, τ_0 is the rotational correlation time of the entire chain, and $\tau_{int}(i)$ is that part of the effective correlation time which can be associated with internal motion of the ith carbon. We

Table 11.1 Spin-lattice relaxation times and effective correlation times for interior carbons of some alkanes at 39 °C[5]

n-Alkane	T_1/s	τ/ps
C-10	5·1	4·6
C-13	2·2	10·7
C-15	1·6	15·2
C-18	1·0	25·1
C-20	0·8	30·2

can define a difference correlation time $\tau(i, j)$ for any two carbon atoms i and j by

$$[\tau(i, j)]^{-1} = [\tau_{eff}(i)]^{-1} - [\tau_{eff}(j)]^{-1} \tag{4}$$

$$= [\tau_{int}(i)]^{-1} - [\tau_{int}(j)]^{-1} \tag{5}$$

The difference correlation times are independent of the overall rotational correlation time of the chains, and so are independent of chain length. Variations in the difference correlation times will, therefore, provide information about the indirect influence of distant units in the chain on the T_1's of interior carbons.

The difference correlation times for a given carbon position shown in Table 11.2 (using the end methyl-carbon effective correlation time as a reference)

Table 11.2 Difference correlation times corresponding to differences in rates of internal rotation for some alkanes at 39 °C[5]

n-Alkane	$\tau(1, 2)/ps$	$\tau(1, 3)/ps$	$\tau(1, 4)/ps$	$\tau(1, \text{interior})/ps$
C-10	7·0	5·4	4·5	4·2
C-13	6·5	5·3	4·7	3·6
C-15	6·9	5·4	4·9	4·4
C-18	6·5	5·3	4·8	4·3
C-20	7·3	6·1	5·4	5·0

are indeed independent of chain-length. A comparison of difference correlation times for different carbon positions of a given chain shows a decrease in the relative flexibility of the chain with increasing length; that is, motion is more restricted near the centre of the chain, presumably because of the substantial mass of the carbons on either side. (Note that since the short correlation time of the end-methyl-carbon has been used as a reference, and equation (4) involves inverses, a small *difference* correlation time actually denotes slow motion associated with a long correlation time.)

The fact that the influence of the end of the chain on the interior persists for at least five carbons suggests that, in general, the segmental motion which determines the correlation time of a given carbon in a polymer involves some 5–6 carbons on either side. This is a relatively long-range effect. If extrapolated to vinyl polymers, this kind of long-range effect means that the T_1 of a given carbon should be thought of in terms of pentads or heptads of tactic placements in the chain. From our earlier discussion of spin-lattice relaxation for poly-styrene (and from similar discussions of other polymer systems[6]), we know, however, that the observed T_1's of high-molecular weight polymers simply do not show this kind of long-range dependence on configuration. This does not mean that segmental motion in the vinyl polymers does not somehow involve cooperative motions of many carbons. Apparently, *unlike* the situation for cooperative motions within the preferred conformations of the interiors of vinyl chains long enough to be free of end effects, the physical interactions controlling the segmental motions near the ends of the short-chain alkanes are considerably different from one carbon position to the next, for at least the first five or so carbons from the ends. The net result is an influence on the local segmental motion of relatively distant interior carbons by the ends of the chain.

It should be pointed out that a more elaborate theoretical analysis than that of equations (3)–(5) has been made of the segmental motion of short-chain alkanes.[7] In effect, this analysis does not assume that the overall tumbling motion of the linear chain is isotropic, but rather employs an anisotropic ellipsoidal model. The picture emerging from this model is still one of increasing motion as one moves out from the centre of inertia towards the ends of the chains, but the differences between effective correlation times of the interior carbons are far less pronounced; that is, the ellipsoidal model attributes much of the apparent long-range effect of the ends of chain on the T_1's of the C_3 to C_5 interior carbons not to an influence on the degree of internal freedom of the interior carbons, but rather to what happens to be the geometrical location of the interior carbons in an unsymmetrical molecule undergoing overall asymmetric rotational tumbling. Unfortunately, because of the relatively large errors involved in the computations of the theoretical analysis for chains much longer than C_6, a clear choice between the two interpretations is not yet possible. However, the issue can be settled, it would appear, from relaxation data from specifically labelled carbons well within the interior of relatively long chains.

11.2.1.3 Elastomers

A simple analysis of spin-lattice relaxation is not valid for many elastomeric polymers.[8] For these polymers the spin-lattice relaxation times are not simply related to the number of nearby- and directly-bonded-protons, and equation (1) (or some reasonably simply modified version) is not an adequate description. This situation occurs for poly(propylene oxide), and is due to the distinctly anisotropic segmental motion of the chain. The anisotropy, whose description requires the specification of several correlation times, destroys the simple inverse proportionality between the spin-lattice relaxation time of a poly-(propylene oxide) main-chain carbon and the number of directly bonded protons.

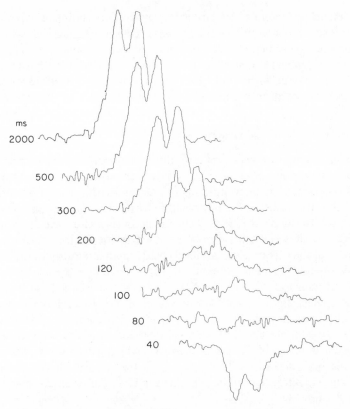

ms
2000
500
300
200
120
100
80
40

Figure 11.1 Partially relaxed ^{13}C n.m.r. spectra of the methine and methylene carbons of a solid poly(propylene oxide). Delay times, 40–2000 ms, are indicated.

The partially relaxed Fourier transform spectra of the main-chain carbons of a solid poly(propylene oxide) illustrate this point (Figure 11.1). Two lines are observed that can be attributed to the methine and methylene carbons. The two lines are about 50 Hz apart and are fairly well resolved, but they are just about indistinguishable in their response to a two-pulse 180°-t-90° sequence. This is also true for dilute solutions of the same high-molecular-weight polymer, as well as for a low-molecular-weight version of poly(propylene oxide).

A similar situation holds for the solid elastomer, cis-polybutadiene $+CH_2CH=CHCH_2+_n$. At 35 °C the methine- and methylene-carbon spin-lattice relaxation times are 600 and 400 ms, respectively. Main-chain rotational motions in cis-polybutadiene have been studied theoretically in some detail.[9] The energetic considerations for rotations about CH_2-CH_2 bonds are quite different from those about $CH-CH_2$ bonds. The overall rotational reorientations are therefore clearly anisotropic and cannot be described by a single correlation time. In addition, the nature of the anisotropy does not lend itself to a straightforward decomposition into an isotropic part and a part associated with complete internal freedom, as is the case for methyl rotation in polyisoprene.

We will defer the problem of describing the involved relaxation in polybutadiene and other elastomers to the end of the next section, which deals with nuclear Overhauser enhancements. The Overhauser enhancements of elastomeric polymers are themselves dependent on the nature of the main-chain segmental motion, and provide important clues as to a suitable general approach for the analysis of relaxation in these systems.

11.2.2 Nuclear Overheauser enhancements

Carbon-13 n.m.r. experiments are often performed with carbon-proton spin-spin interactions removed by scalar decoupling. This decoupling is achieved by some kind of continuous ^{1}H r.f. irradiation of the protons. As a result of this irradiation, the Boltzmann distribution of populations of the Zeeman levels of the protons is altered, which, in turn alters the distribution of populations of those carbons dipolar-coupled to the protons.[2] Changing the populations of the carbon Zeeman levels leads to an alteration of the observed intensities of the ^{13}C n.m.r. transitions. The change in integrated intensity of a carbon resonance due to the irradiation of protons is called the ^{13}C nuclear Overhauser effect. Because the Overhauser effect involves relative populations of Zeeman spin-energy levels (that is, steady-state magnetizations), it seems reasonable to suppose that a theoretical description of the Overhauser effect can be given in terms of the same quantities which were used to describe the ^{13}C spin-lattice relaxation. This is, in fact, the case.[1,2] In the presence of ^{13}C–^{1}H dipolar interactions, the ^{13}C nuclear Overhauser enhancement factor (NOEF) is given by

$$\frac{C_z - C_0}{C_0} = \frac{-f(\omega_H - \omega_C) + 6f(\omega_H + \omega_C)}{f(\omega_H - \omega_C) + 3f(\omega_C) + 6f(\omega_H + \omega_C)} \frac{\gamma_H}{\gamma_C} \tag{6}$$

where the ratio γ_H/γ_C is 3·976 and C_z and C_0 are the steady-state ^{13}C magnetizations with and without decoupling, respectively. The nuclear Overhauser enhancement (NOE) is $1 + $ NOEF, and is the ratio of observed integrated intensities with and without ^{1}H decoupling.

The ratio of f-functions on the right-hand side of equation (6) is referred to as ρ_{CH}, and is a measure of the effectiveness of the dipolar coupling between the ^{13}C and ^{1}H spins. Figure 11.2 shows ρ_{CH} as a function of the product of the ^{13}C Larmor frequency and the segmental correlation time. For any observing ^{13}C n.m.r. frequency, the coupling and hence the Overhauser enhancement are reduced for sufficiently large values of the correlation time. In other words, even in the presence of a completely saturating ^{1}H decoupling field, the Overhauser enhancement is less than the theoretical maximum for those polymer chains that undergo slow motions associated with long correlation times. Motions of this type fail to satisfy the extreme narrowing condition, $\omega_C \ll 1$. Of course, for any given polymer, the correlation time for which this condition is no longer satisfied depends on the observing frequency, and so it will be

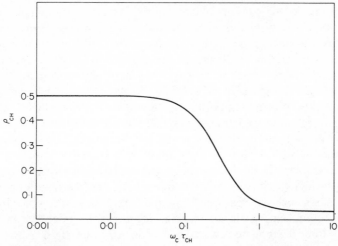

Figure 11.2 A plot of ρ_{CH}, a measure of the effectiveness of the dynamic dipolar coupling between C and H spins, as a function of the product of the ^{13}C frequency and correlation time, $\omega_C \tau$

different for those measurements performed at 15 MHz with iron magnets, relative to those performed at 60 MHz with superconducting solenoids. When the extreme narrowing condition is satisfied, the maximum NOEF and NOE are observed,

$$\rho_{CH} = \tfrac{1}{2} \tag{7}$$

$$\frac{C_z - C_0}{C_0} = 1.988 \tag{8}$$

$$\frac{C_z}{C_0} = 2.988 \tag{9}$$

that is, when the extreme narrowing condition is satisfied, NOEF = 2 and NOE = 3.

Experimental ^{13}C nuclear Overhauser enhancements for solutions of polymers near room temperature are not, in general, the theoretical maximum value.[1] For example, the ratio of the integrated intensities of the spin-decoupled to spin-coupled quaternary carbon line of isotactic polystyrene in o-dichloro-benzene is less than 2. The Overhauser enhancement is reduced for polystyrene because, in terms of a simple one-parameter description of relaxation, the effective correlation time does not satisfy the extreme narrowing condition. Thus, ρ_{CH} of Figure 11.2 is only about one-half of its maximum value of 0.5 for polystyrene. For more flexible polymers such as poly(ethylene oxide) as well as solutions of many polymers at elevated temperatures, the observed Overhauser enhancements are substantially larger since the increased segmental mobility results in a shorter effective correlation time that more nearly satisfies the extreme narrowing condition.

Even though the room-temperature ^{13}C Overhauser enhancements are reduced for many polymers, they do not vary, in general, from one type of carbon to another in the same polymer. The integrated intensity of the quaternary-carbon line of isotactic polystyrene is identical to that of the methine-carbon line as well as to that of the methylene-carbon line.[1] Basically, this is because the effective or average correlation time that best describes those segmental motions important to the Overhauser enhancement is about the same for each carbon in polystyrene. This is also true for elastomers such as polyisoprene, poly(propylene oxide) and polybutadiene,[8] despite the fact that spin-lattice relaxation time measurements on the latter two polymers clearly indicate anisotropy in the main-chain segmental motion. Apparently, given the present level of experimental error of about 10 %, the Overhauser enhancement is simply a less sensitive measure of this kind of motional anisotropy. In practical problems, complications in comparing ^{13}C intensities might be expected for relatively long flexible side-chains connected to more rigid main chains, or for long flexible blocks flanked by more rigid blocks as in ABA-type block copolymers. But, in the majority of cases, especially for synthetic polymers, it is possible to use the integrated intensities of a ^{13}C n.m.r. spectrum of a given polymer solution obtained at any presently available frequency to count the number of carbons contributing to the signal to obtain both structural and steric information.

The average or effective correlation time that one obtains from the observed Overhauser enhancement for solutions of polystyrene at room temperature (using Figure 11.2) is about 10^{-9} s. This value is in reasonable agreement with that obtained from a single-correlation-time analysis of the methine- and methylene-carbon spin-lattice relaxation times.[1] This kind of agreement is not typical. For cis-polybutadiene, for example, spin-lattice relaxation times for the methine and methylene main-chain carbons are of the order of 0·5 s at room temperature, which is comparable to what is observed for dipolar-dominated relaxation in solutions of modest-sized molecules. However, unlike the full Overhauser enhancement of 3·0 expected for such reasonably mobile molecular systems, the enhancement for polybutadiene is 2·4.

The presence of this discrepancy for polybutadiene can be illustrated by the results[8] of a transient NOE experiment. In this experiment, the ^{13}C magnetization, I_z, is measured as a function of time after the ^{1}H saturating field has been turned either on or off. When the ^{13}C magnetization is sampled some time (comparable to the ^{13}C T_1) after the decoupler is turned on, one obtains a spin-decoupled spectrum whose intensity depends on the time delay, t, according to[10]

$$[I_z(t)] = I_0 + 3.976\rho_{CH}I_0(1 - e^{-t/T_1}) \tag{10}$$

with the initial conditions

$$[I_z(0)] = I_0 \tag{11}$$

$$[S_z(t)] = S_z(0) = 0 \tag{12}$$

where S_z is the ^{1}H magnetization.

Thus, a semi-log plot of $[\{I_z(\infty)\} - \{I_z(t)\}]/I_0$ versus time is a straight line whose slope is the ^{13}C spin-lattice relaxation time and whose intercept is the NOEF. In the case when the ^{13}C magnetization is sampled some time after the decoupler is turned off, one obtains a spin-coupled spectrum whose intensity depends on the delay time as a sum of exponentials, which involves the proton as well as the carbon spin-lattice relaxation times. The results of transient NOE experiments on the methylene carbon of *cis*-polybutadiene are shown in Figure 11.3. The solid circles show the experimental data. The solid lines are

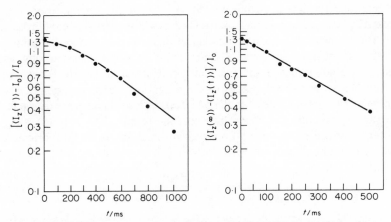

Figure 11.3 Results of transient nuclear Overhauser enhancement experiments for the methylene carbon of solid *cis*-polybutadiene at 35 °C. The plot on the left is for intensity measurements made in the absence of decoupling, and that on the right in the presence of decoupling. Circles show experimental points, and solid curves the theoretical dependence

not fitted to the data points but were independently calculated with equation (10) (and its two-exponential counterpart),[10] using the observed ^{13}C spin-lattice relaxation time, the observed NOEF from steady-state experiments, and the literature estimate for the ^{1}H spin-lattice relaxation time for 90 MHz and 35 °C, 300 ms. The agreement between the observed and calculated values is well within experimental error. The Overhauser enhancement for the main-chain methylene carbon of *cis*-polybutadiene is less than the theoretical maximum, even though the spin-lattice relaxation time is of the order of 0·5 s.

It is clear that this deviation from the theoretical maximum cannot be accounted for by the presence of some very long correlation times (which may determine the linewidth, or T_2), since the Overhauser enhancement depends primarily on T_1-like processes. However the correlation time associated with the observed T_1 is too small to be consistent with the observed Overhauser enhancement. For systems in which dipolar relaxation is dominant, slightly reduced Overhauser enhancements can, in general, be traced to correlation times which are reasonably short but still fail to satisfy the extreme narrowing condition. Thus, the deviation appears to require the presence of some motions associated with correlation times that are long compared to those determining T_1 but short compared to those determining T_2.

This situation seems to suggest that a description of the ^{13}C n.m.r. relaxation phenomena in viscous polymeric systems should be in terms of a broad distribution of correlation times. This is not an unreasonable conclusion. Many studies of polymers and glass-forming materials by dielectric and mechanical loss methods, as well as by n.m.r. techniques, have been interpreted by assuming that the molecular motions involved can be described (at least in an approximate way) by a distribution of correlation times.[11] Actually, some authors[12] prefer to describe relaxation in these multiple-correlation time systems in terms of what are known as non-exponential autocorrelation functions, while still others[13] use models involving a distribution of populations among energy wells associated with the various configurations and conformations of the polymer. Regardless of the details of the model or analytical technique which one prefers, the physical basis for all these descriptions remains the characteristic cooperativity inherent in a macromolecular system.

In terms of a distribution of correlation times for isotropic reorientations, the f-function of equation (2) becomes[8]

$$f(\omega_i) = \int_0^\infty \frac{G(\tau_l)\tau_l \, d\tau_l}{1 + \omega_i^2 \tau_l^2} \tag{13}$$

where now a density function must be specified. Physical variables in complicated systems are frequently assumed to be normally distributed. In fact, normal distributions have been used to describe multiple correlation times in viscous media of various kinds.[11] For polymers, however, it seems reasonable to suppose that seemingly unlikely long-range cooperative motions do occur. Thus, a density function based on a normal distribution, in effect, discriminates against these kinds of cooperative motions, many of which necessarily have long correlation times. Thus, we suspect that the density function must represent a distribution which, while not altogether dissimilar from a normal distribution, is asymmetric, having greater density for very large correlation times than for very small ones. A versatile distribution which fits this description is the χ^2 distribution given by

$$\tilde{G}(\tau_c; p) \, d\tau_c = \frac{1}{\Gamma(p)} (p\tau_c)^{p-1} e^{-p\tau_c} p \, d\tau_c \tag{14}$$

where $\tau_0 = \tau_c/\bar{\tau}$ and where the tilde over the G indicates that this is a trial density function and not necessarily the function to be used in equation (13). The gamma function, $\Gamma(p)$, normalizes $\tilde{G}(\tau_c; p)$ to unity. The distribution is defined by its width and mean. The mean is $\bar{\tau}$. The width of the distribution is characterized

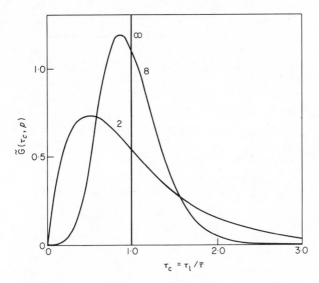

Figure 11.4 A chi-square distribution of correlation times for three values of the width parameter, p, equal to 2, 8 and infinity

by p. The larger p becomes, the narrower the distribution becomes eventually approaching a δ function for very large p (Figure 11.4).

The literature on statistics refers to p as a measure of the number of degrees of freedom, and this terminology is also used and justified in physical discussions.[14] It arises from the fact that the sum of two or more χ^2 distributions is also a χ^2 distribution but of higher order—that is, having a larger p. As applied to a polymer chain, if half of the motions of a unit of the chain are independent of the other half and if the distribution of correlation times associated with the motions of each half is described by the same χ^2 distribution of order p, then a χ^2 distribution of order $2p$ describes the overall distribution. Eventually, as all the motions of any unit in the chain become totally independent, a χ^2 distribution of extremely high order is required to describe the distribution of correlation times. This description is then indistinguishable from one using a single average correlation time.

Correlation times for motions in polymers can differ by many orders of magnitude. In order that the density function be sufficiently wide to account for this variation (which the trial function of equation (14) cannot do), a logarithmic time scale is adopted. We define a $\log$–χ^2 density function by assuming that the functional form of the χ^2 distribution (equation (14)) remains applicable, and now replace τ_c by a logarithmic function of τ_c. The effect of the choice of a $\log$–χ^2 density function is to produce a highly asymmetric distribution of correlation times with a pronounced tail in the long-correlation-time region. One can now calculate T_1, T_2, and the NOEF as a function of p and $\bar{\tau}$.[15] Two

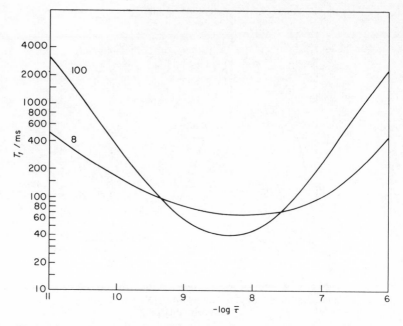

Figure 11.5 The calculated spin-decoupled methine-carbon spin-lattice relaxation time (at 22·6 MHz) for two values, 8 and 100, of the width parameter, p, of a log-χ^2 distribution of correlation times

examples of the T_1 calculation are shown[8] in Figure 11.5. (In these calculations, non-bonded carbon-proton interactions have been ignored.) For a narrow distribution ($p = 100$), the calculation is essentially identical with the prediction of the Bloembergen–Purcell–Pound theory (equations (1) and (2)) with a correlation time $\bar{\tau}$. For a wider distribution ($p = 8$), the T_1 versus $-\log \bar{\tau}$ plot is shallower, flatter, and asymmetric. The asymmetry results in a shift of the T_1 minimum of about 0·3 log units. Examples of the NOEF calculation are shown in Figure 11.6. For $p < 20$ and $\bar{\tau} < 10^{-9}$ s, the NOEF is not a strong function of $\bar{\tau}$ and can be substantially less than the 2·0 theoretical maximum. For the most part, these results are due to the width and pronounced asymmetry of the χ^2 distribution.

As is shown in Table 11.3, the T_1, T_2, and NOEF experimental ^{13}C n.m.r. data for polystyrene and some elastomer systems can be fitted by using a log-χ^2 distribution of correlation time.[8] Furthermore, for each polymer, the fit of three independent data points (at a single temperature) is in terms of only two essential parameter, the average correlation time $\bar{\tau}$ and the width of the distribution, which is related to p. In the spirit of this theoretical analysis, the parameter p can be related to the number of degrees of freedom in the distribution of correlation times and hence to motional cooperativity. By this

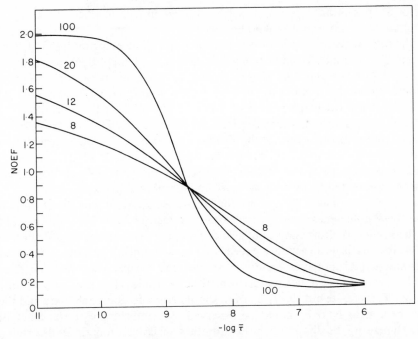

Figure 11.6 The calculated ^{13}C Overhauser enhancement factor (at 22·6 MHz) for four values, 8–100, of the width parameter, p, of a log-χ^2 distribution of correlation times

interpretation, the ratios of p values shown in Table 11.3 suggest that long-range cooperative motions are more likely for the flexible polybutadiene relative to the more rigid solid polyisoprene. This further suggests that as a rigid chain is made flexible (by removal of side groups), long-range cooperative motions become possible that would otherwise have been hampered. Only after considerably more freedom is achieved (by the presence of a solvent) can these long-range

Table 11.3 Observed (22 MHz, ^{1}H; $\gamma H_2/2\pi = 2$ kHz, 35 °C) and calculated (log-χ^2 distribution of Debye oscillators, $b = 1000$) carbon-13 n.m.r. parameters for some main-chain methine carbons[8]

Polymer	State	Observed			Calculated				
		T_1/ms	T_2/ms	NOEF	T_1/ms	T_2/ms	NOEF	$-\log \bar{\tau}$	p
Polystyrene (isotactic)	o-Dichloro-benzene solution	65	26	0·8	68	27	1·0	9·0	18
cis-Polyisoprene	Solid	95	29	1·2	98	32	1·2	9·4	14
cis-Polybutadiene	Solid	600	48	1·4	550	42	1·4	11·0	9

cooperative motions apparently be eliminated by the incoherent averaging resulting from more complete liquid-like behaviour.

The pronounced deviation from 2 of the ratio of the experimental T_1 values for methine and methylene carbon indicates that any kind of rigorous quantitative description of segmental motion in polybutadiene must take the pronounced anisotropy of the main-chain motion into account. This was not done in arriving at the calculated values of Table 11.3. Undoubtedly we could get a better fit using distributed anisotropic rotors as a model, and this would probably mean a more realistic value for the average main-chain segmental reorientation correlation time, as well. Nevertheless, we feel that most of the important qualitative aspects of the physics determining the relaxation in elastomers can be represented by an analysis involving distributed isotropic rotors.

It seems reasonable to suppose that increased freedom can be introduced into these polymer systems by increasing the temperature. This not only will have the effect of decreasing $\bar{\tau}$ but also of changing the shape of the distribution by increasing p. We therefore expect that measurements performed on solid elastomers and polymer solutions at elevated temperatures will produce qualitatively different behaviour from that of Table 11.3; that is, not only should T_1 increase, but the T_1/T_2 ratio should begin to approach unity and the full theoretical NOEF should be observed. An unfortunate consequence of this change in the shape of the distribution of correlation times describing segmental motion may be the inapplicability of simple physical interpretations of the temperature dependence of spin-lattice relaxation times for polymers. In other words, extracting activation energies from the temperature dependence of the spin-lattice relaxation time is complicated by the fact that the distribution of correlation times describing the relaxation is itself temperature dependent. Thus the limiting slopes of a log T_1 versus $1/T$ plot below and above the T_1 minimum[16] are not simply related to some kind of activation energy for the average or effective segmental motion of the polymer, at least in any rigorous quantitative sense. The temperature dependence of the distribution of correlation times describing the segmental motion of the polymer must be established before meaningful information about activation energies can be obtained.

11.3 RELAXATION EXPERIMENTS ON RIGID SOLID POLYMERS BELOW T_g

11.3.1 Spin-lattice relaxation times and nuclear Overhauser enhancements

The natural abundance ^{13}C n.m.r. line-widths of solid polymers well below the glass transition temperature are of the order of tens of kHz, and so virtually impossible to observe. In these systems, however, much of the line broadening is due to ^{13}C–^1H dipolar interactions associated with long correlation times. This broadening can be removed by *strong* proton decoupling[17] (referred to here as dipolar decoupling). Experimentally, dipolar decoupling requires an r.f. field comparable to the ^1H line-width, which for most solid polymers below

T_g is of the order of 5–15 G. This means that about 100 times more power must be used for dipolar decoupling than is used in routine scalar decoupling. The strong decoupling removes near-static dipolar interactions in the same way that conventional weaker decoupling removes weaker scalar interactions, by rapid radio-frequency stirring of the ^{1}H spins,[17] thereby producing what amounts to a high-resolution spectrum of a solid.[18]

Some dipolar-decoupled spectra of solid poly(methyl methacrylate) at 30 °C are shown[19] in Figure 11.7. As the strength of the decoupling field approaches

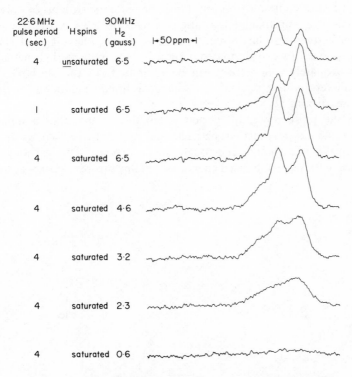

Figure 11.7 Dipolar-decoupled ^{13}C n.m.r. spectra of solid atactic poly(methyl methacrylate) at room temperature under a variety of pulse and decoupling conditions. The saturation of the ^{1}H spin system was achieved with an 0·1 G c.w. decoupling field. The dipolar decoupling field was gated

the ^{1}H line-width of poly(methyl methacrylate) of about 4·5 G, three lines are observed. (We will discuss somewhat later, improvements in the resolution and characterization of these lines made possible by cross-polarization experiments.) The three lines are assigned to the three protonated carbons of the repeating unit, namely the methylene carbon, the ester-methyl carbon, and the α-methyl carbon, in order of increasing magnetic field, respectively. (The non-protonated carbons have much longer spin-lattice relaxation times and so are not observed in an experiment with a pulse repetition period of only a few seconds.) The shape and width of the three lines are unchanged (that is, there is no substantial additional narrowing) as the dipolar decoupling field is increased from 5 G to about 8 G.

The removal of dipolar broadening from ^{13}C n.m.r. spectra often reveals the chemical shift anisotropy which characterizes the electron density distribution about various kinds of carbons.[18] The observation of a chemical shift aniso-tropy is unmistakable since the experimental line-shape has a unique asym-metry. The most commonly observed line-shape for amorphous or polycrystal-line materials is approximately triangular, with the apex of the triangle near one of the extremes of the resonance for those carbons having axial symmetry. The fact that all three protonated-carbon lines in poly(methyl methacrylate) are approximately symmetrical can be attributed, for the most part, to the small inherent anisotropy of the chemical bonding around methyl and methylene carbons.[18]

When the pulse repetition period is lengthened to about a minute, the spectrum of solid poly(methyl methacrylate) shows the resonances due to the slowly relaxing carbonyl and quaternary carbons. The carbonyl-carbon resonance, with its pronounced chemical shift anisotropy, appears at low field,

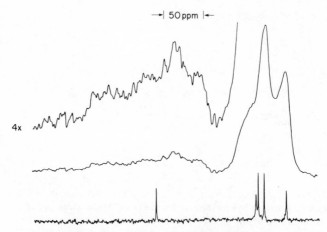

Figure 11.8 The dipolar-decoupled ^{13}C n.m.r. spectrum (upper curves) of solid atactic poly(methyl methacrylate) (pulse period 60 s), compared to the scalar-decoupled spectrum (lower curve) of a solution of highly syndiotactic poly(methyl methacrylate). The solution spectrum is actually a composite of spectra of the polymer in methylene chloride and in ethylene chloride, with the solvent lines omitted for clarity of presentation. From left to right, in order of increas-ing magnetic field, the lines are assigned to the carboxyl, methylene, ester-methyl, quaternary, and α-methyl carbons, respectively.

while the quaternary-carbon resonance is apparent in the change in intensity and centre of the second of the high-field lines, known to have a chemical shift approximately equal to that of the quaternary carbon.[20] The dipolar-decoupled spectrum is compared to a scalar-decoupled high-resolution spectrum of a solution of a stereoregular poly(methyl methacrylate)[19] in Figure 11.8. The carbonyl-carbon resonance of the polymer in solution appears at what amounts to the centre of gravity of the asymmetric carbonyl-carbon resonance in the solid. There are no significant net shifts between the average values of carbon resonances in the solid compared with those in solution.

Since the observed relative intensities of the methyl- and methylene-carbon lines depend upon the pulse repetition time (compare the second and third spectra from the top of Figure 11.7), each line is clearly associated with its own characteristic spin-lattice relaxation time. There is no loss of information due to the averaging of spin-lattice relaxation times by spin diffusion as is the situation for ^{1}H n.m.r. experiments on the same polymer.[21] Furthermore, the ^{13}C relaxation experiment is not confused by the presence of relatively minor impurities (such as residual monomer or absorbed solvent) dominating the observed signal, since all carbon lines can be identified by their characteristic chemical shifts.

From a series of partially relaxed spectra,[19] it can be shown that the shortest T_1 (which is less than 100 ms) belongs to the high-field, α-methyl carbon line, while the ester-methyl and methylene carbon T_1's are substantially longer, having values of the order of 1 s. The T_1's of the two non-protonated carbons in poly(methyl methacrylate) are another order of magnitude longer, with values of about 10 s. Two T_1's can be identified with the most intense line of Figure 11.7. The identification of a short-T_1 component with the ester-methyl carbon and a long-T_1 component with the quaternary carbon is unmistakable because of a shift in the centre of that symmetric line from low field to high field as the longer-T_1 component is allowed to contribute. (The ester-methyl carbon resonance is about 6 ppm to the low-field side of that of the quaternary carbon.[20]) Thus, although the ester-methyl and quaternary-carbon lines are not resolved in the spectrum of the solid, their individual T_1 behavior can still be clearly identified and characterized.[19] The very short T_1 for the α-methyl carbon is consistent with ^{1}H relaxation data which show that, at room temperature, poly(methyl methacrylate) is near a T_1 minimum (which can reasonably be associated with the internal motion of the main-chain methyl group), but far to the high-temperature side of a minimum for the relatively unhindered internal rotation of the side-chain methyl group.[21]

Since the methylene-carbon T_1 is much shorter than the quaternary-carbon T_1, it is likely determined predominantly by dipolar interactions of the methylene carbon with its directly bonded protons, rather than with non-bonded methyl protons. If the latter determined T_1, one would expect a much longer relaxation time, reflecting the several Å internuclear distances involved, and the inverse sixth-power dependence on internuclear separations of these dipolar interactions. Although it is true that each methylene carbon has two nearby α-methyl groups, while a quaternary carbon has only one, an unrealistically large contribution to relaxation from inter-chain dipolar interactions must still be assumed in order to account for the near order-of-magnitude difference in T_1's for the two kinds of main-chain carbons if the methyl protons were primarily responsible for relaxation. Such strong inter-chain interactions are unlikely in view of the probable loose packing of mobile main-chains in a glassy polymer characterized by relatively large intermediate and high-frequency dielectric and mechanical loss figures.[22] In addition, we have observed T_1's of a few seconds for main-chain carbons of other solid polymers at room temperature, for example with polyacrylonitrile and poly(vinyl chloride), where there is no possibility of contributions to spin-lattice relaxation by nearby, mobile side-groups. Thus, the importance of main-chain protons seems clear in the spin-lattice relaxation of directly bonded carbons in solid, glassy polymers in general, and in poly(methyl methacrylate) in particular.

The nuclear Overhauser enhancement (NOE) for all three protonated carbons (including the main-chain methylene carbon) is about 2·0, which is less than the theoretical maximum but still far from the minimum value one might have expected for a rigid system apparently well to the low-temperature side of the T_1 minimum (Figure 11.2). The NOE is readily determined by comparison of the intensities of the fully decoupled spectra obtained with and without pre-saturation of the ^1H spins (the first and third spectra of Figure 11.7).

We attribute the observation of substantial, approximately equal NOE's for both main-chain and side-chain carbons (each relaxed predominantly by its own directly bonded protons) to the presence of a broad distribution of correlation times associated with a variety of segmental rotational and torsional motions. The presence of a distribution of correlation times tends to level the NOE's of most carbons to about the same value (Figure 11.6), even though these carbons may still have significantly different T_1's. Unlike the situation for polymer systems well above T_g, however, the distribution of correlation times for solid poly(methyl methacrylate) probably has a tail in the direction of short correlation times, opposite to that shown in Figure 11.4. In other words, below the glass transition temperature, long-range cooperative motions (necessarily associated with long correlation times) are discriminated against, while many short-range, high-frequency (10^4–10^6 Hz) torsional motions can apparently still persist, even for the main chain, and even well below T_g. Thus, deviations from a symmetrical, narrow distribution of correlation times are most likely to be toward the short-correlation time region. The net result is that a substantial NOE can still exist for carbons in a solid, glassy polymer. Naturally the high-frequency motions resulting in the NOE are also effective contributors to the spin-lattice relaxation times of the various carbons. This is the reason for a methylene-carbon T_1 in the glassy polymer as short as a second, rather than the longer values one might expect for ^{13}C T_1's in highly crystalline materials.[18]

Comparable ^1H T_1 data (where T_1 is averaged over all types of protons in the polymer by spin diffusion) do not, however, suggest the presence of any main-chain segmental mobility for poly(methyl methacrylate) at room temperature.[20] This apparent discrepancy is undoubtedly due to the low activation energy associated with the main-chain motions of loosely packed chains, and therefore the failure of these motions to have a pronounced, observable effect on an average proton T_1 versus temperature plot, even though they have substantial effects on various ^{13}C relaxation parameters determined at a single temperature. Mechanical measurements[22] on poly(methyl methacrylate) have been interpreted in terms of high-frequency motions of the main chain, even at temperatures around 35 °C. Although these interpretations have been considered essentially speculative, they do, in fact, appear to be consistent with the direct evidence on main-chain motion provided by the ^{13}C n.m.r. measurements.

The presence of a variety of segmental motions in poly(methyl methacrylate) has an additional influence on the ^{13}C experiment. Whatever chemical shift anisotropies are present will tend to be obscured by residual dipolar broadening. This broadening is associated with those motions having correlation frequencies relatively large compared to the dipolar decoupling field, $\gamma H_2/2\pi$, and so is not removed by the usual strong decoupling techniques. Of course, motions responsible for broadening need not be as fast as those determining the spin-lattice relaxation in order to have correlation frequencies greater than $\gamma H_2/2\pi$. In fact, intermediate-frequency motions will have the most substantial influence since their broadening is the least averaged by the motion itself. In part, the presence of such motions explains why the small, but significant, chemical shift anisotropies of the methyl and methylene carbons are not apparent.

The chemical shift anisotropy of the carbonyl carbon is, on the other hand, of the order of several kHz, which is sufficiently large to ensure that the anisotropy remains visible despite the presence of some dipolar broadening associated with intermediate-frequency segmental motions. Naturally the fact that the carbonyl carbon has no directly bonded proton tends to reduce any such broadening, relative to that experienced by the methyl and methylene carbons. The presence of the approximately full chemical shift anisotropy, and general polycrystalline asymmetric lineshape,[18] for the carbonyl carbon of poly(methyl methacrylate) at room temperature is a direct proof that the ester side-group as a whole does not engage in extensive internal rotational reorientations, comparable to the internal motions of the ester-methyl (or ester methoxy) and α-methyl carbons. This conclusion had been suspected on the basis of proton n.m.r. relaxation experiments,[23,24] but direct evidence one way or the other had been lacking.

11.3.2 Cross-polarization

Although a resonance from the carbonyl carbon in poly(methyl methacrylate) can be observed, the experiment is made difficult by the long carbonyl-carbon T_1 and necessary delays between sampling pulses. By using cross-polarization, however, we can eliminate waiting the usual several ^{13}C T_1's between samplings of the ^{13}C magnetization (which may be of the order of minutes), and instead are required to wait only for the repolarization of the protons (which for most solid polymers is relatively efficient and occurs in substantially less than a second). The repolarization of the carbon spins then also occurs rapidly because in a solid, cross-polarization of the carbons by the protons occurs through a rapid spin-spin process, rather than by a slow spin-lattice process.[25]

One version of cross-polarization involves producing two strong rotating magnetic fields $(H_1)_{carbon}$ and $(H_1)_{proton}$ tuned to resonance such that

$$\gamma_{carbon}(H_1)_{carbon} = \gamma_{proton}(H_1)_{proton}$$

When this condition (called the Hartmann–Hahn condition) is satisfied, the carbon and proton spin systems are strongly coupled, even though in the static field, H_0, they are not coupled because of their widely different frequencies. The coupling occurs at the Hartmann–Hahn condition since the precession of the proton spins about their H_1 causes the component of the dipolar field along the direction of the static field H_0 to oscillate at an angular frequency $\gamma_{proton}(H_1)_{proton}$. This frequency is just right to induce transitions of the carbon spins relative to their rotating field.[26]

The two-rotating field version of the cross-polarization experiment is performed in four parts.[18,25] First, the proton spins are polarized in a high field. Then, they are placed in the rotating frame by a 90° pulse followed by a 90° phase shift and continuous application of strong ^{1}H r.f. The third part of the experiment is to establish ^{13}C–^{1}H contact for some variable time t by placing the ^{13}C spins into the rotating frame as well. The final step is to sample the ^{13}C magnetization while dipolar decoupling the ^{1}H spins. The maximum polarization of the carbon spins is approximately $\gamma_{proton}/\gamma_{carbon}$, which is larger than what the carbon spins could achieve without cross-polarization, in a time comparable to the carbon T_1, even with the presence of a maximum Overhauser enhancement.

The results of a series of some two-rotating field, cross-relaxation experiments performed on solid poly(methyl methacrylate) are shown[19] in Figure 11.9. The single contact time between the ^{13}C and ^{1}H spins has been varied from 7 ms to 50 μs, with the most favourable contact time appearing to be about 1 ms. The optimum contact time depends upon the cross-relaxation times in the rotating frame, the T_{CH}'s (which in turn, are related to the strengths of the static dipolar interactions between carbons and protons) and the length of time the proton magnetization can be held in the rotating frame and so contribute to the polarization of the carbon spins.[27] The latter time is called the proton $T_{1\rho}$, and for poly(methyl methacrylate) around room temperature, is of the order of 5 ms. The enormous advantage of performing a cross-polarization relative to a simple dipolar-decoupling experiment, is made clear by the observation of strong signals from both of the non-protonated carbons of poly(methyl methacrylate), even with sampling of the ^{13}C magnetization as rapidly as once a second (Figure 11.9). In a standard dipolar-decoupling experiment with a ^{13}C pulse repetition period of 1 s, no signal is observed for either the carbonyl or quaternary carbons (Figure 11.7).

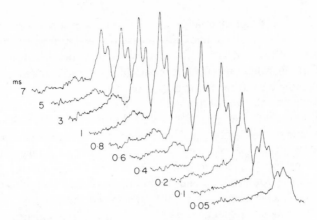

ms
7
5
3
1
0·8
0·6
0·4
0·2
0·1
0·05

Figure 11.9 Dipolar-decoupled ^{13}C n.m.r. spectra of solid atactic poly(methyl methacrylate) obtained following a single Hartmann–Hahn cross-polarization contact of variable length, 0·05–7 ms. The ^{1}H and ^{13}C r.f. fields were approximately 4·5 and 18 G, respectively, and the sequence repetition time was 1 s. The line assignments are those given in Figure 11.8

A cross-polarization experiment also provides new information contained in the differences between cross-relaxation rates. For example, in poly(methyl methacrylate) it is clear from Figure 11.9 that the methylene-carbon line (the lowest-field line in the high-field grouping) has the fastest cross-relaxation rate. The methylene-carbon line remains intense even for very short contact times. (Since the other lines are substantially reduced in intensity for contact times of the order of 100 μs, cross-polarization can obviously serve as a resolution enhancing technique, thereby revealing the shape and position of some lines more clearly than in a standard dipolar-decoupling experiment.) The short methylene-carbon T_{CH} is undoubtedly due to near-static dipolar interactions between the methylene carbon and its two directly bonded protons. Since these protons are also engaged in high-frequency motions responsible for the sizeable NOE and short T_1 of the methylene carbon, and in intermediate-frequency motions responsible for residual dipolar-broadening effects, this information about low-frequency motions of the methylene protons from the cross-polarization experiments supports the idea of a *broad* distribution of correlation times associated with the segmental motions of the polymer main chain in the solid state.

11.3.3 Magic-angle spinning

The dipolar-decoupled ^{13}C n.m.r. spectra of most solid polymers are not as easy to deal with as that of poly(methyl methacrylate), mostly because the chemical shifts are generally not as well separated. For these polymers, chemical shift anisotropy and residual dipolar broadening of carbon resonances having relatively small average chemical shift differences result in disappointingly broad and poorly resolved lines. Such spectra are not well suited for the determination of the various kinds of relaxation parameters discussed above. In order to simplify these spectra, the broadening can be removed by high-speed mechanical sample rotation at the magic angle.[28]

In this experiment, the solid polymer is fashioned into a rotor[29] which is placed in the magnetic field in such a way that the spinning axis of the rotor makes an angle of 54·7° with the field. (A polymer powder can also be placed inside a hollow rotor made of a suitable material such as a rigid fluoropolymer.[29]) The rotor is then spun by means of a gas jet impinging upon small vanes which are a part of the rotor body. The magic angle of 54·7° reduces to zero the $(3 \cos^2 \theta - 1)$ dependence of some kinds of residual static dipolar interactions, as well as the angular dependence of the chemical shift anisotropy. The net result can be a genuine high-resolution spectrum of a solid. The magic-angle spinning experiment becomes useful for ^{13}C studies only when used in conjunction with dipolar decoupling. The reason for this limitation is that it is not experimentally feasible mechanically to spin a rotor at speeds high enough to remove the full static $^{13}C-^1H$ dipolar interaction encountered in solids. Mechanical spinning speeds of 2–4 kHz are relatively simple to achieve, but

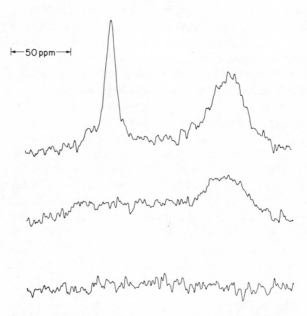

Figure 11.10 Some ^{13}C n.m.r. spectra of a magic-angle rotor made of polystyrene. Lower spectrum: scalar decoupling and magic-angle spinning; centre spectrum: dipolar decoupling only; upper spectrum: dipolar decoupling and magic-angle spinning

narrowing of the complete static dipolar interaction would require speeds at least 5 times as great, which simply is not practical. By removing the bulk of the dipolar interactions by dipolar decoupling, however, the magic-angle spinning need only cope with line broadening effects of a few kHz.

The results of a combination dipolar-decoupling and magic-angle spinning experiment performed on a rotor made from atactic polystyrene are shown[19] in Figure 11.10. The aromatic-carbon resonances are well enough resolved by spinning at 2 kHz that the low-field C_1 quaternary-carbon line is distinguishable, as a shoulder, from the remaining aromatic-carbon lines. The broadening of the aromatic-carbon lines in the simple dipolar decoupling experiment is primarily, but not exclusively, due to chemical shift anisotropy. A long-term time-averaging experiment performed on the polystyrene, in fact, reveals the aromatic-carbon line-shape as asymmetric, but not displaying a clear chemical shift anisotropy. Since the T_1's of the phenyl carbons as well as main-chain carbons can be shown to be less than a few seconds, we feel that substantial torsional motion of the phenyl side group must be occurring, although nothing like free rotation. This torsional motion results in residual dipolar broadening of the overlapping chemical shift anisotropies of the phenyl carbons. In addition, some further dipolar broadening can be expected since the 8-G decoupling field used in these experiments is only comparable to the polystyrene proton line-width. However, as seen in Figure 11.10, some of the dipolar line broadening and the chemical shift anisotropy are readily removed by magic-angle spinning thereby producing a spectrum with which one can readily perform T_1, $T_{1\rho}$, and cross-relaxation and T_{CH} experiments.

11.4 REFERENCES

1. J. Schaefer and D. F. S. Natusch, *Macromolecules*, **5**, 416, (1972).
2. A. Abragam, *The Principles of Nuclear Magnetism*, Oxford University Press, London, 1961.
3. D. Doddrell, V. Glushko and A. Allerhand, *J. Chem. Phys.*, **56**, 3683, (1972).
4. Y. Fujiwara and P. J. Flory, *Macromolecules*, **3**, 43, (1970).
5. J. R. Lyerla, Jr., H. M. McIntyre and D. A. Torchia, *Macromolecules*, **7**, 11, (1974).
6. Y. Inoue, A. Nishioka and R. Chujo, *J. Polymer Sci.*, *Polym. Phys.*, **11**, 2237, (1973).
7. Y. K. Levine, N. J. M. Birdsall, A. G. Lee, J. C. Metcalfe, P. Partington and G. C. K. Roberts, *J. Chem. Phys.*, **60**, 2890, (1974).
8. J. Schaefer, *Macromolecules*, **6**, 882, (1973).
9. See, for example, J. E. Mark, *J. Amer. Chem. Soc.*, **88**, 3454, (1966).
10. J. H. Noggle and R. E. Schirmer, *The Nuclear Overhauser Effect*, Academic, New York, 1971, pp. 113–118.
11. T. M. Connor, *Trans. Faraday Soc.*, **60**, 1574, (1964).
12. Y. Martin-Borret, J. P. Cohen-Addad and J. P. Messa, *J. Chem. Phys.*, **58**, 1700, (1973).
13. J. E. Anderson, *J. Magnetic Reson.*, **11**, 398, (1973).
14. C. E. Porter and R. G. Thomas, *Phys. Rev.*, **104**, 483, (1956).
15. A. choice of b, the base of the log scale, must also be made, but this is not critical to our discussion here. Comparable results are obtained for values of b of 10^1, 10^2, and 10^3. See Ref. 8 for details.

226

16. T. M. Connor, in *N.M.R., Basic Principles and Progress*, P. Diehl, E. Fluck and R. Kosfeld, ed., Springer-Verlag, New York, 1971, Vol. 4.
17. F. Bloch, *Phys. Rev.*, **111**, 841, (1958).
18. A. Pines, M. G. Gibby and J. S. Waugh, *J. Chem. Phys.*, **59**, 569, (1973).
19. J. Schaefer, E. O. Stejskal and R. Buchdahl, *Macromolecules*, **8**, 291, (1975).
20. L. F. Johnson, F. Heatley and F. A. Bovey, *Macromolecules*, **3**, 175, (1970).
21. J. G. Powles, B. I. Hunt and D. J. H. Sandiford, *Polymer*, **5**, 505, (1964).
22. See, for example, N. G. McCrum, B. E. Read and G. Williams, *Anelastic and Dielectric Effects in Polymeric Solids*, John Wiley, London, 1967, p. 250.
23. T. M. Connor and A. Hartland, *Physics Letters*, **23**, 662, (1966).
24. D. W. McCall, *Accounts Chem. Research*, **4**, 223, (1971).
25. S. R. Hartmann and E. L. Hahn, *Phys. Rev.*, **128**, 2042, (1962).
26. F. M. Lurie and C. P. Slichter, *Phys. Rev.*, **133**, A1108, (1964).
27. D. A. McArthur, E. L. Hahn and R. E. Walstedt, *Phys. Rev.*, **188**, 609, (1969).
28. E. R. Andrew, *Progress in N.M.R. Spectroscopy*, **8**, 1, (1971).
29. E. O. Stejskal, J. Schaefer, J. M. S. Henis and M. K. Tripodi, *J. Chem. Phys.*, **61**, 2351, (1974).

12
¹³C N.M.R. Studies of
α-Methylstyrene-Alkane Copolymers

A. V. Cunliffe, P. E. Fuller

Explosives Research and Development Establishment, Waltham Abbey

and R. A. Pethrick

University of Strathclyde

12.1 INTRODUCTION

Over the last fifteen years, n.m.r. spectroscopy has proved to be a very important tool for the investigation of polymer structure.[1,2] Applications range from straightforward identification, compositional analysis and determination of chemical groupings in polymers, to detailed studies of microstructure and sequence distributions. In principle, n.m.r. can also be used to study conformations of macromolecules in solution, although to date this has usually been restricted to model compound studies. N.m.r. is particularly suitable for the analysis of the composition of copolymers, where fairly accurate values can be obtained rapidly and without the need for calibration. It is probably the best general method for the study of the microstructure of macromolecules, well known examples being tacticity measurements in vinyl polymers and the proportions of 1,2-, 1,4- and 3,4- structures in polydienes. Until recently, with the exception of a few fluorine-containing polymers, studies had been restricted to proton (¹H) n.m.r. However, within the last few years commercial spectrometers have become available capable of producing reasonable quality carbon-13 (¹³C) n.m.r. spectra on a routine basis,[3,4] and a rapidly growing literature has demonstrated that ¹³C n.m.r. is a very powerful tool for the investigation of polymer structure, in many cases giving more detailed information than ¹H n.m.r. It is particularly useful for the elucidation of paraffinic structures, providing a chemical shift range roughly forty times greater than proton n.m.r. In the future, it seems likely that ¹H and ¹³C n.m.r. will prove to be very useful complementary techniques. In the present work, a study of a series of hydrocarbon copolymers is described, and it will be seen that this is an example where ¹³C n.m.r. gives a lot of information about sequence distribution, in a case where very little information was obtained from ¹H n.m.r.

In a series of papers,[5-9] Richards, Scilly and co-workers have shown that it is possible to prepare regular copolymers by the following general reaction scheme. A vinyl[5,6] or diene[5,7-9] monomer M (2 moles), capable of polymerization by an anionic mechanism, is reacted with lithium or sodium metal in tetrahydrofuran (THF) solution, in the presence of a difunctional reagent RX_2 (1 mole), such as a dibromoalkane. It is readily apparent, from the facts that the solution remains colourless and that the red or yellow colour characteristic of anions produced from M is confined to the metal surface, that the reaction occurs exclusively close to the metal surface, and it appears that the following general mechanism applies. The monomer M reacts with the metal to form radical anions, which then dimerize to produce the dimer dianion; see reaction 1. Copolymers are formed by the reaction of dimer dianions with the dihalide RX_2, according to reactions 2 and 3.

$$2M \xrightarrow[\text{Na. Li}]{} 2 \cdot M^- \rightarrow {}^-M{-}M^- \tag{1}$$

$$^-M{-}M^- + RX_2 \rightarrow {}^-M{-}M{-}R{-}X + X^- \tag{2}$$

$$^-M{-}M{-}R{-}X \xrightarrow[{}^-M{-}M^-]{RX_2} \quad \text{+}M{-}M{-}R\text{+} \tag{3}$$

For a vinyl monomer $CH_2{=}CR_1R_2$, such as styrene or α-methylstyrene, the dimer dianion has the tail-to-tail linked structure (1).[10,11]

$$^-\underset{\underset{R_2}{|}}{\overset{\overset{R_1}{|}}{C}}{-}CH_2{-}CH_2{-}\underset{\underset{R_2}{|}}{\overset{\overset{R_1}{|}}{C}}{}^-$$

(1)

Thus, for the case of α-methylstyrene as monomer and the normal dibromo-alkanes $Br(CH_2)_nBr$, n = 3–10, as linking agents, which is the system described in the present paper, if this mechanism is correct the copolymers will have the structure (2).

$$\left[\text{+}\underset{\underset{Ph}{|}}{\overset{\overset{Me}{|}}{C}}{-}CH_2{-}CH_2{-}\underset{\underset{Ph}{|}}{\overset{\overset{Me}{|}}{C}}{-}(CH_2)_n\text{+} \right]_m$$

(2)

It is clear that there are also a number of other reactions which could occur in this system. The dianion (1) may react with another monomer molecule, according to reaction 4, to give the species (3) which contains the extra α-methyl-styrene unit linked in a head-to-tail fashion.

$$^-\underset{\underset{Ph}{|}}{\overset{\overset{Me}{|}}{C}}{-}CH_2{-}CH_2{-}\underset{\underset{Ph}{|}}{\overset{\overset{Me}{|}}{C}}{}^- \xrightarrow{M} {}^-\underset{\underset{Ph}{|}}{\overset{\overset{Me}{|}}{C}}{-}CH_2{-}CH_2{-}\underset{\underset{Ph}{|}}{\overset{\overset{Me}{|}}{C}}{-}CH_2{-}\underset{\underset{Ph}{|}}{\overset{\overset{Me}{|}}{C}}{}^- \tag{4}$$

(3)

Presumably this can occur relatively easily at each end of the dimer to form a tetrameric dianion (**3a**) further addition in the case of α-methylstyrene being much more difficult.

$$
\begin{array}{c}
\text{Ph} \\
|\\
{}^{-}\text{C}-\text{CH}_2-\text{CH}_2-\overset{\displaystyle\text{Ph}}{\underset{\displaystyle\text{Me}}{\text{C}}}-\text{CH}_2-\overset{\displaystyle\text{Ph}}{\underset{\displaystyle\text{Me}}{\text{C}}}{}^{-} \xrightarrow[\text{RX}_2]{\text{M}}
\end{array}
\quad
\left[\begin{array}{c}
\overset{\displaystyle\text{Ph}}{\underset{\displaystyle\text{Me}}{\text{C}}}-\text{CH}_2-\overset{\displaystyle\text{Ph}}{\underset{\displaystyle\text{Me}}{\text{C}}}-\text{CH}_2-\text{CH}_2-\overset{\displaystyle\text{Ph}}{\underset{\displaystyle\text{Me}}{\text{C}}}-\text{CH}_2-\overset{\displaystyle\text{Ph}}{\underset{\displaystyle\text{Me}}{\text{C}}}
\end{array}\right]
$$

$$\qquad\qquad\qquad\qquad (3) \qquad\qquad\qquad\qquad\qquad\qquad\qquad (3a)$$

In addition, a number of structures are possible from reactions of the dihalide rather than monomer directly with the metal, as in reactions 5 to 8.

$$\text{BrRBr} \xrightarrow{\text{Li}} \text{BrRLi} \tag{5}$$

$$\text{BrRLi} \xrightarrow{\text{BrRBr}} \text{Br}-\text{R}-\text{R}-\text{Br} \tag{6}$$

$$(4)$$

$$\text{BrRLi} \xrightarrow{\text{M}} \text{BrR}-\text{MLi} \xrightarrow[\text{etc.}]{\text{BrRBr}} +\text{R}-\text{M}-\text{R}+ \tag{7}$$

$$(5)$$

$$
\text{BrR}-\text{MLi} \xrightarrow{\text{M}} \text{BrR}-\text{M}-\text{MLi} \rightarrow
\left[\begin{array}{c}
\text{R}-\text{CH}_2-\overset{\displaystyle\text{Me}}{\underset{\displaystyle\text{Ph}}{\text{C}}}-\text{CH}_2-\overset{\displaystyle\text{Me}}{\underset{\displaystyle\text{Ph}}{\text{C}}}
\end{array}\right]
\tag{8}
$$

$$(6)$$

Thus, we may expect that, in addition to the desired structure (**2**), a variety of other structures (**3**) to (**6**) may occur.

In addition to the sequence distribution in the copolymers, the n.m.r. spectra of these materials are likely to be sensitive to the stereochemical arrangement of the α-methylstyrene units. If we assume that the main structure is structure (**2**), the problem of the stereoregularity of these materials is a particularly simple one, provided that the value of n in (**2**) is sufficiently large for interactions between adjacent dimer units to be neglected. In this case, there are only two possible tactic forms, corresponding to the two possible arrangements of α-methylstyrene units in the dimeric units in structure (**2**).

With these considerations in mind, we may identify three problems which we may attempt to solve in a ^{13}C n.m.r. study of these materials. These are:

(a) to show that the copolymers have predominantly the tail-to-tail linked structure (**2**);

(b) to determine what proportion of other structures such as (**3**) to (**6**) occur;

(c) to determine the stereochemical arrangements of the α-methylstyrene units in the copolymers.

Prior to this work, there was an appreciable amount of evidence to suggest that the materials were predominantly of structure (**2**).[5,6,12,13] The ratio of the aliphatic to aromatic areas in the proton n.m.r. showed that the ratio of monomer to linking group R in the materials was close to 2 to 1, as in the initial reactant stoichiometry, in accordance with the proposed structure. Moreover, the absence of the characteristic high-field methyl absorptions and the low-field resonance of a CH_2 group between two $C-Ph$ groups suggested that the main structure did not consist of head-to-tail linked monomer units. However, it was not possible to obtain more detailed information from the proton n.m.r., since all the aliphatic hydrogens overlapped to give a broad band with little resolved structure. Support for the proposed structure also arose from model compound studies. A very useful model for the copolymer system is obtained by carrying out an identical reaction to the copolymerization, except that 2 moles of monohalide HRX are used instead of 1 mole of RX_2.[11,12,13] If reactions 1 to 3 are correct for the copolymerization, then we expect reaction 1 to be followed by reaction 9 for the model system.

$$^-M-M^- \xrightarrow[2HRX]{} HR-\overset{\overset{\displaystyle Me}{|}}{\underset{\underset{\displaystyle Ph}{|}}{C}}-CH_2-CH_2-\overset{\overset{\displaystyle Me}{|}}{\underset{\underset{\displaystyle Ph}{|}}{C}}-RH \qquad (9)$$

$$(7)$$

It is found that, when HRX is a normal alkyl bromide, over 90% of the tail-to-tail dimer (7) is obtained, supporting the hypothesis that the copolymer consists predominantly of structure (2). However, although there were a number of indications as to the structure of these materials, there was, prior to the ^{13}C n.m.r. study, no technique available which gave detailed structural information on the copolymers themselves.

12.2 EXPERIMENTAL

Full experimental details for the preparation of the copolymers are given in previous papers.[5,6] In a typical reaction about 1·0 mole of lithium in 500 ml THF was stirred under nitrogen in a one litre three-necked flask. A mixture of α-methylstyrene (0·4 moles) and dihalide (0·2 moles) in THF was added, and the solution was stirred for several hours, the reaction temperature being maintained at 0 °C. The polymer was precipitated from methanol, purified by reprecipitation and dried in a vacuum oven at 60 °C.

Proton-noise-decoupled ^{13}C n.m.r. spectra were recorded at 25·15 MHz on a JEOL PFT 100 instrument with a JEC-6 computer in the Fourier Transform mode. Approximately 25% w/v solutions in $CDCl_3$ were employed, in 8 mm o.d. sample tubes, the $CDCl_3$ solvent being used to provide the internal ^2D lock signal. The transients were stored as 4K data points, transforming to a 2K data point spectrum. For the polymer samples, 2500 transients, with a pulse width of 10 μs (45° pulse) and a repetition rate of 5 s, were accumulated. For the dimers, 250 accumulations were sufficient. Spectra were recorded at ambient temperature (circa 30 °C). Chemical shifts were measured relative to tetramethylsilane (TMS) internal reference.

12.3 ¹³C N.M.R. SPECTRA OF α-METHYLSTYRENE/ALKANE DIMERS

We first looked at the ^{13}C spectra of the dimers of structure (7), with RH = H, Me, Et, Pr, Bu, pentyl, in order to determine the chemical shifts to be expected in structure (2). The dimers also gave strong indications as to the stereochemical configuration of the α-methylstyrene units in the copolymers. The aromatic region consisted of the typical pattern to be expected from a monosubstituted alkylbenzene, with peaks at 147·7, 128·1, 126·5 and 125·5 ppm relative to TMS, with areas very roughly in the ratio 1:2:2:1. The low-field peak, corresponding to the phenyl carbon C_1 attached to the aliphatic chain, was clearly split into two peaks of practically equal intensity. Studies on polystyrene[14,15] and poly(α-methylstyrene)[15] have shown that C_1 is sensitive to stereochemical effects, and it seems likely, therefore, that the two peaks correspond to the two diastereoisomers. The aliphatic region for the propyl dimer is shown in Figure 12.1. It is clear that a number of peaks are split into doublets, and from the

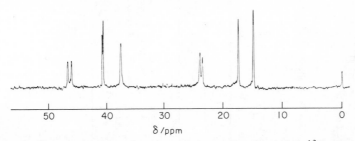

δ /ppm

Figure 12.1 The aliphatic region of the proton-decoupled ^{13}C n.m.r. spectrum of the α-methylstyrene/propyl dimer (7, RH = C_3H_7). The peak at highest field is TMS internal reference, and the solvent $CDCl_3$ appears as three small peaks (folded) near the centre of the spectrum. (The scale in all the Figures is in ppm downfield from TMS)

assignments given in Table 12.1 we see that the large splittings occur for the CH_2 group C_4, the methyl group C_3, and to a lesser extent the quaternary carbon C_2, using the numbering system in Table 12.1. It is thus reasonable to assume that the doublet splitting is due to the two diastereoisomers arising from the two possible arrangements of the asymmetric quaternary carbons. This was confirmed in the case of the ethyl dimer, which readily separated into a solid form and a liquid form, each isomer showing one of the two doublet peaks. Because of the close similarity between the preparation of the dimers and the copolymers, the essentially equal amounts of the two isomers of the dimers suggest that the stereochemical arrangement of the α-methylstyrene units in structure (2) in the copolymers will be essentially random.

The other main interest in the ^{13}C spectra of the dimers lies in the chemical shifts of the various carbon atoms, since it should be possible to use them to

Table 12.1 α-Methylstyrene dimers: calculated and observed chemical shifts in ppm relative to TMS

$$
\begin{array}{c}
\overset{3}{}CH_3 \\
| \\
-\underset{1}{CH_2}-\underset{2}{C}-\underset{4}{CH_2}\cdots\underset{8}{CH_3} \\
| \\
Ph
\end{array}
$$

Group attached to carbon 2	C_1		C_2		C_3		C_4		C_5		C_6		C_7		C_8	
	Calc.	Obs.	Calc.	Obs.	Calc.	Obs.	Calc.	Obs.	Calc.	Obs.	Calc.	Obs.	Calc.	Obs.	Calc.	Obs.
H	37·0	36·3	42·2	39·8	22·3	22·4										
Methyl	39·2	39·2	43·9	37·4	29·5	29·2	29·5	29·2								
Ethyl	37·0	37·1	46·5	42·8	27·0	23·2	34·8	35·7	9·4	8·8						
Propyl	37·3	37·5	44·0	40·6	27·3	23·8	44·2	46·7	18·5	18·2	14·2	14·9				
Butyl	37·4	37·5	44·4	40·8	27·4	24·0	41·6	43·3	27·9	26·4	23·3	23·5	14·0	14·2		
Pentyl	37·3	37·5	44·5	40·5	27·4	24·0	41·9	43·4	25·3	23·5	32·6	32·7	23·0	22·9	13·9	14·1

predict the expected shifts in the copolymers. We wish to assign the various aliphatic peaks to the carbon atoms in structure (7). This can be partly achieved from the undecoupled spectra. This readily allows identification of the quaternary carbon C_2, which is a singlet, and also distinguishes the CH_2 carbons from the two CH_3 groups. The quaternary carbon is also readily identified by its long spin lattice relaxation time T_1. However, in order to distinguish between the various CH_2 carbons a scheme for predicting the chemical shifts of the carbon atoms is required. Such a scheme has been given by Grant and Paul.[16] They have shown that, for a number of saturated hydrocarbons, the carbon chemical shifts are given by a simple additive equation

$$\delta_c^i = B + \sum_j A_j n_{ij} \qquad (10)$$

In this equation, δ_c^i is the chemical shift of the ith carbon atom, in ppm relative to TMS. The coefficient A_j has characteristic values for carbon atoms, $\alpha, \beta, \gamma, \delta$ or ε relative to the ith carbon atom, and n_{ij} is the number of carbon atoms with coefficient A_j. Thus, the shift of the ith carbon atom is made up of a series of additive contributions from the adjacent carbon atoms. The constant B has a value of -2.6 ppm, which is very close to the chemical shift of methane. The scheme is easily extended to include contributions from phenyl groups. The values of the coefficients A_j are shown in Table 12.2. With these values,

Table 12.2 Calculation of ^{13}C shifts in terms of $\alpha, \beta, \gamma, \delta, \varepsilon$ effects.

δ_c^i for a particular carbon atom i is given by $\delta_c^i = B + \sum_j A_j n_{ij}$ where δ_c^i is the shift in ppm relative to TMS, $B = -2.6$, and A_j takes the following values for carbon atoms $\alpha, \beta, \gamma, \delta$ or ε to carbon atom i.

	A_j (Carbon)	A_j (Phenyl)
α	$+9.1$	$+23$
β	$+9.4$	$+9.5$
γ	-2.5	-2
δ	$+0.3$	
ε	$+0.1$	

the calculated chemical shifts given in Table 12.2 are obtained. By comparison with the observed values, it is clear that the agreement is relatively poor for the quaternary carbon C_2. However, since this is readily identified by other means, this is not serious. The fit is relatively poor generally for the carbon atoms close to the quaternary carbon C_2, but assuming that we can identify C_2 from other considerations, the agreement is good enough to assign unambiguously all the carbon atoms in the six dimer molecules. The additive scheme has, in addition, a property which is very useful in predicting the shifts in the copolymers. It implies that, neglecting the very small contributions from δ and ε carbons, the chemical shift of a given carbon atom depends only on

234

atoms or groups α, β or γ to it. Thus, the chemical shifts of the α-methylstyrene units and the CH_2 groups of the alkane units adjacent to the quaternary carbon atom in structure (2) should have the same shifts as the equivalent carbon atoms in the pentyl dimer. This should be a useful method for predicting the chemical shifts in the copolymers.

12.4 ^{13}C N.M.R. SPECTRA OF α-METHYLSTYRENE/ALKANE COPOLYMERS

The copolymers described in the present work are restricted to those with $n = 3$ or greater in structure (2), that is with the linking agent dibromopropane up to dibromodecane. The products from the reactions with dibromoethane and dibromomethane as linking agent give much more complicated ^{13}C n.m.r. spectra, which are still being assigned. This is partly to be expected from the fact that, with the shorter linking aliphatic chains, interactions between adjacent dimer units become important, and the possible stereochemical arrangements become much more complicated. In addition, previous work[5,6] has shown that dibromoethane behaves differently from other dibromides, tending to give an elimination reaction rather than the substitution reaction between the dianion and the dihalide. Compared with the proton spectra, the ^{13}C spectra of these materials show a great deal of detail, and it is hoped that they will provide information on their structure.

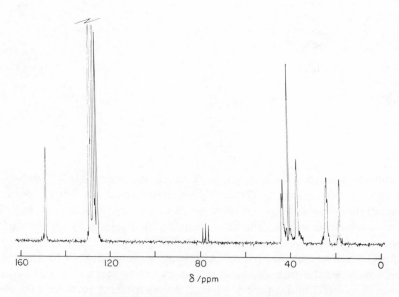

Figure 12.2 The proton-decoupled ^{13}C n.m.r. spectrum of the α-methyl-styrene/propane copolymer ($n = 3$ in structure 2)

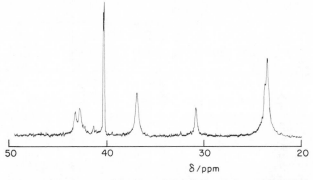

Figure 12.3 The aliphatic region of the proton-decoupled
^{13}C n.m.r. spectrum of the α-methylstyrene/pentane copolymer ($n = 5$ in structure **2**)

Figure 12.2 shows the spectrum of the α-methylstyrene/propane copolymer. The aromatic region at low field is very similar to that of dimers. The quaternary carbon atom consists of a doublet, as for the dimers. Figure 12.3 shows the aliphatic region for the α-methylstyrene/pentane copolymers. It is seen that the low-field peaks are very similar to those for the dimers. As in the dimer case, the C_4 carbon (CH_2 group adjacent to the dimer unit) and the quaternary Carbon C_2 are split into doublets of approximately equal intensity (the methyl peaks overlap with C_5) so that the stereochemical arrangement of the two α-methylstyrene units in structure (**2**) is essentially random. Interactions in the ^{13}C n.m.r. spectra between adjacent dimer units in (**2**) are apparently negligible for $n > 3$. Neglecting a few small peaks which will be discussed later, the observed chemical shifts of the main peaks are shown in Table 12.3. The experimental values are compared with values for structure (**2**) calculated by two methods. The first set of values is calculated using the additive scheme of Grant and Paul,[16] using the values of Table 12.2. The second set of values is calculated from the shifts of the pentyl dimer, making small corrections where appropriate for the extra atoms and phenyl groups in the copolymers, using the values of Table 12.2. It is seen that the latter values are very close to the observed values. Thus, all the major peaks in the spectra can be predicted from the dimer chemical shifts in terms of structure (**2**), confirming that this is the predominant structure in the copolymers. Although the major peaks are explained in terms of structure (**2**), it is clear that there are a number of small peaks which must arise from other structures.

In addition to the main structure, two other structures were identified. No evidence was found for structure (**4**) in any of the copolymers. For the propyl and butyl copolymers, this would lead to hexyl and octyl linking groups, which should be readily observed since the central carbon atom should be well removed from other peaks in the spectra. However, the other two structures which might be expected from the above considerations, namely head-to-tail linked and monomeric α-methylstyrene units are both observed. The most

Table 12.3 α-Methylstyrene-alkane copolymers: calculated and observed shifts in ppm relative to TMS
(i) Calculated from scheme in Table 12.2. (ii) Estimated from pentyl dimer

$$
\begin{array}{c}
\text{Me} \qquad\qquad \underset{3}{\text{Me}} \\
-\underset{}{\overset{|}{\text{C}}}-\text{CH}_2-\underset{1}{\text{CH}_2}-\underset{2}{\overset{|}{\text{C}}}-\underset{4}{\text{CH}_2}-\underset{5}{\text{CH}_2}-\underset{6}{\text{CH}_2}-\underset{7}{\text{CH}_2}-\underset{8}{\text{CH}_2}- \\
\overset{|}{\text{Ph}} \qquad\qquad \overset{|}{\text{Ph}}
\end{array}
$$

n in structure **2**	C_1			C_2			C_3			C_4			C_5		C_6		C_7		C_8	
	(i)	(ii)	Obs.	(i)	(ii)	Obs.	(i)	(ii)	Obs.	(i)	(ii)	Obs.	(i)	Obs.	(i)	Obs.	(i)	Obs.	(i)	Obs.
10	37.4	37.5	37.4	44.5	40.5	40.6	27.4	23.9	23.7	41.9	43.5	43.5	25.7	24.1	30.5	30.4	30.3	30.0	30.2	30.0
6	37.4	37.5	37.4	44.5	40.5	40.6	27.4	23.9	24.2	41.9	43.5	43.5	25.7	24.2	30.5	30.3				
5	37.4	37.5	37.1	44.5	40.5	40.4	27.4	23.9	23.9	41.9	43.5	43.5	25.8	23.9	30.9	31.1				
4	37.4	37.5	37.0	44.5	40.5	40.4	27.4	23.9	23.7	42.0	43.6	43.6	26.0	24.7						
3	37.4	37.5	37.3	44.6	40.6	40.7	27.4	23.9	24.2	42.2	43.7	43.7	21.2	19.1						

intense extra peak in the spectrum occurs at 41·5 ppm, 1·0 ppm to low field of the quaternary carbon in structure (2), and the relatively long T_1 value shows that it is a quaternary carbon atom. The identity of this peak was determined in the following way.

In order to prepare the regular copolymers, the monomer and dihalide are reacted together in the ratio 2:1. However, if the reaction is carried out with a considerable excess of monomer, for instance 2 moles of monomer to $\frac{1}{16}$ mole of dibromide,[17,18] then the product contains a considerable excess of monomer. We expect these extra units to be incorporated in a head-to-tail fashion, and the variation in the spectra with increasing excess of α-methylstyrene suggests that the extra units are incorporated mainly as structures (3) and (3a).

Figure 12.4 compares the 2:1 α-methylstyrene/butane copolymer with the 2:$\frac{1}{16}$ copolymer. The 2:$\frac{1}{16}$ material has a composition of 1 mole of α-methylstyrene to 0·19 moles of butane, from the proton n.m.r. With this composition, we expect the extra head-to-tail units to be predominantly in tetrameric units

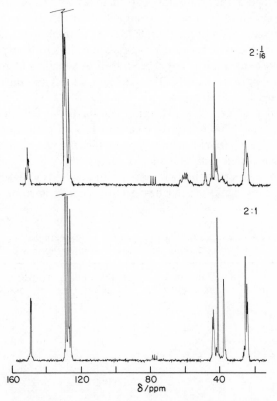

Figure 12.4 Comparison of the proton-decoupled
^{13}C n.m.r. spectra of α-methylstyrene/butane copolymers produced with initial reactant stoichiometries of 2:1 (lower trace) and 2:$\frac{1}{16}$ (upper trace)

as in structure (**3a**), although there must also be some units with more head-to-tail units added on to (**3a**). The spectrum shows a number of peaks in addition to those in the 2:1 copolymer, associated with the extra α-methylstyrene units. Apart from the sharp, intense peak at 41·5 ppm, which is discussed below, the spectrum shows a complicated multiplet at about 60 ppm, which is readily shown to be due to CH_2 groups, and has approximately the same chemical shift as the methylene groups in normal head-to-tail poly(α-methylstyrene).[15] The complexity is due to stereochemical factors, and also to the different environments as successive monomer units are added in a head-to-tail fashion first to the tail-to-tail dimer to form (**3a**), and then to (**3a**) to form longer α-methylstyrene sequences. The singlet at 43·1 ppm, which is readily shown to be a quaternary carbon, has the same chemical shift as the quaternary carbon in the homopolymer,[15] and presumably corresponds to the central carbon of three head-to-tail monomer units, as distinct from the quaternary carbons in structure (**3a**), which are not in the same environment as the homopolymer.

The most prominent peak in the spectrum is a sharp, intense peak at 41·5 ppm, 1·0 ppm to low field of the quaternary carbon in (**2**), which is clearly shown to be a quaternary peak, and is therefore assumed to correspond to the quaternary carbons in the head-to-tail linked units (**3a**). For small quantities of structure (**3**) or (**3a**) in the otherwise regular 2:1 copolymer, it is to be expected that this peak will be most easily observed. The shift of this peak is identical to that of the most intense extra peak in the 2:1 spectrum, so it is assumed that the latter is due to the head-to-tail structure.

A similar extra peak occurs at 40·0 ppm, 0·5 ppm to high field of the quaternary carbon in structure. This quaternary carbon has an identical chemical shift to the quaternary carbon in 5-methyl-5-phenyldecane. We assume that it arises from monomeric α-methylstyrene units (structure (**5**)).

It therefore appears that the two most prominent extra peaks in the spectra are the quaternary peaks of the heat-to-tail α-methylstyrene units and the

Table 12.4 Composition of a series of α-methylstyrene/alkane copolymers[a] having the basic tail-to-tail structures **2**

n in structure **2**	Tail-to-tail dimer units	Head-to-tail units	Monomer units
3	94·5	4·5	1·0
4	86·4	10·0	3·6
4	90·6	9·4	<0·5
4	93·0	5·6	1·4
5	94·6	3·9	1·5
6	82·7	14·2	3·1
10	83·3	13·9	2·8
10	72·0	26·9	1·1

[a] These polymers were all made under conditions of 2:1 reactant stoichiometry (see text).

monomeric α-methylstyrene units. The amounts of these structures are measured from the relative areas of the corresponding peaks. The results are shown in Table 12.4. The three values for the α-methylstyrene/butane copolymers ($n = 4$) refer to different samples, and given an indication of the extent of variation in composition in samples prepared at different times. Even allowing for this variation, it appears that the regularity of the structure decreases as the length of the alkane chain increases, the longer dibromoalkanes giving a considerable number of head-to-tail linkages. However, for the shorter dibromoalkanes, it appears that the proportion of these irregular structures is low, the copolymers consisting predominantly of structure (2). The stereochemical arrangement of the α-methylstyrene in the chain units is essentially random, consisting of approximately equal amounts of the two possible arrangements. Effects in the ^{13}C spectrum between adjacent dimer units are negligible for $n \geqslant 3$ in structure (2).

12.5 REFERENCES

1. F. A. Bovey, *Progress in Polymer Science*, **3**, 1, (1971).
2. F. A. Bovey, *High Resolution N.M.R. of Macromolecules*, Academic Press, New York, 1972.
3. J. B. Stothers, *Carbon*-13 *N.M.R. Spectroscopy*, Academic Press, New York, 1972.
4. T. C. Farrar and E. D. Becker, *Pulse and Fourier Transform N.M.R.*, Academic Press, New York, 1971.
5. D. H. Richards, N. F. Scilly and F. Williams, *Polymer*, **10**, 603, (1969).
6. D. H. Richards, N. F. Scilly and S. M. Hutchison, *Polymer*, **10**, 611, (1969).
7. M. G. Pemberton, D. H. Richards and N. F. Scilly, *European Polymer J.*, **6**, 1083, (1970).
8. A. Davis, D. H. Richards and N. F. Scilly, *European Polymer J.*, **6**, 1293, (1970).
9. A. V. Cunliffe, R. J. Pace, D. H. Richards and N. F. Scilly, *European Polymer J.*, **9**, 125, (1973).
10. D. R. Weyenberg, *J. Org. Chem.*, **30**, 3236, (1965).
11. D. H. Richards and N. F. Scilly, *J. Chem. Soc. C*, 55, (1969).
12. A. Davis, D. H. Richards and N. F. Scilly, *Makromol. Chem.*, **152**, 121, (1972).
13. A. Davis, D. H. Richards and N. F. Scilly, *Makromol. Chem.*, **152**, 133, (1972).
14. L. F. Johnson, F. Heatley and F. A. Bovey, *Macromolecules*, **3**, 175, (1970).
15. Y. Inoue, A. Nishioka and R. Chujo, *Makromol. Chem.*, **156**, 207, (1972).
16. D. M. Grant and E. G. Paul, *J. Amer. Chem. Soc.*, **86**, 2984, (1964).
17. A. V. Cunliffe, W. J. Hubbert and D. H. Richards, *Makromol. Chem.*, **157**, 23, (1972).
18. A. V. Cunliffe, W. J. Hubbert and D. H. Richards, *Makromol. Chem.*, **157**, 39, (1972).

13

The Characterization of Diene Polymers by High Resolution Proton Magnetic Resonance

J. R. Ebdon

University of Lancaster

13.1 INTRODUCTION

Following the pioneering work of Bovey and Tiers[1] which demonstrated that high resolution proton magnetic resonance spectroscopy (p.m.r.) could be used to determine the tacticity of poly(methyl methacrylate), the interest in the use of p.m.r. for the elucidation of polymer microstructure has increased rapidly. Today p.m.r. ranks alongside infrared spectroscopy as a standard technique for polymer characterization.

In recent years, the initial impetus has been maintained through the development of p.m.r. spectrometers equipped with superconducting magnets and capable of operating at frequencies of 220, 250 or even 300 MHz. These instruments together with, in some instances, the judicious use of selective deuterium labelling, spin decoupling or signal enhancement techniques, have enabled the detailed microstructures of many polymers, both synthetic and natural, to be determined.

It is probable that in the future the use of p.m.r. for polymer structure determination will to some extent be overshadowed by the more recently developed technique of Fourier-transform carbon-13 nuclear magnetic resonance (C-13 n.m.r.) discussed elsewhere in this volume. However, it must be remembered that obtaining a satisfactory C-13 n.m.r. spectrum may take hours rather than the minutes usually required to obtain a p.m.r. spectrum, especially if reliable quantitative information is needed and if only materials containing the natural abundance of C-13 are available ($1 \cdot 1 \%$ of the total carbon). Thus, where the same information is available from the application of either technique, p.m.r. will be at an advantage.

In view of the commercial importance of diene homopolymers and copolymers as synthetic rubbers, it is not surprising that there should be an interest in the determination of their microstructures. The information thus obtained can lead to a better understanding of the relationships between the properties of the polymers and their structures, and also can often provide a deeper insight into the mechanisms of the polymerizations.

The aim of this chapter is to illustrate the type of study which has been carried out and the range of information which is currently available through the use of p.m.r. concerning the microstructure of diene polymers. Much of the work in this area performed before 1971 has already been very lucidly reviewed.[2] Consequently this chapter deals mostly with work published since that date.

13.2 HOMOPOLYMERS

Butadiene may be polymerized with a variety of initiators to give polymers containing 1,4-(1) and 1,2-(2) monomer units.

$$-CH_2-CH=CH-CH_2- \qquad -CH_2-CH-$$
$$(1) \qquad\qquad\qquad\qquad CH$$
$$\qquad\qquad\qquad\qquad\quad || $$
$$\qquad\qquad\qquad\qquad\quad CH_2$$
$$\qquad\qquad\qquad\qquad\quad (2)$$

The 1,4-units may exist either as *cis*- or *trans*-isomers. 1,2-Units contain an asymmetric carbon atom and thus long sequences of such units will give rise to chains or portions of chains of a particular tacticity. In addition, 1,2-units may be linked in a head-to-head, head-to-tail, or tail-to-tail formation with one another or with a 1,4-unit. With 2-substituted dienes, such as isoprene and chloroprene, the situation becomes more complex as a 3,4-unit is now distinguishable from a 1,2-unit and sequences of 1,4-units may contain head-to-head, head-to-tail and tail-to-tail linkages.

1-Substituted dienes such as 1,3-pentadiene can give rise to polymers containing *cis*-isotactic, *cis*-syndiotactic, *trans*-isotactic, *trans*-syndiotactic and atactic structures, *cis*- and *trans*-3,4-units and a variety of head-to-head, head-to-tail and tail-to-tail linkages. Increasing the complexity of the substitution pattern gives rise to an increased number of isomeric possibilities, most of which cannot, as yet, be identified by proton magnetic resonance or indeed by any other technique.

13.2.1 Butadiene

In the p.m.r. spectrum of a polybutadiene, 1,2-units may be recognized by the presence of peaks arising from the olefinic methylene protons at about 4·8 δ. However, the resonances of the methylene and olefinic methine protons in *trans*-1,4-units are barely distinguishable from those of *cis*-1,4-units at all normal observing frequencies unless spin decoupling is employed. Only at 300 MHz are these resonances clearly resolved into peaks characteristic of *cis*- and *trans*-units.[3] With *o*-dichlorobenzene as solvent, the *trans*-1,4-methylene peak is centred at 1·98 δ whilst that of the *cis*-1,4-unit is at 2·03 δ. The *trans*- and *cis*-olefinic methine peaks occur at 5·37 δ and 5·32 δ respectively.

Hatada *et al.*[4] have analyzed the spin-decoupled spectra of U.V. isomerized 1,4-polybutadienes at 100 MHz using CDCl$_3$ as solvent. Under these conditions, the decoupled methylene resonance occurs as three peaks. The peak at lowest field is assigned to the central methylene groups in *cis-cis* dyads whilst that at highest field is assigned to *trans-trans* dyads. The central peak is assigned to *cis-trans* and *trans-cis* dyads. The decoupled olefinic methine resonance is split into two peaks characteristic of *cis-* and *trans*-units respectively. Thus, not only does such an experiment allow the measurement of the relative proportions of *cis*-1,4- and *trans*-1,4-units, but gives some insight also into the sequence of such units along the polymer chain. When a similar decoupling experiment is performed at 300 MHz,[5] the olefinic methine resonance of 1,4-units is split still further into a total of six peaks as shown in Figure 13.1. In order of

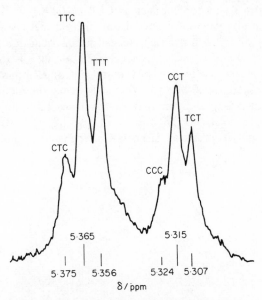

Figure 13.1 Decoupled olefinic methine resonances at 300 MHz of a 1,4-polybutadiene containing *cis*-(C) and *trans*-(T) units. Triad assignments are indicated (Reproduced from Santee *et al.*[5])

increasing field strength these have been assigned to the olefinic protons of the central butadiene units in *cis-trans-cis, trans-trans-cis, trans-trans-trans, cis-cis-cis, cis-cis-trans,* and *trans-cis-trans* triads respectively. Julémont *et al.*,[6] have recently employed spin-decoupled p.m.r. at 250 MHz to establish that the so-called equibinary 1,4-polybutadiene prepared in the presence of bis(h[3]-allylnickel trifluoroacetate) contains an essentially random sequence of *cis*- and *trans*-units. A similar conclusion has been reached by Hatada *et al.*[7] from 100 MHz spectra.

The 100 MHz spectra of isotactic, syndiotactic and atactic 1,2-polybutadienes have been described by Zymonas *et al.*[8] In the syndiotactic polymer, the methylene protons are all magnetically equivalent and so the methylene resonance occurs as a simple triplet due to coupling to the neighbouring methine protons. In the isotactic polymer, the methylene groups contain non-equivalent protons and the resonance is more complex. The methine patterns of all the polymers are very similar and contain six peaks centred approximately at 2·2 δ.

13.2.2 Isoprene

The relative proportions of *cis*- and *trans*-units in a 1,4-polyisoprene may be obtained quite simply from the intensities of the methyl peaks at about 1·7 δ (*cis*) and 1·6 δ (*trans*) in a solvent such as CCl_4 or $CDCl_3$.[9,10] The presence of 1,2- and/or 3,4-units can be inferred from olefinic methylene resonances at about 4·7–5·0 δ and 1,2-units may be estimated quantitatively from their methyl resonance at 1·05 δ. In a polyisoprene containing 1,4- and 3,4-structures, the direct quantitative estimation of the amount of 3,4-units is less easy but can be accomplished at 220 MHz at which frequency the methyl peak of these units is partially resolved from that of the *trans*-1,4-units[11] (Figure 13.2). It is interesting to note also that the pattern of the methylene resonances in an all-*trans*-1,4-polyisoprene at 220 MHz in $CDCl_3$ is significantly different

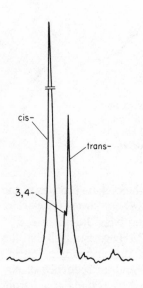

Figure 13.2 Methyl resonances at 220 MHz in $CDCl_3$ of a polyisoprene containing *cis*-1,4-, *trans*-1,4- and 3,4-units

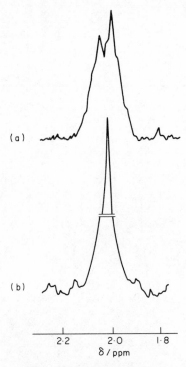

2·2 2·0 1·8

δ / ppm

Figure 13.3 Methylene resonances at 220 MHz in CDCl₃ of
(a) an all-*trans*-1,4-polyisoprene
and (b) an all-*cis*-1,4-polyiso-
prene

from that in all-*cis*-1,4-polyisoprene[11] (Figure 13.3). This indicates that the
difference in shielding between the protons in the 1 and 4 methylene groups is
rather greater in the *trans*-polymer than in the *cis*-polymer.

P.m.r. thus appears to give no information on the proportions of head-to-
head, head-to-tail and tail-to-tail linkages in 1,4-polyisoprenes, although all
three types of linkage have been detected in a variety of polyisoprenes by
pyrolysis and ozonolysis.[12]

13.2.3 Chloroprene

Ferguson[13] studied the 100 MHz p.m.r. spectra of predominantly *trans*- and
cis-1,4-polychloroprenes. The spectra do not differ markedly one from another,
but are interesting in that they reveal the presence of head-to-head and tail-to-
tail linkages in addition to the normal head-to-tail sequences. The proportions
of such linkages may be estimated from their methylene peaks. These peaks are
particularly well resolved at 220 MHz in the spectrum of an approximately
90%-*trans*-1,4-polychloroprene[11] (Figure 13.4).

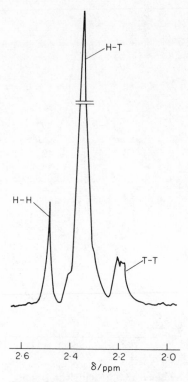

Figure 13.4 Methylene resonances at 220 MHz in CDCl$_3$ of an approximately 90%-*trans*-1,4-polychloroprene showing presence of head-to-head (H-H), head-to-tail (H-T) and tail-to-tail (T-T) linkages

13.2.4 Other dienes

Ohno *et al.*[14] have discussed the 100 MHz spectra of a number of 2-alkyl substituted polybutadienes prepared with anionic and coordination catalysts. In all the polymers, the *cis*-1,4- and *trans*-1,4-units give rise to methyl resonance patterns composed of two closely overlapping triplets. In 2-ethyl-, 2-isopropyl- and 2-*n*-butyl-butadiene, the *cis*-triplet is at slightly lower field than the *trans*-triplet, whereas in 2-*n*-propylbutadiene the reverse occurs. 3,4-Units can be identified from their characteristic olefinic resonances.

The microstructure of poly-1,3-pentadiene has been studied most recently by Denisova *et al.*[15] They have investigated the cationic polymerization of *cis*- and *trans*-1,3-pentadiene and characterized the resulting polymers at 60 MHz. The linear unsaturation of the polymer chains was found to consist of *trans*-1,4- and 1,2-units regardless of the particular pentadiene isomer chosen. The polymers also contained substituted cyclohexene type structures.

220 MHz spectra of some poly-2,3-dimethylbutadienes have been obtained by Blondin *et al.*[16] using chlorobenzene as solvent at 100 °C. The polymers were prepared using radical and anionic initiators and were found to have predominantly 1,4-structures. The methylene resonances are split into three peaks and have been assigned to the adjacent methylene groups of *cis-cis* (2·24 δ), *cis-trans* (2·20 δ) and *trans-trans* (2·16 δ) dyads. The methyl resonances are grouped into two sets of three peaks. The more shielded group has been assigned to the *cis*-units at the centre of *trans-cis-trans* (1·70 δ), *trans-cis-cis* (1·73 δ) and *cis-cis-cis* (1·74 δ) triads. The low-field group (1·77, 1·78 and 1·80 δ) is assigned to *trans-trans-trans*, *cis-trans-trans* and *cis-trans-cis* triads. On the basis of the measured peak areas, Blondin *et al.* have found that Bernoullian trial statistics are followed both by the radical and by the anionic polymerizations. In the former polymerizations the probability of *cis*-placement is 0·24 whilst in the latter it is 0·40.

13.3 COPOLYMERS

In addition to providing the type of information available for homopolymers, p.m.r. of copolymers can be expected to yield information also about overall composition and, in favourable cases, about the average distribution of the monomer units along the copolymer chains. Measurements of monomer sequence distributions are particularly important since to a large extent this distribution determines the physical and chemical properties of the copolymer. Moreover, such measurements can often allow one to distinguish between various possible copolymerization mechanisms. For example, if in a copolymer composed of monomer units A and B the relative proportions of AB, AA and BB dyads are known, then a pair of simple reactivity ratios r_A and r_B may be obtained for the two types of chain centre ending in an A-unit and a B-unit respectively. If, however, triad fractions can be measured (i.e. AAA, AAB, BAB, BBA etc.) then four reactivity ratios r_{AA}, r_{BA}, r_{BB} and r_{AB} may be calculated and the possible effects of penultimate monomer units on the reactivities of the growing centres can be determined. Similarly, a knowledge of tetrad fractions can allow antepenultimate effects to be investigated. Sequence measurements can also help to determine whether propagation reactions involving monomer-monomer complexes are important.

13.3.1 Methyl methacrylate-diene copolymers

Bevington and Ebdon[17-19] have investigated the microstructures of some essentially random methyl methacrylate-butadiene, methyl methacrylate-isoprene and methyl methacrylate-chloroprene copolymers by 220 MHz p.m.r. spectroscopy. All the copolymers were prepared by radical polymerization at 60 °C at an initial overall monomer concentration of 5·0 mol litre^{-1} using benzene as solvent. The p.m.r. spectra were recorded in $CDCl_3$ and either C_6D_6 or C_5D_5N at ambient temperatures. Proportions of 1,2-diene units in the butadiene-containing copolymers and 1,2- and 3,4-units in the isoprene and chloroprene copolymers can be determined from the olefinic methylene resonances at around 5 δ. Methacrylate-centred triad fractions can be obtained from the resonances of the methacrylate α-methyl protons which occur between 0·7 and 1·3 δ in CDCl$_3$. The α-methyl peaks for a range of methyl methacrylate-chloroprene[19] copolymers are shown in Figure 13.5. Similar peaks are observed

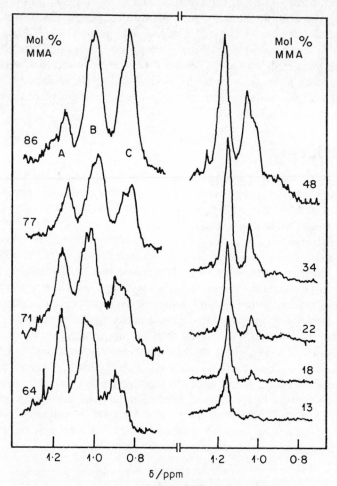

Figure 13.5 α-Methyl resonances at 220 MHz in CDCl$_3$ for a range of methyl methacrylate-chloroprene copolymers. Methyl methacrylate contents are shown. (Reproduced from Ebdon[19] by permission of IPC Business Press Ltd.)

in the corresponding spectra of methyl methacrylate-butadiene[18] and methyl methacrylate-isoprene[17] copolymers although the resolution is less good. In the methyl methacrylate-chloroprene spectra (Figure 13.5), the α-methyl peaks within group A (1·3–1·1 δ) are assigned to the methacrylate units at the centre of CMC triads, isotactic MMM triads and MMC and CMM triads in which the constituent MM dyad is in a meso (dd or ll) configuration (M = methyl methacrylate, C = chloroprene). α-Methyl peaks in group B (1·1–0·95 δ) are assigned to heterotactic MMM triads and MMC and CMM triads in which the constituent MM dyad is in a racemic (dl or ld) configuration. Group C (0·95–0·7 δ) is assigned solely to syndiotactic MMM triads. Analogous assignments can be made for the corresponding α-methyl peaks in the spectra of the methyl methacrylate-butadiene and the methyl methacrylate-isoprene copolymers. In the methyl methacrylate-chloroprene system, the relative amounts of the various methacrylate-centred triads obtained from p.m.r. spectra are in good agreement with those calculated from the reactivity ratios $r_{MM} = 0·107$, $r_{CM} = 0·057$ and $r_C = 6·7$. In other words, the measured triad fractions are consistent with a copolymerization in which the reactivity of the methacrylate-ended radicals is influenced to a slight extent by the nature of the penultimate unit (i.e. whether methyl methacrylate or chloroprene). A similar penultimate effect is not apparent with methyl methacrylate-butadiene and methyl methacrylate-isoprene. However, in these systems the detection of such an effect is made more difficult by the poorer resolution of the α-methyl resonances and the uncertainty of the positions and intensities of peaks arising from triads containing 1,2- and 3,4-diene units.

The α-methyl peaks from methyl methacrylate-butadiene and methyl methacrylate-isoprene copolymers are interesting also in that they provide (at the higher diene contents) a measure of the average configuration of one of the diene units in the diene-methacrylate-diene triad. At high diene contents it is observed that the predominant peak in group A of the α-methyl peaks (assigned to the diene-methacrylate-diene triad) can be resolved into two overlapping components (A1 and A2) the relative areas of which are independent of the copolymer composition. This splitting for one of the methyl methacrylate-butadiene copolymers can be seen in Figure 13.6. The most probable diene unit responsible for the splitting is that attached to the quaternary carbon of the methacrylate unit, and it is suggested that peak A1 arises from triads in which this diene unit is in the *cis*-1,4-configuration and that A2 arises from triads in which this diene unit is in the *trans*-1,4-configuration. On this basis, the ratio of *cis*-to-*trans* enchainment of diene units adjacent to methacrylate units is 36:64 for the butadiene-containing polymers and 25:75 for those containing isoprene. The overall *cis-trans* ratio of a sample of polyisoprene prepared under similar conditions estimated from methyl peak areas is 38:62. Alternating methyl methacrylate-butadiene and methyl methacrylate-isoprene copolymers prepared in the presence of zinc chloride display only a single α-methyl peak at A2 characteristic of copolymers containing only *trans*-1,4-diene units.

The 220 MHz spectra of an alternating methyl methacrylate-isoprene copolymer can be used also to determine the relative orientations of the methacrylate and isoprene units within the chains, i.e. whether the methacrylate-isoprene dyads have structure (3) or structure (4).[11]

250

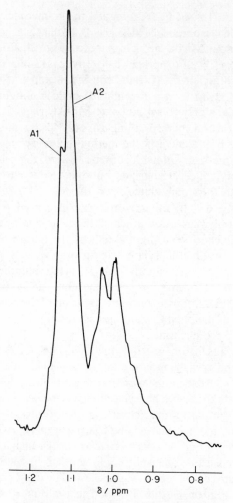

Figure 13.6 α-Methyl resonances at 220 MHz in CDCl₃ for a methyl methacrylate-butadiene copolymer containing 51 mol % MMA showing splitting due to neighbouring *cis*-(A1) and *trans*-(A2) butadiene units

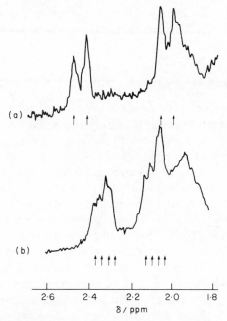

Figure 13.7 Methylene resonances at 220 MHz in CDCl$_3$ of (a) an alternating methyl methacrylate-isoprene copolymer and (b) an alternating methyl methacrylate-butadiene copolymer. Components of AB patterns shown thus: ↑

As can be seen from Figure 13.7a, the methylene peaks for such a copolymer contain a prominent AB quartet-type pattern centred at $2.23\,\delta$ ($J_{AB} = 14\,$Hz, $\Delta\delta_{AB} = 0.43\,$ppm) characteristic of an isolated methylene group containing non-equivalent protons. Such a methylene group occurs only in structure (**3**). In the spectrum of an alternating methyl methacrylate-butadiene copolymer prepared under similar conditions the corresponding AB pattern (Figure 13.7b) is, as expected, further split due to coupling of the corresponding methylene protons with the neighbouring olefinic methine proton ($J = 6\,$Hz). Structure (**4**), if it were to occur, could reasonably be expected to give rise to an AB pattern resembling that in Figure 13.7b.

That the dyads in the alternating copolymer of methyl methacrylate and isoprene are predominantly of structure (**3**) is evidence against there being a connection between this type of copolymerization and the Diels–Alder reaction of methyl methacrylate and isoprene as has been proposed by Oikawa and Yamamoto.[20] In the major Diels–Alder adduct of methyl methacrylate and isoprene (methyl-1,3-dimethyl-3-cyclohexane-1-carboxylate) the relative orientation of the two monomer units is the same as in (**4**).

A structure analogous to (3) has been proposed for the dyads in alternating acrylonitrile-isoprene copolymers by Gatti and Carbonaro[21] from C-13 n.m.r. measurements.

13.3.2 Acrylonitrile-diene copolymers

100 MHz p.m.r. spectra of random copolymers of acrylonitrile (A) and butadiene (B) have been recorded by Kuzay and Kimmer[22] at 80 °C in $CDCl_3$. They observed that the olefinic methine signals of the butadiene units were split into two sets of peaks centred at 5·55 and 5·41 δ respectively. The former set has been assigned to 1,4-butadiene units in AB dyads and the latter to 1,4-units in BB dyads.

Similar polymers have been studied at 220 MHz by Suzuki et al.[23] They observed that in $CDCl_3$ at 60 °C, the acrylonitrile methine resonance was split into three peaks at 3·05, 2·87 and 2·62 δ. These have been assigned to AAA, AAB and BAB triads in order of increasing field. The butadiene methylene signals also give some sequence information; they appear as two peaks centred at 2·12 and 2·30 δ characteristic of 1,4-butadiene units in BB and BA dyads. Suzuki et al.[23] have also prepared some copolymers containing 1,1,4,4-tetradeuteriobutadiene. In these polymers, the olefinic methine protons give rise to eight peaks between 5·29 and 5·68 δ which can be assigned to the various butadiene-centred pentads. On the basis of these pentad measurements, the authors conclude that the radical copolymerization of acrylonitrile and butadiene in bulk at 50 °C follows first-order Markov statistics with $r_B = 0.35$ and $r_A = 0.05$. However, Pham et al.[24] have presented some evidence which suggests that statistics of this type are not obeyed at the extreme ends of the monomer feed range and have proposed that this might be due to the variation in the configurations of the butadiene units with copolymer composition. At high acrylonitrile contents, all the butadiene units appear to be in the trans-1,4-configuration.

Lindsay et al.[25] have reported studies of acrylonitrile-1,1,4,4-tetradeuteriobutadiene copolymers at 300 MHz. From the olefinic methine resonance pattern of alternating copolymers prepared in the presence of $Et_3Al_2Cl_3$ and $VOCl_3$ they conclude that meso and racemic placements of neighbouring acrylonitrile units are equally probable.

P.m.r. spectra of random and alternating copolymers of acrylonitrile and isoprene have been interpreted by Patnaik et al.[26] and Koma et al.[27] Masaki and Yamashita[28] have used p.m.r. to confirm the alternating structures of copolymers of acrylonitrile and chloroprene prepared with catalysts such as $EtAlCl_2 - VOCl_3$.

13.3.3 α-Methylstyrene-butadiene copolymers

Copolymers of α-methylstyrene (S) and butadiene (B) have been prepared by Elgert et al.[29] using butyl lithium in tetrahydrofuran at -75 °C. Their microstructures have been determined at 220 MHz using hexachlorobutadiene at 150 °C as solvent. The butadiene units are found to be predominantly in the 1,2-configuration. Butadiene-centred triad and pentad fractions can be obtained from the olefinic methine and methylene signals and α-methylstyrene-centred triad fractions can be obtained from the α-methyl peaks.

Using 3,4,5-trideuterio-α-methylstyrene it is possible to assign configurational and compositional triads in the resonances of the ortho phenyl protons. Measured compositional triad and pentad fractions are in good agreement with those calculated from the penultimate reactivity ratios:

$$r_{SS} = r_{BS} = 0.025, \qquad r_{BB} = 2.3 \quad \text{and} \quad r_{SB} = 3.7$$

13.3.4 Other copolymers

Other diene copolymers recently studied by p.m.r. include butadiene with methacrylonitrile,[30] propylene[31,32] and acetylene[33]; 1,1,4,4-tetradeuterio-butadiene with methacrylonitrile[34]; 1-chloro-1,3-butadiene with styrene[35]; and various dienes with sulphur dioxide[36–38] and maleic anhydride.[39,40]

13.4 REFERENCES

1. F. A. Bovey and G. V. D. Tiers, *J. Polymer Sci.*, **44**, 173, (1960).
2. F. A. Bovey, *High Resolution N.M.R. of Macromolecules*, Academic Press, New York and London, 1972, Chapter XI.
3. E. R. Santee, R. Chang and M. Morton, *J. Polymer Sci. (Polymer Letters)*, **11**, 449, (1973).
4. K. Hatada, Y. Tanaka, Y. Terawaki and H. Okuda, *Polymer J.*, **5**, 327, (1973).
5. E. R. Santee, V. D. Mochel and M. Morton, *J. Polymer Sci. (Polymer Letters)*, **11**, 453, (1973).
6. M. Julémont, E. Walckiers, R. Warin and P. Teyssié, *Makromol. Chem.*, **175**, 1673, (1974).
7. K. Hatada, Y. Terawaki, H. Okuda, Y. Tanaka and H. Sato, *J. Polymer Sci. (Polymer Letters)*, **12**, 305, (1974).
8. J. Zymonas, E. R. Santee and H. J. Harwood, *Macromolecules*, **6**, 129, (1973).
9. H. Y. Chen, *Anal. Chem.*, **34**, 1134, 1793, (1962).
10. M. A. Golub, S. A. Fuqua and N. S. Bhacca, *J. Amer. Chem. Soc.*, **84**, 4981, (1962).
11. J. R. Ebdon, unpublished work.
12. M. J. Hackathorn and M. J. Brock, *Rubber Chem. Technol.*, **45**, 1295, (1972).
13. R. C. Ferguson, *J. Polymer Sci.*, A2, **11**, 4735, (1964).
14. R. Ohno, M. Kaami and Y. Tanaka, *Polymer J.*, **4**, 49, (1973).
15. T. T. Denisova, I. A. Livshits and E. R. Gershtein, *Vysokomol. Soedinenya*, **16A**, 880, (1974).
16. D. Blondin, J. Regis and J. Prud'homme, *Macromolecules*, **7**, 187, (1974).
17. J. C. Bevington and J. R. Ebdon, *Makromol. Chem.*, **153**, 173, (1972).
18. J. R. Ebdon, *J. Macromol. Sci.-Chem.*, A, **8**, 417, (1974).
19. J. R. Ebdon, *Polymer*, **15**, 782, (1974).
20. E. Oikawa and K. Yamamoto, *Polymer J.*, **1**, 669, (1974).
21. G. Gatti and A. Carbonaro, *Makromol. Chem.*, **175**, 1627, (1974).
22. P. Kuzay and W. Kimmer, *Plaste u. Kaut.*, **18**, 743, (1971).

23. T. Suzuki, Y. Takegami, J. Furukawa, E. Kobayashi and Y. Arai, *Polymer J.*, **4**, 657, (1973).
24. Q. T. Pham, J. Vialle and J. Guillot, *Proceedings of the 2nd European Symposium on Polymer Spectroscopy*, Milan, June 1971, Consiglio Nazionale delle Recherche, Rome, 1973, p. 42.
25. G. A. Lindsay, E. R. Santee and H. J. Harwood, *Amer. Chem. Soc., Polymer Preprints*, **14**, 646, (1973).
26. B. Patnaik, A. Takahashi and N. G. Gaylord, *J. Macromol. Sci.-Chem.*, **A, 4**, 143, (1970).
27. Y. Koma, K. Iimura and M. Takeda, *J. Polymer Sci. (Polymer Letters)*, **10**, 2983, (1972).
28. A. Masaki and I. Yamashita, *J. Macromol. Sci.-Chem.*, **A, 6**, 439, (1972).
29. K. F. Elgert, E. Seiler, G. Puschendorf and H.-J. Cantow, *Makromol. Chem.*, **165**, 245, 261, (1973).
30. R. Schmolke, W. Kimmer, P. Kuzay and W. Hufenreuter, *Plaste u. Kaut.*, **18**, 95, (1971).
31. T. Suzuki, Y. Takegami, J. Furukawa and R. Hirai, *J. Polymer Sci.*, **B, 9**, 931 (1971).
32. J. Furukawa, H. Amano and R. Hirai, *J. Polymer Sci.*, **A-1, 10**, 681, (1972).
33. J. Furukawa, E. Kobayashi and T. Kawagoe, *J. Polymer Sci. (Polymer Letters)*, **11**, 573, (1973).
34. T. L. Ang, R. C. Chang, E. R. Santee and H. J. Harwood, *J. Polymer Sci. (Polymer Letters)*, **10**, 791, (1972).
35. A. Winston and P. W. Wichacheewa, *Macromolecules*, **6**, 200, (1973).
36. M. Iino, M. Matsuda and R. Asami, *J. Polymer Sci.*, **B, 9**, 473, (1971).
37. K. J. Ivin and N. A. Walker, *J. Polymer Sci.*, **B, 9**, 901, (1971).
38. M. Matsuda and Y. Hara, *J. Polymer Sci.*, **A-1, 10**, 837, (1972).
39. N. G. Gaylord, M. Stolka, A. Takahashi and S. Maiti, *J. Macromol. Sci.-Chem.*, **A, 5**, 867, (1971).
40. N. G. Gaylord, M. Stolka and B. K. Patnaik, *J. Macromol. Sci.-Chem.*, **A, 6**, 1435, (1972).

14
Spin Labels and Probes in Dynamic and Structural Studies of Synthetic and Modified Polymers

P. Tormala and J. J. Lindberg

University of Helsinki

14.1 INTRODUCTION

The use of spin labelling and probe techniques in structural and dynamic studies of polymers is based on the analysis of electron spin resonance (e.s.r.) spectra of radicals or paramagnetic ions incorporated in polymer systems. Stable nitroxide radicals have proved to be one of the most important groups of labels and probes so far developed. There is already a considerable literature dealing with the chemistry of nitroxides,[1] and the principles of spin labelling and its application to studies of proteins, lipids, and biological and synthetic membranes.[2-10]

The basic structural unit of stable nitroxide radicals used in labelling is the nitroxide group that is protected by bulky side groups (R and R' in Figure 14.1).

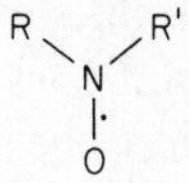

Figure 14.1 The basic unit of nitroxide radicals used as spin labels and probes

The side groups R and R' may be aliphatic or aromatic in nature. Very stable radicals are achieved if tertiary carbons are bonded to the nitrogen so that disproportionation reactions are impossible.[3] Such radicals are e.g. di-*tert*-butylnitroxide[11] (I), 2,2-dimethyl oxazoline derivatives[12] (II), pyrroline (III) and piperidine (IV) nitroxides,[1,13] as shown in Figure 14.2. These radicals are most conveniently prepared by oxidation of the corresponding secondary amines.[1]

In spin labelling, group R of radicals III and IV may include a reactive group by which the radical can be bonded covalently to the studied molecule. Such groups include the hydroxyl-, carboxyl-, chloroformyl, cyano-, isocyanate-, isothiocyanate-, maleimide-, and hydroxyimino-group. In spin probes R is non-reactive, e.g. hydrogen, aliphatic or aromatic hydrocarbon, keto-, ester-, ether- or amide group.

256

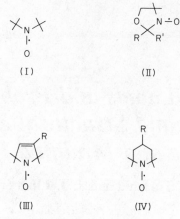

Figure 14.2 Di-*tert*-butylnitroxide
(I), 2,2-dimethyloxazoline nitroxide
(II), 2,2,5,5-tetramethylpyrroline
nitroxide (III) and 2,2,6,6-tetra-
methylpiperidine nitroxide (IV)

The paramagnetic labels have been found to be sensitive indicators of the structure and dynamics of their environment. The rotational rate and the degree of orientation of radicals are calculated from their e.s.r. spectra, and by means of this information the mobility, structure and orientations of the environment can be studied indirectly.

Since label and probe radicals, as a rule, are impurities in the studied systems, their concentration is kept as small as possible to minimize the alteration in properties of the system. The standard amount of radical used in 100 ppm.

We wish here to introduce those calculation methods that are useful for polymer chemists and physicists interested in labelling and probe techniques and also to discuss studies of synthetic polymers made by means of spin probes and labels. The results of these studies are compared with data obtained by other dynamic and relaxation measurement methods. Included in the article are some results of recent spin labelling and probe studies undertaken in our laboratory and not previously published.

14.2 THEORETICAL METHODS

The paramagnetic resonance spectra of nitroxide radicals in an external magnetic field can be expressed in quantitative form by the spin Hamiltonian

$$H = -\beta \cdot \bar{H}_0 \cdot \bar{g} \cdot \bar{S} + \bar{S} \cdot \bar{T} \cdot \bar{I} \tag{1}$$

where β, $\bar{H}_0$, $\bar{g}$, $\bar{S}$, $\bar{T}$, and $\bar{I}$ are the Bohr magneton, the external magnetic field, the $\bar{g}$ tensor, the electron spin operator, the hyperfine tensor, and the nuclear spin operator, respectively. It is assumed here that the concentration of radicals is so low ($\leq 10^{-3}$ M) that the intermolecular dipole-dipole interaction and electron exchange are negligible.

H is composed of an isotropic and anisotropic part:

$$H = H^{iso} + H^{aniso} \tag{2}$$

$$H^{iso} = -\beta \cdot g \cdot \bar{H}_0 \cdot \bar{S} + a \cdot \bar{I} \cdot \bar{S} \tag{3}$$

$$H^{aniso} = -\beta \cdot \bar{H}_0 \cdot \bar{g}' \cdot \bar{S} + \bar{S} \cdot \bar{T}' \cdot \bar{I} \tag{4}$$

The parameters g and a are the isotropic g-value and the isotropic hyperfine coupling constant, respectively. They are defined by

$$g = \tfrac{1}{3}(g_x + g_y + g_z) = \tfrac{1}{3} \operatorname{Tr} \bar{g} \tag{5}$$

$$a = \tfrac{1}{3}(T_x + T_y + T_z) = \tfrac{1}{3} \operatorname{Tr} \bar{T} \tag{6}$$

and $\bar{g}'$ and $\bar{T}'$ are traceless tensors. Their components are related to the total $\bar{g}$ and $\bar{T}$ tensors through the relations

$$g_i = g_i' + g \tag{7}$$

$$T_i = T_i' + a \qquad (i = x, y, z) \tag{8}$$

If the nitroxide radicals are in a solution of low viscosity (e.g. in benzene) at room temperature, they are in rapid rotational Brownian motion. This leads to an averaging of the anisotropies of the $\bar{g}$ and $\bar{T}$ tensors, i.e. $\bar{g}'$ and $\bar{T}'$ approach $\bar{0}$. Then the resultant spectrum consists of three identical peaks (caused by the

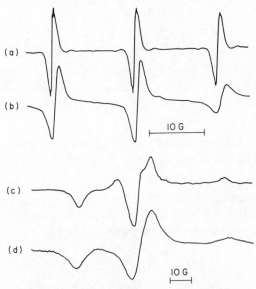

Figure 14.3 E.s.r. spectra of radical III (R = —COCH$_3$) in polyethylene glycol of $\bar{M}_n = 4000$ at (a) 373 K, (b) 323 K, (c) 263 K and (d) 100 K

interaction of the unpaired electron with ^{14}N nucleus, whose spin quantum number is $-1, 0,$ or $+1$) as shown in Figure 14.3a. The rotational correlation time τ of radicals is then $\sim 10^{-11}$ s. τ can be explained as the time required for a nitroxide radical to make one complete rotation about its axis.

In viscous liquids the tumbling rates of the molecules decrease and the averaging of the tensors is far from complete. This leads to a broadening of the hyperfine components in such a way that the high-field peak broadens most (Figure 14.3b). The tumbling rate of isotropically moving nitroxide radicals in viscous solutions can be calculated by means of the known elements of $\bar{g}$ and $\bar{T}$ tensors and the measured line-widths of hyperfine components. When $\tau < 5 \times 10^{-9}$ s, the line-width contributions can be calculated by means of the Redfield relaxation matrix method[14,15] or the method of Kubo and Tomita as applied by Kivelson.[16,17] In the case of nitroxide radicals both of these methods lead to the following expression for the line-widths[18]:

$$\frac{1}{T_2(M)} = \left[\tfrac{1}{20}b^2(3 + 7u) + \tfrac{1}{15}(\Delta\gamma H)^2(\tfrac{4}{3} + u) + \tfrac{1}{8}b^2M^2 \cdot \left(1 - \frac{u}{5}\right)\right.$$
$$\left. + \tfrac{1}{5}b\Delta\gamma HM(\tfrac{4}{3} + u)\right] \cdot \tau \tag{9}$$

where

$$\Delta\gamma = \frac{|\beta|}{\hbar}[g_z - \tfrac{1}{2}(g_x + g_y)] \tag{10}$$

$$b = \tfrac{2}{3}[T_z - \tfrac{1}{2}(T_x + T_y)] \tag{11}$$

$$u = \frac{1}{1 + \omega_0^2\tau^2} \tag{12}$$

ω_0 is the Larmor frequency of the electron in the applied magnetic field H. Equation (9) is valid only when the following conditions are satisfied[19]:

(a) $(T_x + T_y + T_z)/3\omega_0 \ll 1$;
(b) $T_x \simeq T_y$ and $g_x \simeq g_y$ (e.g. the tensors are axially symmetrical)
(d) the rotation is isotropic and
(d) $T_2(M) \gg \tau$.

Combining the equations (9) for different hyperfine components ($M = 0, \pm 1$), and assuming sufficiently slow rotations to make $u = 0$, the following equations for calculation of τ are obtained (cf. Waggoner et al.[20]):

$$\tau(1) = \left(\frac{\Delta H_1}{\Delta H_0} - \frac{\Delta H_{-1}}{\Delta H_0}\right)\left(\frac{15\pi\sqrt{3}\,\Delta H_0}{8Hb\,\Delta\gamma}\right) \tag{13}$$

$$\tau(2) = \left(\frac{\Delta H_1}{\Delta H_0} + \frac{\Delta H_{-1}}{\Delta H_0} - 2\right)\left(\frac{4\pi\sqrt{3}h}{g\beta b^2}\right) \tag{14}$$

where ΔH_i = the peak-to-peak width of the hyperfine component ($M = i$) in gauss and the unit of b is gauss. The equations are valid in the τ range from 5×10^{-11} s to 5×10^{-9} s and in principle they should yield equal values for τ.

The correlation time τ is related to the radius (r) of the tumbling molecule and to the viscosity (η) of the solvent by the simple Stokes expression

$$\tau = \frac{4\pi\eta r^3}{3kT} \tag{15}$$

On the basis of recent studies it is also possible to calculate slow-motional correlation times (5×10^{-9} s $< \tau < 10^{-6}$ s) for nitroxide radicals from their e.s.r. spectra.[21-25]

McCalley et al.[24] have reported a useful relationship between τ and the shift of the position of the high-field peak from its position in a powder spectrum (Δ):

$$\Delta = \text{constant} . \tau^{-\frac{3}{2}} \tag{16}$$

The equation (16) serves for the calculation of the correlation times in the region of about 5×10^{-9}–10^{-7} s, which makes it useful for studying the rotational diffusion of macromolecules.

On the basis of the studies of Freed et al.,[23] Goldman et al.[25] have shown that it is possible to express τ as a function of the parameter $S = A'_z/A_z$, where A'_z is one-half the separation of the outer hyperfine extrema of the e.s.r. spectrum and A_z is the rigid-limit value for the same quantity. Thus

$$\tau = a(1 - S)^b \tag{17}$$

where a and b are dependent upon the diffusion model adopted and the intrinsic line-width of the spectrum. This expression can be used conveniently in the range $10^{-8} < \tau < 10^{-6}$ s.

There are special techniques which can be used to determine τ in the more complex cases where powder spectra seem to be identical because the rotational rate of radicals is very low compared to the frequency of hyperfine coupling. Thus it is possible to determine τ in the range 2×10^{-7} s $< \tau < 2 \times 10^{-5}$ s by using an adiabatic fast passage technique[26] or pulsed e.s.r. measurements of electron spin relaxation times.[27]

14.3 LINEAR POLYMERS

14.3.1 Spin probe studies

In studies of properties of nitroxide radicals, Stryukov et al.[28] and Stryukov and Rozantsev[29] found that some radicals readily diffuse into various polymers. At temperatures above T_g (the glass transition point) of polymer (e.g. polyethylene), the e.s.r. spectrum of radicals consisted of a well-resolved triplet. This indicated that the radicals possessed a high degree of rotational motion in partially amorphous polymers. The fact that the irreversible deformation,

Table 14.1 The rotational activation energies (E_a) of some spin probe radicals in linear polymers

Polymer	Radical	$E_a/\text{kJ mol}^{-1}$	Reference
Polyethylene, $\rho = 0.918\,\text{g cm}^{-3}$	(IV) R $=$ $-$H	44·8	29
Polyethylene, $\rho = 0.950\,\text{g cm}^{-3}$	R $=$ $-$H	43·5	29
Polyisobutene	R $=$ $-$H	49·4	1
Poly(butyl methacrylate)	R $=$ $-$H	50	1
Polystyrene	R $=$ $-$H	94·6	1
Polyisoprene	R $=$ $-$H	22·4	30
Polystyrene	(IV) R $=$ $=$O	76·1	31
Polypropylene, atactic	R $=$ $=$O	78·2	31
Polypropylene, isotactic	R $=$ $=$O	43·9	31
Butadiene rubber	R $=$ $=$O	48·1	31
Natural rubber	R $=$ $=$O	35·6	31
Polyethylene (low density)	(IV) R $=$ $-$OBz	46·0	32
Polyethylene (high density)	R $=$ $-$OBz	46·0	32
Polypropylene	R $=$ $-$OBz	51·5	32
Polystyrene	R $=$ $-$OBz	76·6	32
Polyethylene glycol	(III) R $=$ $-$COCH$_3$	39·7	33

which reduces the number of vibrational degrees of freedom of the polymer molecules, retards also the rotational movements of radicals, showed the close connection of the mobility of radicals with the segmental mobility of polymer molecules.

Numerous studies have shown that in linear polymers and elastomers the rotational correlation times obey Arrhenius law. Thus

$$\tau = \tau_0 \exp(E_a/RT) \tag{18}$$

when $T > T_g$. Hence when $\log \tau$ is drawn as a function of $1/T$ straight lines are achieved. In Table 13.1 are collected the rotational activation energy values (E_a) of some spin probe radicals in different polymers.

It can be seen from Table 14.1 that the functional groups of radicals have an effect on the E_a value; however, the magnitude of E_a is the same for different small nitroxide radicals in the same polymer. Wasserman et al.[31] have shown that the pre-exponential factor (τ_0) has a linear coupling with E_a and have concluded that E_a is an effective value which exceeds the actual activation energy E. Accordingly E is defined by the mobility of segments of a macromolecule and the temperature dependence of E strongly contributes to E_a. The value E_a of radical rotation does not depend on the size of the radical but only on the properties of the polymer matrix.[34]

Comparisons have been made which show that the E_a values found for radicals approximate the activation energies of segmentary motions of the corresponding polymers measured by n.m.r. spectroscopy[31,35] and by dielectric relaxation methods.[35,38] The activation energy of local segmental diffusion of polyethylene glycol as measured by neutron quasi-elastic incoherent scattering[39]

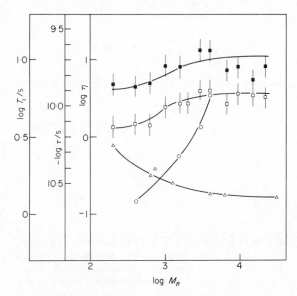

Figure 14.4 The nuclear spin-lattice relaxation time (T_1) of PEG at 423 K ($\triangle$), τ of free ($\square$) and covalently bonded (■) pyrroline nitroxide radicals in PEG at 353 K and the absolute viscosity (η) of PEG at 343 K (○) as a function of log $\overline{M}_n$

is also in agreement with spin probe studies.[33] These observations confirm that the radical rotation frequency is determined by the mobility of the polymer segments. This is further supported by Figure 14.4 where τ-values of radical III (R = $-COCH_3$) in polyethylene glycol[40] (PEG) ($\square$) are compared with spin-lattice relaxation times (T_1) derived from n.m.r. measurements[41] ($\triangle$) and with absolute viscosities[42] (○). τ-values of covalently bonded radicals are also given (■). T_1 and τ reach their limiting values at log $\overline{M}_n$ = 3–4 and τ is only slightly dependent on $\overline{M}_n$. This is in contrast to the behaviour of η, which increases rapidly with increasing $\overline{M}_n$. This difference can be explained by the hypothesis that short-range segmental motions contribute strongly to τ. On the long time-scale of viscosity measurements, transitions between rotational conformers are averaged. On the other hand, fluctuations in the macromolecular end-to-end distance are so slow that they contribute strongly to the bulk viscosity of the polymer.[43] Because τ depends on $\overline{M}_n$ only slightly when $\overline{M}_n \geqslant 200$, it is probable that the polymer segments whose rotations predominantly determine τ are smaller than five monomer units. This hypothesis is supported by the linear correlation between τ of amorphous PEG and dielectric γ-relaxation of PEG in the glass state (Figure 14.5).

The limiting values of dielectric relaxation time, τ_{min} and τ_{max}, calculated by means of the damped torsional oscillator (DTO) model[43,45] for PEG 4000

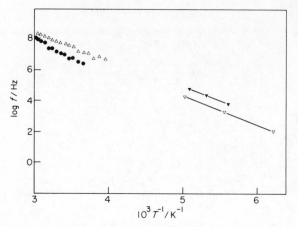

Figure 14.5 log f vs $1/T$ for PEG. Dielectric relaxation.[44]
($\triangledown$) melt crystallized; ($\blacktriangledown$) solution crystallized.
E.s.r.:[36,37,38] ($\triangle$) solution crystallized (probe radicals);
($\bullet$) solution crystallized (label radicals)

at 353 K are $\tau_{min} = 0.03 \times 10^{-19}$ s and $\tau_{max} = 12 \times 10^{-9}$ s. The experimental τ-value (the rotational correlation time of free nitroxide radicals) is 0.20×10^{-9} s: thus $\tau_{min} < \tau < \tau_{max}$. Because the DTO model gives the relaxation times of a segment of about three chain bonds in length, this result also indicates that radicals experience rotations of short polymer segments, perhaps about one monomer unit in length.

Probe radicals can reveal phase transitions of polymers if during phase transition there occurs a change in the short-range segmental motions of polymer molecules. Because probes are located in the amorphous phase[46,47] they can reflect transitions of this phase. There are already some reports which indicate that T_g of partially amorphous polymers can be determined by spin probes because the slope of the $(1/T, \log \tau)$ plot changes at this temperature.[37,40,48,49] In highly crystalline PEG ($\overline{M}_n < 22,000$) spin probes also experience the crystalline melting transition T_m. Evidently this is a consequence of the break up of strain forces induced by crystalline phase and maintained by hydroxyl hydrogen bonds in the amorphous phase; for the esterification of hydroxyl end-groups diminishes the melting transition markedly.[47] In high molecular weight PEG ($M_v \sim 900,000$), where the end-group effect is eliminated the spin probes, accordingly, do not experience the crystalline melting point.[50] In Figure 14.6 is given the $(1/T, 1, \tau)$ plot of radical III ($R = -COCH_3$) in PEG ($M_n = 1500$) where both T_m and T_g transitions are illustrated.

It can be seen that PEG has two glass-transitions (T_{g1} and T_{g2}), as has been reported earlier.[51]

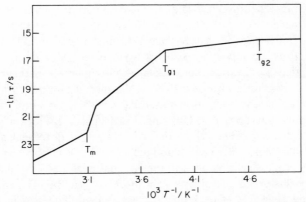

Figure 14.6 $(1/T, \ln \tau)$-plot of radical III $(R = -COCH_3)$ in PEG 1500; $T_m = 325$ K, $T_{g_1} = 262$ K and $T_{g_2} = 213$ K

14.3.2 Spin labelling studies

Although spin labelling studies of biopolymers have clearly shown the potentialities of this method for studies of high-molecular-weight compounds, there has not yet been much work on the spin labelling (covalent bonding of radicals) of synthetic polymers. Griffith *et al.*[52] prepared a nitroxide polymer of $\overline{M}_n = 1050$ by polymerizing methacrylate ester of radical IV. They also labelled maleic anhydride/methyl vinyl ether (1:1) copolymer ($M_w \sim 500,000$) with radical IV ($R = -OH$). The room temperature e.s.r. spectra of solid polymers consisted of a single exchange-narrowed line. The authors suggested that the polymerization of nitroxide monomers may be useful in studying exchange effects and polymer dynamics.

Drefahl *et al.*[53] published a method for preparing spin-labelled polystyrene from mercurated polystyrene *via* a nitroso derivative of polystyrene (Figure 14.7). Dipolar exchange broadened the hyperfine structure of the e.s.r. spectrum

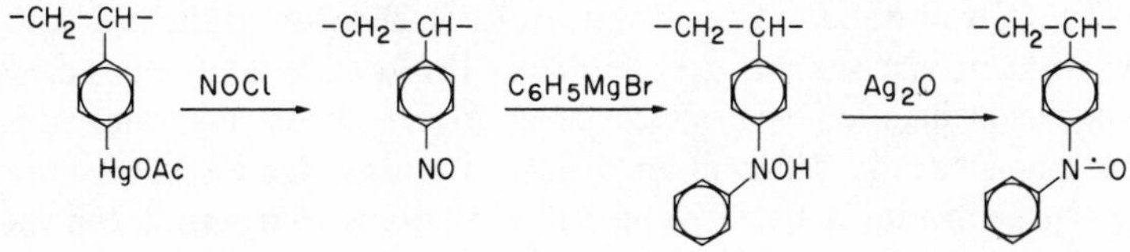

Figure 14.7 Preparation of spin labelled polystyrene

in tetrahydrofuran. Therefore the rotational rate could not be calculated from the line-widths of spectral components and it was impossible to do any structural determinations on the basis of e.s.r. spectral data.

The first spin labelling studies applied to the study of the structural and dynamic properties of synthetic polymers used poly(ethylene glycol),[54] polystyrene[55] and polyamides and polyesters.[55] Cameron *et al.* modified the method of Drefahl to produce polystyrenes with a much lower nitroxide content.[55,58-59] They have also synthesized spin labelled poly(methyl methacrylate).[61] In solutions of polystyrene with a small enough amount of label radicals, dipolar exchange is eliminated and motional modulation is the most important process broadening the hyperfine lines. Utilizing this property, Bullock, Cameron and Smith studied the segmental relaxations of polystyrene in toluene solution and also in solid state below T_g. They found a close fit between the correlation time derived from n.m.r. experiments and the Arrhenius plot of correlation time (τ) of labelled polystyrene in toluene solution.[55] From the Arrhenius plot an activation energy (E_a) of 18.0 kJ mol^{-1} was obtained. This was in good agreement with the value of 20.0 kJ mol^{-1} for poly(-p-chlorostyrene) and poly(p-fluorostyrene) from dielectric relaxation measurements.[62] On the basis of these comparisons Cameron *et al.* associated τ and E_a with some form of segmental motion of the polymer backbone. Their later studies with fractions of polystyrenes of narrow molecular weight distribution ($\overline{M}_n = 2025$–196,000) have confirmed the model of local segmental reorientations.[58] However it was found that the rotational correlation time was dependent on $\overline{M}_n$. Accordingly it was found that at high molecular weights τ was independent of chain length and it seems that a local mode or segmental relaxation process was being observed. This will be characterized by a correlation time τ_{lm}. As the molecular weight decreases, the rotational frequency of the whole macromolecule increases and ultimately the magnitude of the correlation time τ_{eoe} describing this end-over-end rotation becomes comparable to τ_{lm} and contributes significantly to the relaxation process. The activation energy for this rotational mode agreed well with that of viscous flow of the solvent (9.0 kJ mol^{-1}).

Törmälä and Lindberg prepared spin-labelled polyethylene glycol,[33,54] nylon 6,10 and polyesters[56] by Schotten–Baumann condensation of radical III (R = −COCl) to −OH or −NH$_2$ end-groups of polymers. It was found that in pure polymers there is only a comparatively small difference between rotational rates of free (spin probes) and bonded (spin labels) radicals.[33] This indicated that the bonded radicals rotate quite freely around the ultimate C−O bond and that in viscous polymer matrices the rotational frequencies of the whole polymer molecule are very slow compared to the local-mode relaxations. The situation is totally different in dilute solutions. Under these conditions bonded radicals are intimately coupled to the polymer structure. On the other hand, free radicals under these same conditions experience predominantly the solvent environment and only collide randomly with polymer molecules. Therefore the differences in τ values may be considerable.[63,64]

Comparisons of rotational frequencies of spin labels with other relaxational data in solid polystyrene,[59] polyethylene glycol, and nylon 6,10[40] have shown that labels reflect the local γ or δ-relaxations of their environment. There is however some kind of delaying effect to the direction of lower frequencies as compared with the values of corresponding dielectric and mechanical relaxation measurements (cf. Figure 14.5). Probably this shift is caused by the intramolecular slightly restricted rotation around the covalent bond that connects the label radical to the polymer backbone.

Because $\bar{g}$ and $\bar{T}$ parameters are tensors, spin labelling can be used to study orientations of polymers. Such studies have been made e.g. by Sanson and Ptak[65] and Wee and Miller[65] who used poly(benzyl glutamate).

Synthesis of spin labelled polymers have been reported also by Braun and Hauge[67] (poly-N-4-diphenylamine acrylamide with diphenylnitroxide units), Forrester and Hepburn[68] (poly-t-butyl vinylphenyl nitroxide) and Kurosaki et al.[69] (nitroxide polymers from 4-methacryloyl derivatives of radical IV).

In addition some recent work on intramolecular collision frequency has been done by Szwarc and co-workers[70] using radical anions of α-butylnaphthalene as a label in hydrocarbon chains (also see p. 317).

14.3.3 Application potentialities

Because molecular and submolecular motions contribute to the macroscopic behaviour of polymers, spin labelling and probe methods complement the established methods of polymer material research. This was clearly seen in a study concerning the dynamic state of plasticized poly(vinyl chloride) (PVC).[38] In Figure 14.8 are given values of radical III (R = $-CONH_2$) in poly(vinyl chloride)-di(2-ethylhexyl) phthalate (DEHP) plastisols and gels as a function of PVC weight fraction (w_{PVC}).

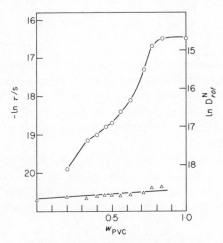

Figure 14.8 τ of radical III (R = $-CONH_2$) in poly(vinyl chloride) (PVC)-di(2-ethylhexyl) phthalate plastisols ($\triangle$) and gels ($\bigcirc$) as a function of PVC weight fraction (w_{PVC})

266

It can be seen that rotations of radicals in plastisols are only slightly dependent on w_{PVC}. This also shows the short-range nature of measurements by the spin probe method. Gelation facilitates the penetration of plasticizer molecules into the polymer. Further, Figure 14.8 shows that τ in gels is strongly dependent on w_{PVC} and a marked increase in τ is observed. The increased frequency of collisions of radicals with the PVC network is probably the main factor retarding the rotational relaxations of radicals.

There is a linear correlation between τ and Young's modulus (G) of plasticized (gelated) PVC (Figure 14.9). This indicates that the mobility of molecules on the submicroscopic scale and the macroscopic elastic properties of gel are intimately connected.

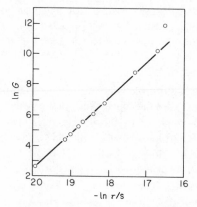

Figure 14.9 The correlation between τ of radical III and the modulus values (G) of PVC gels[71]

Lateral diffusion constants of plasticizer molecules in gelated PVC, calculated from τ values of probe radicals, were at room temperature in the range from 0.2×10^{-8} cm^2 s^{-1} ($w_{PVC} = 0.83$) to 4.6×10^{-8} cm^2 s^{-1} ($w_{PVC} = 0.20$). These values are near the diffusion constants of probe lipids in model and natural membranes.[72] This explains the observation that plasticizers are able to diffuse into lipophilic systems.[73]

Some other areas that lend themselves to study by labels and probes are plasticizer efficiency, gelation, and effects of additives.

Since nitroxide radicals move easily from solution into polymers, it is also possible to study modified polymer materials e.g. synthetic fibres by means of these radicals.[74]

14.4 NETWORK POLYMERS

Network polymers are difficult objects of investigation for polymer chemists because of their complex structure and insolubility. The use of label and probe radicals would appear to be a tempting method of studying the microstructure of heterogeneous networks, because these radicals experience structural features only in their nearest environment. So far, there are only a few studies in this field. Chesnut and Hower[75] measured τ of radical IV (R = —OH) in the aqueous interior of a series of cation ion-exchange resins (Dowex 50 W) at room temperature. There were only minor differences between the values of the various ionic forms (Li$^+$, Na$^+$, K$^+$) of the resins studied but a significant dependence on cross-linking. The values of τ increased with increasing cross-linking up to 8 and 10% cross-linking, but a reduced value was obtained for 12% resin. It was proposed that this effect was probably due to the heterogeneous nature of the resin interior.

Ward and Books[76] labelled amorphous cross-linked polyurethane elastomers with the imidazole derivative of radical III (R = —COOH). The spin label was exclusively attached to the urethane blocks of the network chains. E.s.r. spectra were obtained for the dry unstrained polymer, on swollen unstrained polymer, and on dry polymer in various states of strain. Distinct differences between the spectral parameters of the swollen and dry samples were seen. Rotational correlation times varied between approximately 10^{-9} and 10^{-11} s. In the dry samples held in various strain states, temperature strongly influenced the spectral parameters. In the neighbourhood of 240 K a spectral change was observed at low strains. Strain-temperature-spectra correlations were made. The spin-label technique showed promise of being a powerful microstructure determinant particularly in polymers of heterogeneous character.

In studies on constituents of wood we have examined the structure of thioglycolic acid lignin (TGL) by spin labelling and probe methods.[63,64] The analytical characteristics of TGL are given in Table 14.2.

Table 14.2 Thioglycolic acid lignin from spruce (Betula verrucosa)

Molecular weight, $\overline{M}_n$	$3000 \pm {20\% \atop 5\%}$	(GPC)
COOH-groups	163	g . equiv.$^{-1}$
OCH$_3$-groups	17·74	per cent
Yield of TGL from dry wood	16·1	per cent by weight
Intrinsic viscosity [η]		
In pyridine	0·0752	dl g^{-1} at 302 K
In DMSO − 1% (w/w) H$_2$O	0·0708	dl g^{-1} at 302 K

The spin labelling of TGL was performed by Schotten–Baumann condensation of radical III (R = —COCl) with lignin. Studies with model compounds[7] indicated that both alcoholic and phenolic hydroxyl groups of lignin can be labelled by this radical. According to the e.s.r. spectra, the rotations of radicals around the ester bond are in both cases so free that the rotational rates in solution are inseparable. The spin probe radical III (R = —CONH$_2$) that we used did not react with TGL under the conditions employed. The correlation time of probe radicals in lignin solutions (dioxan, pyridine or dimethylsulphoxide-water) decreased continuously with temperature but increased with increasing lignin concentration. The influence of lignin concentration was taken into account using the specific correlation time, τ_{sp}

$$\tau_{sp} = \frac{\tau_{solution} - \tau_{solvent}}{\tau_{solvent}} \tag{19}$$

At 325 K, τ_{sp} of 15 % TGL solution showed the following trend: DMSO—10 wt % H_2O ($\tau_{sp} = 0.34$) < pyridine (0.39) < DMSO—1 wt % H_2O (0.56) ≪ dioxan (3.09). From the data it was concluded that the collision frequency between radicals and lignin molecules is most effective in dioxan where the solvation of lignin molecules is highest.

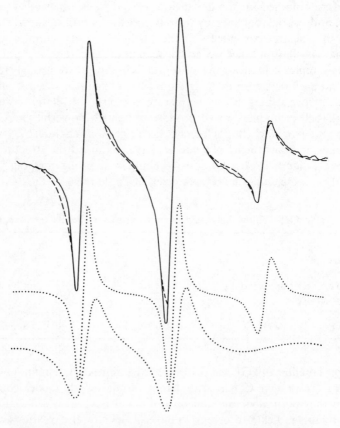

Figure 14.10 The experimental and simulated e.s.r. spectra of 15 % (w/w) spin labelled lignin in DMSO-1 % (w/w) H_2O at 326 K; full line, the experimental spectrum; dashed line, the simulated spectrum; dotted lines, the components of the simulated spectrum

Although the e.s.r. spectra of probe radicals were homogeneous, the label radicals in TGL experienced two environments. The correlation times, activation energies and the hydrodynamic radii of rotating frames were solved by simulation of the superposition spectra. In Figure 14.10 are given the experimental and the simulated spectra of spin labelled TGL in DMSO-1 wt % H_2O at 326 K (15% w/w solution).

The results of the measurements of the correlation times of the two environments (τ_1 and τ_2) and the corresponding activation energies E_1 and E_2 at 303–343 K are given in Table 14.3. For comparison the τ value of probe radical III (R = $-CONH_2$) in TGL-pyridine is also given.

Table 14.3 Correlation times τ and activation energies of radical-TGL complex and free radical III-TGL in pyridine or DMSO-1% (w/w) H_2O at 303–343 K. Concentration of TGL: 15% (w/w)

Solvent	Temperature K	τ_1 ns	E_1 kJ mol^{-1}	τ_2 ns	E_2 kJ mol^{-1}
		Radical-TGL complex			
Pyridine	303	0·75		6·5	
	313	0·69	7·2	6·0	8·5
	323	0·63		5·4	
	333	0·58		4·8	
		±0·025		±0·25	
DMSO-H_2O	313	0·70		6·8	
(1% w/w H_2O)	326	0·62	7·8	6·0	9·9
	333	0·58		5·4	
	343	0·54		4·9	
		±0·025		±0·25	
		Free radical III-TGL			
Pyridine	325			0·029	
				±0·003	

From the data it is evident that two major correlation times are observed in spin-labelled TGL, i.e. about 0·7 ns and about 6 ns as compared to about 0·03 ns for the free radical in TGL pyridine.

The existence of two major correlation time regions indicated that these radicals experience two different major environments, whose local degrees of motional freedom differ from each other. E_a values, which are very similar, indicate that the rotational activation mechanism is determined principally by the solvent medium.

Since label radicals are bonded covalently to lignin molecules, a logical explanation of the existence of two correlation time regions is that some radicals are coupled tightly in the inner stiff network structure of lignin molecules (τ_2) and others to the more loosely arranged surface regions (τ_1).

The corresponding effective radii calculated from τ values by means of the Stokes–Einstein equation had the values $r_2 \sim 2$ nm and $r_1 \sim 1$ nm at ambient temperatures.

14.5 CONCLUSIONS

Spin labelling and probe techniques afford a new way of studying static structural features and rapid time-dependent processes in synthetic polymers, polymer solutions and polymer additive mixtures. The techniques are indirect in nature and therefore cannot replace more traditional methods but only complement them. They have great application potentialities in local relaxation studies of complex polymer, copolymer and polymer network systems because label radicals give information only about their nearest environment. Selective labelling and simulation of superposition spectra makes possible the study of complex relaxation problems in non-homogeneous polymer systems.

14.6 REFERENCES

1. E. G. Rozantsev, *Free Nitroxyl Radicals*, Plenum, New York, 1970.
2. C. L. Hamilton and H. M. McConnell, in *Structural Chemistry and Molecular Biology*, ed. A. Rich and N. Davidson, Freeman, San Francisco, 115, 1968.
3. J. D. Ingham, in *Reviews in Macromolecular Chemistry*, vol. 3, ed. G. B. Butler and K. F. O'Driscoll, Marcel Dekker, New York, 279, 1968.
4. O. H. Griffith and A. S. Waggoner, *Accounts Chem. Res.*, **2**, 17, (1969).
5. H. M. McConnell and B. G. McFarland, *Quart. Rev. Biophysics*, **3**, 91, (1970).
6. P. Törmälä, J. Brotherus and J. J. Lindberg, *Finska Kemistamf. Medd.*, **81**, 11, (1972).
7. P. Törmälä, J. J. Lindberg and L. Koivu, *Paper and Timber*, **54**, 159, (1972).
8. J. Brotherus and P. Törmälä, *Finska Kemistsamf. Medd.*, **81**, 49, (1972).
9. S. Schreier-Muccillo and I. C. P. Smith, in *Progress in Surface and Membrane Science*. vol. 9, ed. J. F. Danielli, M. D. Rosenberg and D. A. Cadenhead, Academic Press, New York, in press, 1973.
10. L. J. Berliner, *Spin Labeling. Theory and Applications.* Academic Press. New York, 1974.
11. A. K. Hoffman, *J. Amer. Chem. Soc.*, **83**, 4671, (1961).
12. J. F. W. Keana, S. B. Keana and D. Beetham, *J. Amer. Chem. Soc.*, **89**, 3055, (1967).
13. R. Brière, H. Lemaire and A. Rassat, *Bull. Soc. Chim. Fr.*, 327 , (1965).
14. J. H. Freed and G. K. Fraenkel, *J. Chem. Phys.*, **39**, 326, (1963).
15. J. H. Freed and G. K. Fraenkel, *J. Chem. Phys.*, **40**, 1815, (1964).
16. D. Kivelson, *J. Chem. Phys.*, **27**, 1087, (1957).
17. D. Kivelson, *J. Chem. Phys.*, **33**, 1094, (1960).
18. G. Poggi and S. Johnson, *J. Magn. Reson.*, **3**, 436, (1970).
19. T. J. Stone, T. Buckman, P. C. Nordio and H. M. McConnell, *Proc. Natl. Acad. Sci. U.S.*, **54**, 1010, (1965).
20. A. S. Waggoner, O. H. Griffith and C. R. Christensen, *Proc. Natl. Acad. Sci. U.S.*, **57**, 1198, (1967).
21. N. N. Korst and A. V. Lazarev, *Mol. Phys.*, **17**, 481, (1969).
22. I. V. Alexandrov, A. N. Ivanova, N. N. Korst, A. V. Lazarev, A. J. Prikhozenko and V. B. Stryukov, *Mol. Phys.*, **18**, 681, (1970).
23. J. H. Freed, G. U. Bruno and C. F. Polnasek, *J. Phys. Chem.*, **75**, 3385, (1971).
24. R. C. McCalley, E. J. Shimshick and H. M. McConnell, *Chem. Phys. Letters*, **13**, 115, (1972).

25. S. A. Goldman, G. V. Bruno and J. H. Freed, *J. Phys. Chem.*, **76**, 1858, (1972).
26. J. S. Hyde and L. Dalton, *Chem. Phys. Letters*, **16**, 568, (1972).
27. I. M. Brown, *Chem. Phys. Letters*, **17**, 404, (1972).
28. V. B. Stryukov, Yu. S. Karimov and E. G. Rozantsev, *Vysokomol. Soedinenya, Ser. B*, **7**, 493, (1967).
29. V. B. Stryukov and E. G. Rozantsev, *Vysokomol. Soedinenya, Ser. A*, **10**, 626, (1968).
30. A. Rousseau and R. Lenk, *Mol. Phys.*, **15**, 425, (1968).
31. A. M. Wasserman, A. L. Buchachenko, A. L. Kovarskii and M. B. Neiman, *Vysokomol. Soedinenya, Ser. A*, **10**, 1930, (1968).
32. G. P. Rabold, *J. Polymer Sci.*, *A*-1, **7**, 1203, (1969).
33. P. Törmälä, H. Lättilä and J. J. Lindberg, *Polymer*, **14**, 481, (1973).
34. A. L. Kovarskii, A. M. Wasserman and A. L. Buchachenko, *Vysokomol. Soedinenya, Ser. A*, **13**, 1647, (1971).
35. A. M. Wasserman, A. L. Buchachenko, A. L. Kovarskii and M. B. Neiman, *European Polymer J.*, **5**, 473, (1969).
36. P. Törmälä, *Polymer*, **15**, 124, (1974).
37. P. Törmälä, *European Polymer J.*, **10**, 519, (1974).
38. P. Törmälä, *Angew Makromol. Chem.*, **37**, 135, (1974).
39. G. Allen, J. S. Higgins and C. J. Wright, *J. Chem. Soc. Faraday Trans. II*, **70**, 348, (1974).
40. P. Törmälä, *Thesis*, Helsinki, (1973).
41. G. Allen, T. M. Connor and H. Pursey, *Trans. Faraday Soc.*, **59**, 1525, (1963).
42. J. J. Lindberg, P. Törmälä and M. Lähteenmäki, *Suomen Kemistilehti*, **B**, **45**, 295, (1972).
43. A. V. Tobolsky and D. B. DuPré, *Adv. Polymer Sci.*, **6**, 103, (1969).
44. N. G. McCrum, B. E. Read and G. Williams, *Anelastic and Dielectric Effects in Polymeric Solids*, Wiley, Bristol, p. 421, 1967.
45. A. V. Tobolsky and J. J. Aklonis, *J. Phys. Chem.*, **68**, 1970, (1964).
46. A. L. Kovarskii, A. M. Wasserman and A. L. Buchachenko, *J. Magn. Reson.*, **7**, 225, (1972).
47. P. Törmälä and J. Tulikoura, *Polymer*, **15**, 248, (1974).
48. S. C. Gross, *J. Polymer Sci.*, *A*-1, **9**, 3327, (1971).
49. A. Savolainen and P. Törmälä, *J. Polymer Sci., Polymer Physics Edition*, **12**, 1251, (1974).
50. P. Törmälä and L. Mannila, unpublished results.
51. R. F. Boyer, *Plastics and Polymers*, **41**, 71, (1973).
52. O. H. Griffith, J. F. W. Keana, S. Rottschaefer and T. A. Warlick, *J. Amer. Chem. Soc.*, **89**, 5072, (1967).
53. G. Drefahl, H. H. Hörhold and K. D. Hofmann, *J. Prakt. Chem.*, **4**, 137, (1968).
54. P. Törmälä, J. Martinmaa, K. Silvennoinen and K. Vaahtera, *Acta Chem. Scand.*, **24**, 3066, (1970).
55. A. T. Bullock, J. H. Butterworth and G. G. Cameron, *European Polymer J.*, **7**, 445, (1971).
56. P. Törmälä, K. Silvennoinen and J. J. Lindberg, *Acta Chem. Scand.*, **25**, 2659, (1971).
57. A. T. Bullock, G. G. Cameron and P. Smith, *Polymer*, **13**, 89, (1972).
58. A. T. Bullock, G. G. Cameron and P. M. Smith, *J. Phys. Chem.*, **77**, 1635, (1973).
59. A. T. Bullock, G. G. Cameron and P. M. Smith, *J. Polymer Sci., Polymer Physics Edition*, **11**, 1263, (1973).
60. A. T. Bullock, G. G. Cameron and P. M. Smith, *Polymer*, **14**, 525, (1973).
61. A. T. Bullock, G. G. Cameron and J. M. Elsom, *Polymer*, **15**, 74, (1974).
62. W. H. Stockmayer, *Pure Appl. Chem.*, **15**, 539, (1967).
63. J. J. Lindberg, I. Bulla and P. Törmälä, *J. Polymer Sci., Polymer Symposia Edition*, **51**, (1975).
64. P. Törmälä, J. J. Lindberg and S. Lehtinen, *Paper and Timber*, **57**, (1975).
65. A. Sanson and M. Ptak, *C.R. Acad. Sc. Paris, D*, 1319, (1970).

66. E. L. Wee and W. G. Miller, *J. Phys. Chem.*, **77**, 182, (1973).
67. D. Braun and S. Hauge, *Makromol. Chem.*, **150**, 57, (1971).
68. A. R. Forrester and S. P. Hepburn, *J. Chem. Soc.*, *C*, 3322, (1971).
69. T. Kurosaki, K. W. Lee and M. Okawara, *J. Polymer Sci.*, *Polymer Chemistry Edition*, **10**, 3295, (1972).
70. H. D. Connor, K. Shimada and M. Szwarc, *Macromolecules*, **5**, 801, (1972).
71. A. T. Walter, *J. Polymer Sci.*, **13**, 207, (1954).
72. C. J. Scandella, P. Devaux and H. M. McConnell, *Proc. Natl. Acad. Sci. U.S.*, **69**, 2056, (1972).
73. R. L. Jaeger and R. J. Rubin, *Science*, **170**, 460, (1970).
74. P. Törmälä, J. Sundquist, T. Penttila and J. J. Lindberg, unpublished results.
75. D. B. Chesnut and J. F. Hower, *J. Phys. Chem.*, **75**, 907, (1971).
76. T. C. Ward and I. T. Books, *Macromolecules*, **7**, 207, (1974).

15
E.s.r. Studies of Spin-Labelled Synthetic Polymers

A. T. Bullock and G. G. Cameron

University of Aberdeen

15.1 INTRODUCTION

In the field of biological macromolecules experiments on the technique of spin-labelling were first reported in 1965.[1] It is now reasonably well established and is the subject of several review articles.[2] A *spin-label* is a stable radical covalently bonded to a macromolecule. The electron spin resonance (e.s.r.) spectrum of the label is sensitive to its environment and in favourable circumstances may be analysed to yield information on the dynamics and orientation of the labelled polymer molecule. Stable radicals are also employed as *spin probes* when they are not attached chemically to a macromolecule but simply dispersed in the polymeric or other matrix. The results obtained by the label and probe techniques are often complementary. In this review we are concerned only with the former and in particular with its comparatively recent application to synthetic polymers where it has been shown that spin-labelling can usefully complement studies by other techniques such as dielectric relaxation, fluorescence depolarization, nuclear magnetic relaxation etc. The study of the dynamics of synthetic polymers was the primary aim of this work but in the process of investigating methods of attaching spin-labels to polymers some additional information on chemical reactions of polymers came to light; this latter aspect of the work is discussed in the last section.

One of the advantages of the spin-labelling technique is that, provided the position of the label on the polymer is known, the part of the polymer chain being examined is unambiguously defined. This of course assumes that the motions of the label and the polymer chains are cooperative. Therefore, in theory at least, it should be possible to study the motions of polymer chain-ends as distinct from groups or segments within the chain.

A variety of relatively stable radicals may be attached to polymer molecules[3,4] but for dynamic and structural studies on polymers only nitroxide (nitroxyl) radicals have been employed successfully. These radicals are particularly useful as spin-labels since, if they carry suitable substituents, they are quite stable and are fairly easy to synthesize. Furthermore, they usually possess well-defined anisotropic g tensors and hyperfine coupling tensors to the ^{14}N nucleus, and these are fundamental requirements for the nitroxide-labelled polymer to yield quantitative dynamic data from a line-width analysis of its e.s.r. spectrum. In order to avoid line-broadening effects from spin exchange the polymer must be very lightly labelled, typically one labelled monomer unit in 200–600 normal units.

Thus, the two basic essentials of a spin-labelling study are first to synthesize the polymer with nitroxide groups at known points on the chain, and second to analyse the e.s.r. spectrum so as to obtain the correlation time for rotational diffusive motion of the radical. The practical and theoretical aspects of these requirements in relation to the work of the present authors are discussed in the next two sections.

15.2 PREPARATION AND CHARACTERIZATION OF SPIN-LABELLED POLYMERS

The most direct means of attaching a nitroxide label to a polymer is to employ a small nitroxide carrying a suitable functional group which is capable of reacting with an appropriate group on the polymer. For example, a hydroxyl-terminated polyether may be labelled at the chain end in the condensation reaction:

$$\sim\!\!-CH_2OCH_2CH_2OH + \quad\text{(1)}\quad \longrightarrow \quad \sim\!\!-CH_2OCH_2CH_2O-\overset{O}{\underset{\|}{C}}$$

Other examples of this approach are described elsewhere in this volume[5] and several functional nitroxides similar to (1) above are now commercially available. Unfortunately many addition polymers do not possess suitably reactive functional groups and these have to be incorporated in the polymer either by copolymerization or by chemical modification. Attempts to prepare labelled addition polymers by direct polymerization or copolymerization of nitroxide-containing monomers such as

$$CH_2\!\!=\!\!\underset{\underset{O}{\underset{\|}{C=O}}}{\overset{}{C}}\!\!-CH_3$$

have not been conspicuously successful.[6,7] Usually only oligomers or very low polymers are obtainable because the propagating radical or ion reacts readily with the nitroxide group. In other cases the monomeric nitroxide is unstable. However, labelled polymers and copolymers have been prepared by direct polymerization of the monomeric amine precursor

$$
\begin{array}{c}
\text{CH}_3 \\
| \\
\text{CH}_2\!=\!\text{C} \qquad \text{X = NH, O} \\
| \\
\text{C}\!=\!\text{O} \\
| \\
\text{X}
\end{array}
$$

followed by oxidation of the polymer to yield piperidine N-oxide units in the side chain.[8]

In the work discussed here the direct reaction, that is using a small functional nitroxide similar to (1), was used only to a limited extent in labelling poly(acrylic esters). Most of the polymers were labelled indirectly by some variation of the reaction sequence

$$
\text{R}\!-\!\text{NO} + \text{R}'\text{M} \rightarrow \text{R}\!-\!\underset{\underset{\text{R}'}{|}}{\text{N}}\!-\!\text{O}^-\text{M}^+
$$

$$\downarrow \text{hydrolysis/oxidation}$$

$$\text{R}\!-\!\underset{\underset{\text{R}'}{|}}{\text{N}}\!-\!\text{O}\cdot$$

where R—NO is a nitroso derivative and R'—M an organometallic reagent. A more detailed account of the labelling process adopted for individual polymers follows.

15.2.1 Polystyrene

Polystyrene was labelled with phenyl nitroxide groups in the *para* position[9] by a variation of the method of Drefahl *et al.*[10] the first step of which involves mercuration. The mercurated units are then converted with nitrosyl chloride to nitroso groups which yield the phenyl hydroxylamine on treatment with phenyl magnesium bromide.

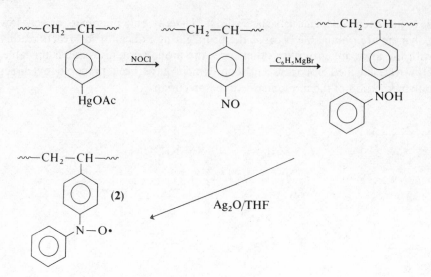

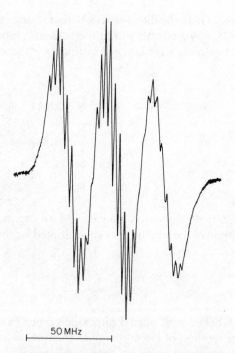

Figure 15.1 E.s.r. spectrum of polystyrene ($\overline{M}_n \sim 1000$) labelled with p-phenyl nitroxide groups (**2**). (1·0% in toluene at 40 °C)

Table 15.1 Characterization of spin-labelled polymers

Polymer	Molecular weight	Mode of polymerization	Type of label	g_{iso}	Coupling constants[e] MHz	Ref.
Polystyrene	~1000 and 115,000[a]	Radical	p-phenyl nitroxide	—	a_N28.4	9
Polystyrene	1950–193,000[d]	—	p-t-butyl nitroxide	2.00577	a_N35.4, $a_{H(2,6)}$2.48, $a_{H(3,5)}$5.56, a_{t-Bu}0.21	13, 14
Polystyrene	126,000[a]	Radical	m-t-butyl nitroxide	2.00589	a_N35.6, $a_{H(2,4,6)}$5.0, $a_{H(5)}$2.5	14
Polystyrene	30,000[b]	Anionic Na Naphth./THF/−70°C	t-butyl nitroxide end label	2.0059	a_N41.50, a_H9.27	18
Poly(α-methyl styrene)	104,000[a]	Anionic BuLi/THF/−80°C	p-t-butyl nitroxide	2.00576	a_N34.8, $a_{H(2,6)}$2.48, $a_{H(3,5)}$3.56	18
PMMA	2300[a], 25,000[c], 81,000[c]	Anionic BuLi/THF/−70°C	t-butyl nitroxide end label	2.00599	a_N41.91	19, 21
PMMA[f]	550,000[c] (See Table 5)	Radical	2,2,6,6-tetramethyl piperidine-1-oxyl	2.006079	a_N43.36, $a_{(\gamma-CH)}$0.23, $a_{(\alpha-CH_3,axial)}$0, $a_{(\alpha-CH_3,equat.)}$ − 1.21, $a_{(\beta-CH_2,axial)}$ − 0.87, $a_{(\beta-CH_2,equat.)}$ − 1.35	16, 21

[a] Number average. [b] From GPC. [c] Viscosity average. [d] Series of GPC calibration fractions. [e] Toluene solution (1–5%) at room temperature. [f] Coupling constants from data in ref. 45.

In the final step the hydroxylamine is oxidized to a nitroxide group by shaking in THF solution with silver oxide. The original procedure of Drefahl *et al.* gives a rather heavily labelled polymer showing broad, poorly resolved lines in the e.s.r. spectrum. By reducing the degree of mercuration in the first step a lightly labelled polymer, showing aromatic proton hyperfine splitting in the e.s.r. spectrum, was obtained[9] (Figure 15.1). Spectral data for this and other labelled polymers are summarized in Table 15.1. If t-butyl magnesium chloride is substituted for phenyl magnesium bromide in the above synthesis t-butyl nitroxide labels result[11]:

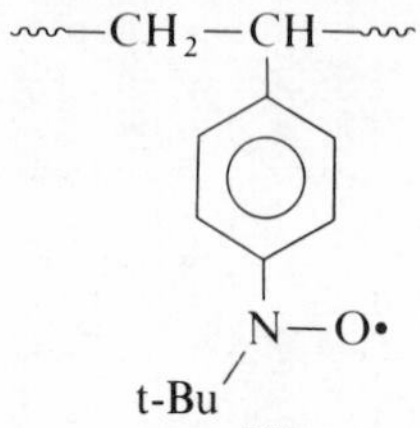

$$\sim\!\!\sim\!\!-CH_2-CH-\!\!\sim\!\!\sim$$

N—O•

t-Bu

(3)

In many ways this is a more convenient label as its e.s.r. spectrum is simpler (Figure 15.2) and is readily simulated.

Apart from preliminary experiments with phenyl nitroxide labels, t-butyl nitroxide has been the label of choice with styrene polymers. However, the above synthesis suffers the drawbacks that the mercurated polymer tends to crosslink and the nitroso intermediate is unstable. A more satisfactory synthetic route employs *para*-lithiated polystyrene as the reactive intermediate. This is readily prepared by Braun's method[12] and can be reacted with 2-methyl-2-nitrosopropane (MNP) to yield t-butyl nitroxide labels. The iodine in the *para* position after the first stage is exchanged for lithium using *n*-BuLi in benzene solution.

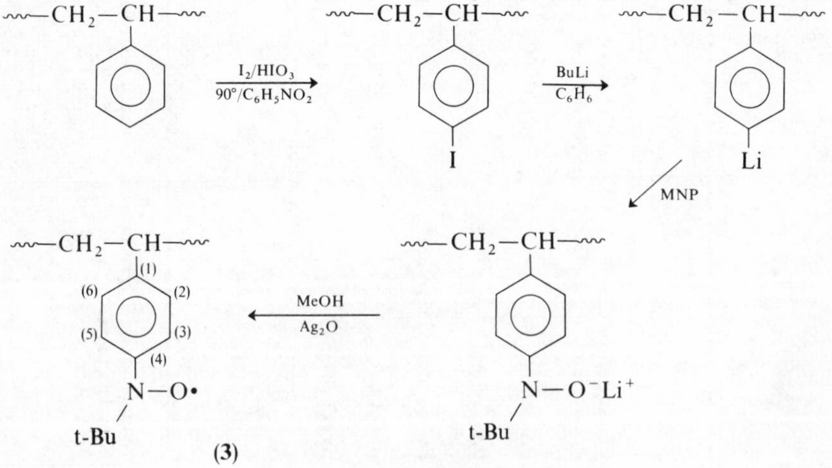

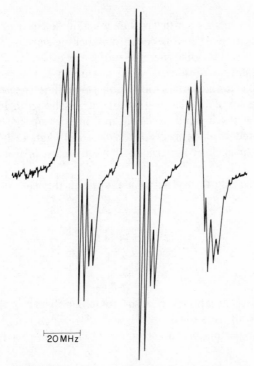

20 MHz

Figure 15.2 E.s.r. spectrum of a polystyrene fraction ($\overline{M}_n = 1950$) labelled with *p*-t-butyl nitroxide groups (**3**). (1 % in toluene at room temperature).[14] (Reproduced by permission of the American Chemical Society)

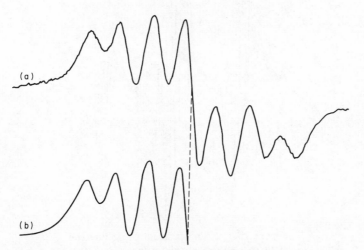

Figure 15.3 Recorded (a) and simulated (b) spectra of the central multiplet ($m_I = 0$) of Figure 15.2.[14] (Reproduced by permission of the American Chemical Society)

280

None of the steps in this synthesis induce chain scission or cross-linking as side reactions and it is therefore ideal for labelling narrow fractions of polystyrene.[13,14] The results of dynamic studies on a series of labelled narrow fractions of polystyrene are described in section 15.4. The density of labelling can be controlled at the initial iodination stage. The spectrum of a narrow fraction of polystyrene labelled in this manner is shown in Figure 15.2, while Figure 15.3 shows the expanded central multiplet compared with a simulated multiplet computed from the spectral parameters in Table 15.1.[14] Each multiplet consists of seven lines, due to interaction with aromatic protons, in the intensity ratio 1:2:3:4:3:2:1.

Polystyrene carrying t-butyl nitroxide labels in the *meta*-position

$$\text{\textasciitilde\textasciitilde}-CH_2-CH-\text{\textasciitilde\textasciitilde}$$

(4)

can be prepared by treating the *meta*-lithiated polymer in a like manner.[15] The e.s.r. spectrum is shown in Figure 15.4. Each line of the nitrogen triplet shows a hyperfine proton structure of eight lines in the intensity ratio

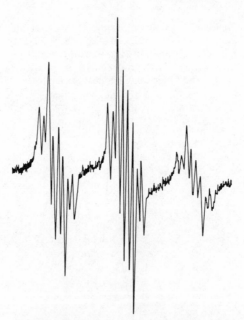

Figure 15.4 E.s.r. spectrum of polystyrene with *m*-t-butyl nitroxide groups (4). (5% in toluene at 52 °C).[15] (Reproduced by permission of I.P.C. Science and Technology Press Ltd.)

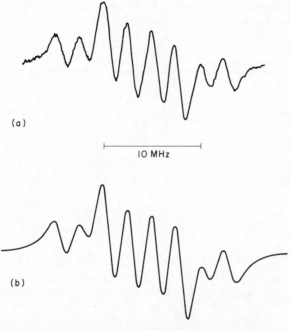

(a)

|———————| 10 MHz

(b)

Figure 15.5 (a) Expanded central multiplet from Figure 15.4; (b) computer-simulated multiplet.[15] (Reproduced by permission of I.P.C. Science and Technology Press Ltd.)

1:1:3:3:3:3:1:1. Figure 15.5 shows a comparison of the central multiplet with a simulated version computed from the coupling constants in Table 15.1.[15] The synthesis of this polymer involved direct lithiation of polystyrene with n-BuLi complexed with N,N,N',N'-tetramethylethylene diamine (TMEDA). This is not an ideal synthesis because some cross-linking occurs. Polystyrene labelled in the same position can also be prepared from a copolymer of styrene and m-bromostyrene by lithium-halogen exchange with n-BuLi/TMEDA.[16] Experiments on the *meta*-lithiation of polystyrene revealed some interesting aspects of the chemical modification of polystyrene which are discussed in section 15.5.

It should be noted that in the polymers carrying t-butyl nitroxide labels the contributions of the t-butyl protons do not show as resolvable splittings. Nevertheless, the unpaired electron does couple to a small extent with these protons and the resultant inhomogeneous line broadening has to be taken into account in simulating the spectrum.[14] The importance of adequate spectral simulation in precise calculations of correlation times is discussed in more detail in sections 15.3 and 15.4.

Polystyrene carrying chain-end *t*-butyl nitroxide labels **(5)** may be prepared by 'killing' the living polymer with MNP[17]:

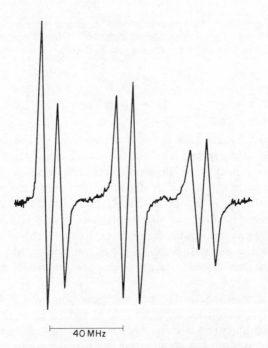

By choosing the appropriate initiating system end-labelled polystyrene carrying one nitroxide group per chain (BuLi) or two nitroxide groups per chain (sodium naphthalene) may be produced.[18] The e.s.r. spectrum of end-labelled polystyrene

40 MHz

Figure 15.6 Polystyrene end-labelled with t-butyl nitroxide groups **(5)** from anionic polymerization with Na naphthalene in THF.[18] (1% in toluene at 23 °C)

shows the expected triplet of doublets (Figure 15.6) due to the effect of the α-proton.[17,18] This polymer has not yet been used for dynamic studies.

The synthetic procedures for labelled polystyrene also apply to poly(α-methyl-styrene) which, when labelled in the aromatic ring, yields spectra very similar to those from the corresponding polystyrene samples.[18]

15.2.2 End-labelled poly(methyl methacrylate)

Poly(methyl methacrylate) (PMMA) carrying terminal t-butyl nitroxide groups can be synthesized in the same manner as end-labelled polystyrene.[19] However, the 'living' anionic polymerization of MMA is subject to several side reactions which can affect the labelling efficiency and indeed the point of attachment of the label. The reaction conditions and choice of initiator are therefore of critical importance. This was not appreciated in early attempts at preparing labelled PMMA.[17,20] Low temperatures (*ca.* −70 °C) appear to be most suitable and under these conditions with THF as solvent *n*-BuLi, 9-fluorenyl lithium and naphthyl sodium all give a labelled polymer with the three-line spectrum shown in Figure 15.7.[19] This spectrum is consistent with the structure

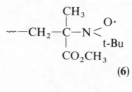

$$
\sim\!\!-CH_2-\overset{\overset{\displaystyle CH_3}{|}}{\underset{\underset{\displaystyle CO_2CH_3}{|}}{C}}-N\!\!<^{\textstyle O\cdot}_{\textstyle t\text{-}Bu}
$$

(6)

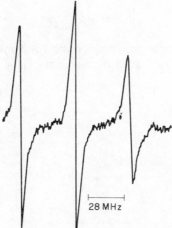

28 MHz

Figure 15.7 E.s.r. spectrum of PMMA end-labelled with t-butyl nitroxide groups (**6**). (5% in toluene at room temperature).[19] (Reproduced by permission of I.P.C. Science and Technology Press Ltd.)

When butyl lithium is employed as initiator the three-line spectrum is obtained only if the polymerization and termination are conducted at $-70\,°C$. If the temperature at which MNP is added is allowed to reach $0\,°C$ or above side reactions set in and a more complex spectrum is obtained.[19] The chemistry underlying these complications is discussed in section 15.5.

High molecular weight PMMA (i.e. $\overline{M}_n > 100{,}000$) labelled at chain ends has not yet been prepared. Attempts at this have so far given a polymer showing no detectable e.s.r. signal. The spectrum of PMMA with one end-label per 1000 monomer units should be detectable and this failure, despite elaborate precautions, seems to indicate that living ends are lost either prior to or during addition of MNP. It is relevant to note that the efficiency of the reaction between MNP and carbanions is not known at present. Several PMMA samples of rather low molecular weight labelled at chain ends have been the subject of dynamic studies (see section 15.4).

15.2.3 Polyacrylates labelled on ester groups

Polyacrylates and polymethacrylates carrying labelled ester groups can be prepared readily[16] by the direct reaction of the appropriate acid chloride copolymer with the functional nitroxide:

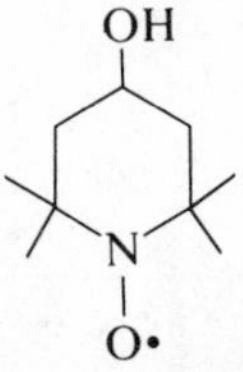

(7)

Reaction of (7) with a copolymer (from radical polymerization) of MMA and methacryloyl chloride yields labelled units in the chain:

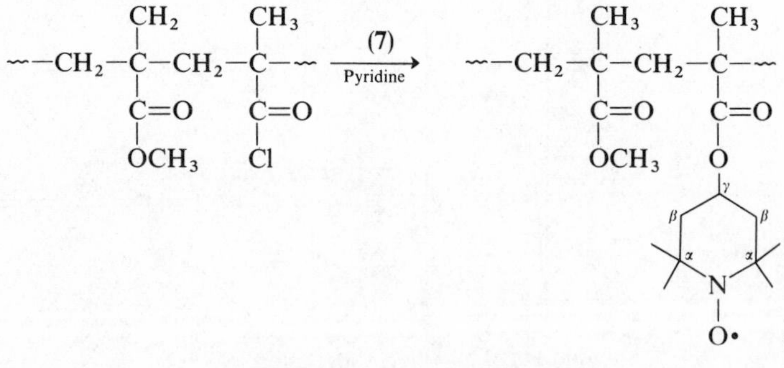

Precipitation of the resulting labelled polymer in methanol converts any unreacted acid chloride groups to methyl methacrylate units. The e.s.r. spectrum of a solution of PMMA labelled in this manner is shown in Figure 15.8.[21] Each line of the nitrogen triplet shows some fine structure due to interaction of

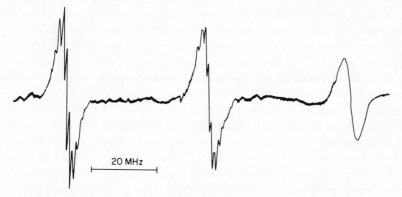

Figure 15.8 E.s.r. spectrum of PMMA labelled at the ester group with 2,2,6,6-tetramethyl piperidine-1-oxyl labels (**8**).[21] (5% in toluene at 21 °C)

the unpaired electron with the protons of the piperidine ring. Figure 15.9 shows a comparison of a computer-simulated multiplet and the observed low-field multiplet (from Figure 15.8).[21] The coupling constants used for this simulation are tabulated in Table 15.1. Because of the many weak proton interactions

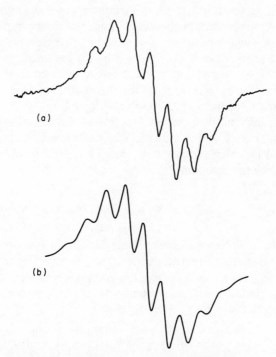

Figure 15.9 (a) Expanded low-field multiplet from Figure 15.8; (b) computer-simulated multiplet from coupling constants in Table 15.1[21]

this is a particularly difficult spectrum to simulate adequately. As will be shown later, however, this is also a good example for illustrating the importance of the choice of simulation model for accurate line-width measurements.

Radical (7) has also been used to label poly(n-butyl methacrylate) and poly(methyl acrylate) by the analogous synthetic route to that above.[16,21] The e.s.r. spectra of the resulting polymers are essentially the same as Figure 15.8.

This is a versatile route to labelled polyacrylates. The proportion of acid chloride units in the chain, and hence the spin concentration, can be controlled by suitable adjustment of the monomer molar ratios during radical polymerization. Furthermore, the average molecular weight of the product can be varied by adjusting the usual parameters—initiator concentration, polymerization temperature, amount of transfer agent etc. Its main disadvantage is that the polymer has a fairly wide molecular weight distribution ($\overline{M}_w/\overline{M}_n = 1.5$–$2.0$) and should narrow fractions be required a lengthy fractionation process would have to be devised.

15.3 THEORY AND METHOD OF LINE-WIDTH ANALYSIS

Elsewhere in this volume[5] are given several equations which relate the line-width parameter T_2 to τ_c, the rotational correlation time of the spin-label and to several parameters in the spin Hamiltonian for the nitroxide. However, we have encountered certain theoretical and practical complications in line-width analyses[14] which have been neglected by other workers with one notable exception.[22] For this reason, the theory is briefly reviewed below.

Figure 15.2, showing the e.s.r. spectrum of *para*-labelled polystyrene (3), provides a convenient starting point for the discussion. Both the ^{14}N nucleus and the aromatic protons show resolvable couplings to the electron. These are all anisotropic and the rotational motion of the radical will thus modulate the couplings. This, together with the effect of an anisotropic g tensor, gives rise to line widths which depend on the nuclear quantum numbers m_i and which are expressed by the equation[23]:

$$T_2^{-1}(m_1, m_2, \ldots, m_n) = A + \sum_{i=1}^{n} B_i m_i + \sum_{i=1}^{n} C_i m_i^2 + \sum_{i \neq j = 1}^{n} E_{ij} m_i m_j \qquad (1)$$

Fortunately, it is often possible to select lines in the e.s.r. spectrum for which the resultant ^{1}H spin quantum number, i.e. $\sum m_H$, is zero. Under these conditions, the line widths depend only on m_I, the component of the ^{14}N nuclear spin, various spin Hamiltonian parameters, the applied magnetic field and the rotational correlation time τ_c. The expression[22] is

$$T_2(m_I)^{-1} = \left[\frac{b^2}{20}(3 + 7u) + \frac{1}{15}(\Delta\gamma B_0)^2 \left(\frac{4}{3} + u \right) + \frac{1}{8}b^2 m_I^2 \left(1 - \frac{u}{5} \right) \right.$$
$$\left. - \frac{1}{5}b \, \Delta\gamma \, B_0 m_I \left(\frac{4}{3} + u \right) \right] \tau_c + X \qquad (2)$$

where $b = (4\pi/3)[T_{zz} - \frac{1}{2}(T_{xx} + T_{yy})]$,

$$\Delta\gamma = -\frac{|\beta|}{\hbar}[g_{zz} - \frac{1}{2}(g_{xx} + g_{yy})] \quad \text{and} \quad u = 1/(1 + \omega_0^2\tau_c^2)$$

T_{ii} and g_{jj} are the diagonal components of the hyperfine coupling and g tensors respectively (the former in Hz), B_0 is the applied magnetic field, ω_0 the corresponding electronic Larmor angular frequency and X the broadening by other mechanisms independent of m_I. The life-time broadening, or nonsecular, contribution to the line widths represented by u has been found to be negligible in all cases so far studied in this laboratory since τ_c has always been greater than 10^{-10} s and the microwave frequency is $2\pi \times 9\cdot27 \times 10^9$ rad s^{-1}. Neglecting the nonsecular terms, equation (2) becomes:

$$T_2(m_I)^{-1} = \left[\frac{3b^2}{20} + \frac{4}{45}(\Delta\gamma\, B_0)^2 + \frac{b^2m_I^2}{8} - \frac{4}{15}b\,\Delta\gamma\, B_0 m_I\right]\tau_c + X \tag{3}$$

Equation (3) is readily modified[1] to eliminate the parameter X, thus:

$$T_2(0)/T_2(m_I) = 1 - \frac{4}{15}\tau_c b\,\Delta\gamma\, B_0 T_2(0)m_I + \frac{1}{8}\tau_c b^2 T_2(0)m_I^2 \tag{4}$$

Using $R_\pm$ to represent $T_2(0)/T_2(\pm1)$ it follows that

$$R_+ + R_- - 2 = (1/4)\tau_c b^2 T_2(0) \tag{5}$$

The most accurate method of determining $R_\pm$ is to measure the peak-to-peak intensities, Y, of the relevant lines since:

$$R_\pm = [Y(0)/Y(\pm1)]^{\frac{1}{2}} \tag{6}$$

The methods used in deriving the parameters b and $T_2(0)$ will be described shortly. However, it is first necessary to discuss an assumption implicit in deriving equation (4). This is that X, the broadening parameter independent of m_I, represents a source of homogeneous broadening.[24] In most spin-labelled polymers there is a source of inhomogeneous broadening namely, unresolved hyperfine couplings to various magnetic nuclei. In particular, the t-butyl protons present when MNP is used in the labelling process, whilst having a rather small isotropic coupling constant to the electron (*ca.* 0·2–0·3 MHz), nonetheless make a marked contribution to the observed line widths. This is largely due to the fact that, for the t-butyl group, $\sum m_H$ takes values between the limits $-9/2$ and $+9/2$. The reason why this broadening may not be included in X is simply that the overall observed line width is not a linear combination of the natural line width given in the equations (1)–(4) and the contribution from the unresolved hyperfine couplings. This point was first discussed by Poggi and Johnson[22] and will be considered later in this article.

The anisotropy parameter b may be determined from the powder spectrum of the solid, labelled polymer (shown in Figure 15.10) together with a measurement of a_N, the isotropic coupling constant.[9] It may be shown[25] that the separation between the extrema in this spectrum is equal to $2T_{zz}$ *providing that*

288

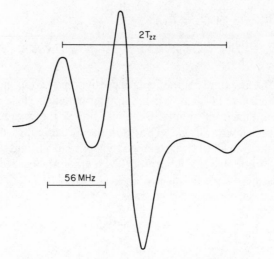

Figure 15.10 Solid state powder e.s.r. spectrum of
polystyrene labelled with *p*-t-butyl nitroxide (**3**).[50]
(Reproduced by permission of John Wiley and Sons,
Inc.)

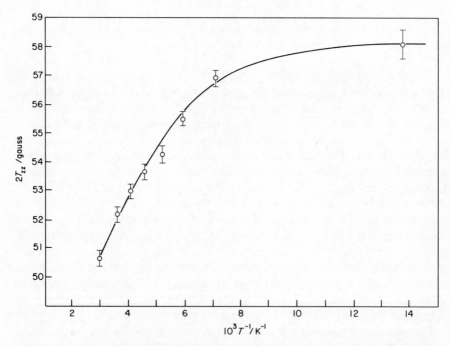

Figure 15.11 Dependence of the field separation of the spectral extrema upon $1/T$
showing the asymptotic approach to the rigid lattice value $2T_{zz}$ (1 gauss = 2·81 MHz).[50]
(Reproduced by permission of John Wiley and Sons, Inc.)

the randomly oriented radicals are at their rigid limit, i.e. $\tau_c \rightarrow \infty$. The safest method of ensuring this is to determine the extrema separation as a function of temperature. Figure 15.11 shows that the rigid limit value is reached asymptotically. $T_2(0)$ is determined simply by measuring the peak-to-peak width of the central line of the derivative spectrum ($m_I = 0$). If this width is Δv Hz, then $T_2 = (\pi\sqrt{3}\,\Delta v)^{-1}$.

This theoretical section may be concluded by returning to the problem of inhomogeneous broadening. To allow for this feature we resort to computer spectral simulations. The program used has facilities for plotting the simulated spectrum on a graph plotter and includes in its output the coordinates of all the turning points in the first-derivative spectrum. The first facility greatly assists visual evaluation of the adequacy of the simulation whilst the second simplifies the problem of correcting $T_2(0)$ and $R_\pm$ for the effects of overlap and inhomogeneous broadening. Briefly, an experimental spectrum is chosen at random. Simulations are made for a range of all the relevant coupling constants, including those responsible for the inhomogeneous broadening, together with the "true" peak-to-peak line width Δv (Hz) until a satisfactory simulation is obtained. Such a case is shown in Figure 15.3. Simulations are then repeated over a suitable range of values of 'true', or input, line widths. It is thus possible

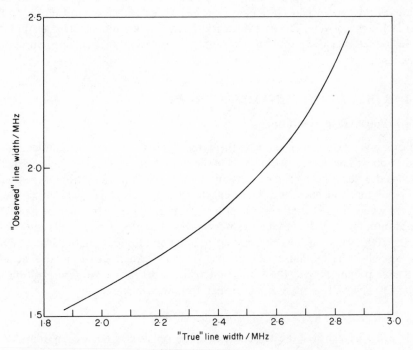

Figure 15.12 Calibration plot of observed *vs.* true peak-to-peak line width.[14] (Reproduced by permission of the American Chemical Society)

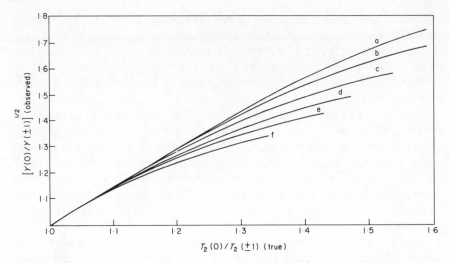

Figure 15.13 Calibration plots of observed $[Y(0)/Y(\pm1)]^{\frac{1}{2}}$ *vs.* $T_2(0)/T_2(\pm1)$ for the following values of Δv $(=[\pi\sqrt{3}T_2(0)]^{-1})$ in MHz: (a) 1·90, (b) 2·00, (c) 2·10, (d) 2·20, (e) 2·30 and (f) 2·40.[14] (Reproduced by permission of the American Chemical Society)

to obtain calibration plots of observed *versus* true line widths and observed values of R_+ *versus* true $T_2(0)/T_2(\pm1)$ ratios. These are shown in Figures 15.12 and 15.13 respectively. The non-linearity shown in these plots is strong supporting evidence that the parameter X cannot include inhomogeneous sources of broadening.

15.4 RESULTS OF DYNAMIC STUDIES

15.4.1 Polystyrene in solution

For a given polymer in solution the dynamics of the various available modes of motion will be determined by intramolecular steric effects, molecular weight, viscous drag exerted by the solvent and intermolecular interactions dependent upon polymer concentration. These effects are considered in turn in this section. Except where noted otherwise, the polymers studied were narrow fraction GPC calibration standards having number-average molecular weights of 1950, 3100, 9700, 19,650, 49,000, 96,200 and 193,000 and are designated PS1 to PS7 respectively. The labelling method has been described earlier in this article and gave polymers with the t-butyl nitroxide label (**3**) in the *para* position at a density always less than one label per 160 monomer units.

15.4.1.1 The dependence of τ_c on molecular weight[14]

Figure 15.14 shows the dependence of τ_c on molecular weight for a 1% solution of labelled polystyrene in toluene at 294·2, 312·6 and 345·2 K. A common feature of these plots is the independence of τ_c on chain length at high molecular

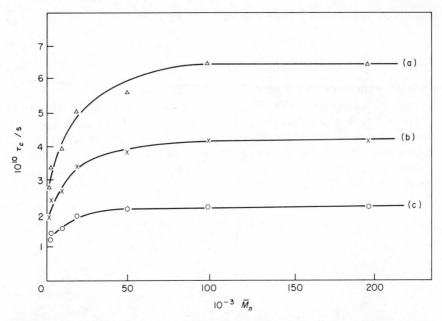

Figure 15.14 Dependence of the rotational correlation time τ_c upon molecular weight at (a) 294·2, (b) 312·6 and (c) 345·2 K.[14] (Reproduced by permission of the American Chemical Society)

weights. This is characteristic of a 'local mode' or segmental relaxation which may be associated with a correlation time τ_{lm}. At lower molecular weights other modes are expected to contribute to the observed correlation time τ_c, in particular the first normal mode which represents 'end-over-end' rotation of the macromolecule. This mode should clearly depend on molecular weight and will dominate the relaxation process for short chain lengths. This model is qualitatively supported by the results summarized in Figure 15.14.

In an attempt to extract more quantitative information and to check certain theoretical estimates of τ_{eoe}, the correlation time characterizing the end-over-end rotation of the polymer molecule, it is assumed that the three correlation times τ_c, τ_{lm} and τ_{eoe} are related thus:

$$1/\tau_c = (1/\tau_{eoe}) + (1/\tau_{lm}) \tag{7}$$

Correlation times in magnetic resonance are normally concerned with the time evolution of certain second-order spherical harmonics of the position coordinates. On the other hand, τ_1, the relaxation time of the first normal mode of a polymer chain, describes the time dependence of the first-order spherical harmonic $\cos\theta$ where θ is the angle between the vector joining the ends of the chain and some vector fixed in an external frame of reference. Using the relationship $\tau_{eoe} = \tau_1/3$ there are two models commonly used for

292

theoretical estimates of τ_1, and hence τ_{eoe}. These are[26]

$$\tau_{eoe} = 0.61M[\eta]\eta_0/RT \quad \text{(free-draining coil)} \tag{8a}$$

$$\tau_{eoe} = 0.42M[\eta]\eta_0/RT \quad \text{(non-draining coil)} \tag{8b}$$

where M is the molecular weight, $[\eta]$ the limiting viscosity number and η_0 the viscosity of the solvent. A detailed discussion of the quantitative agreement between experimental values of τ_{eoe} determined from equation (7) and theoretical estimates from equations (8) will be given later. For the present it is sufficient to note that, provided $[\eta]$ is independent of temperature,[14] both equations (8) predict that a plot of $\log(T\tau_{eoe})$ *versus* $1/T$ should give an activation energy equal to that for viscous flow of the pure solvent. Figure 15.15 shows such a plot

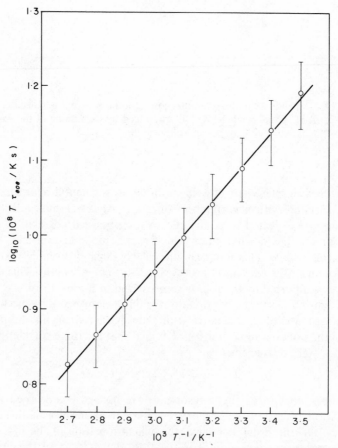

Figure 15.15 A plot of $\log_{10}(10^8 T\tau_{eoe})$ *vs.* $10^3/T$ for labelled polystyrene ($\bar{M}_n = 1950$) in toluene (1%): $E_a = 8.8 \pm 1.7 \, \text{kJ}$ mol^{-1}.[14] (Reproduced by permission of the American Chemical Society)

and the activation energy obtained is $8.8 \pm 1.7 \, \text{kJ mol}^{-1}$. This compares favourably with that for viscous flow in toluene, namely, $9.0 \, \text{kJ mol}^{-1}$.

A criticism sometimes levelled against labelling techniques, notably spin- and fluorescence-labelling, is that the rotational diffusion of the label may not accurately reflect the dynamics of the macromolecular chain to which the label is covalently attached.[27] Further, the motion of the label may be aniso-tropic. In the present case such motions could be rapid rotation about the C(4)—N bond or about the main chain carbon-phenyl carbon bond. If these criticisms have substance then it is to be expected that changing the pendant group bonded to the nitrogen atom would alter the apparent activation energy for the rotational diffusive process determining τ_c for the segmental relaxation. Also, alteration of the position of attachment of the label to the styryl ring (say to the *meta* position) would, at a given temperature, give a value of τ_c different to that obtained in the *para*-labelled polymer. Both aspects have been checked. A heterodisperse sample of polystyrene having an average molecular weight of *ca.* 115,000 was labelled in the *para* position of the styryl rings first[9] with the phenyl nitroxide label (2) and then the t-butyl nitroxide label (3). Figure 15.16 shows a composite Arrhenius plot[14] for the two labelled polymers

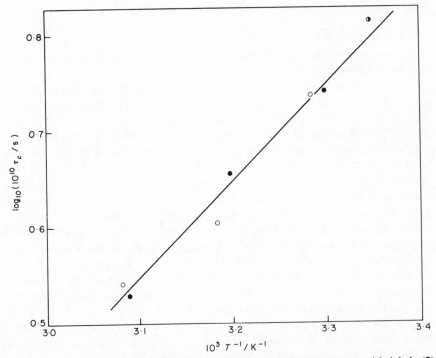

Figure 15.16 Composite Arrhenius plot for heterodisperse polystyrene with labels (2) and (3): ●, label (2), ○, label (3); ◑, n.m.r. from ref. 28.[14] (Reproduced by permission of the American Chemical Society)

together with a point derived from some ^{1}H spin-lattice relaxation time measurements made by McCall and Bovey.[28] The close agreement between the correlation times for the dissimilar labels strongly supports our view that the e.s.r. line-width measurements are accurately reflecting the dynamics of the polystyrene chain. The activation energy from Figure 15.16 is 19·5 kJ mol^{-1}. Further support comes from some preliminary measurements of τ_c for polystyrene labelled in the *meta* position of the styryl ring,[14,15] (4). At 52 °C, 1% toluene solutions of the *para*- and *meta*-labelled polymers have values of τ_c of $(3·2 \pm 0·3) \times 10^{-10}$ s and $(3·9 \pm 0·4) \times 10^{-10}$ s respectively. This agreement to within experimental error, provides final confirmation that the motion responsible for the line width effects in the e.s.r. spectra is essentially isotropic and is a property of the polymer.

Comparison of the results described above with those obtained by other techniques is of interest. An Arrhenius plot of τ_c for the segmental, or local mode, process which obtains at high molecular weights, yields an activation energy for the labelled polymer in toluene solution of $18·0 \pm 0·8$ kJ mol^{-1}. This is close to the values of $18·8 \pm 2·1$ and *ca.* 20 kJ mol^{-1} found by Stockmayer *et al.*[29] for the dielectric relaxation of poly(*p*-fluorostyrene) and poly(*p*-chlorostyrene). We have commented elsewhere that this comparison indicates that the stiffness of the chain is little influenced by the presence of large polar substituents in the *para* position of the styryl ring.[14]

15.4.1.2 Solvent effects[30]

It was noted earlier that the viscous drag exerted by the solvent is expected to affect the dynamic properties of a macromolecular chain in solution. It is of particular interest to determine the relative contributions that intramolecular steric effects and solvent viscosity make to the observed segmental motions. In other words it is important to be able to distinguish and separate the components of the so-called 'internal viscosity'[31] of the chain. To this end, τ_c was determined over a range of temperatures for samples PS1 to PS7 in two more solvents namely, α-chloronaphthalene and cyclohexane. The reasons for these choices were as follows. Toluene and α-chloronaphthalene are equally good solvents, in the thermodynamic sense, for polystyrene over a wide range of molecular weights and temperature.[30] However, the viscosity of α-chloronaphthalene is always several times greater than that of toluene over the range of temperatures studied. In addition its activation energy for viscous flow is approximately twice that for toluene. On the other hand, cyclohexane is a poor solvent for the higher molecular weight samples at temperatures above its theta temperature (34 °C) and a nonsolvent for molecular weights greater than, say, 20,000 at temperatures lower than 34 °C. For PS1 and PS2, however, [η] was the same for all three solvents at 55 °C and so the differences in solvent power are negligible at low molecular weights and high temperatures. Viscosity data on the three solvents are summarized in Table 15.2.

A detailed discussion of the dependence of τ_c upon molecular weight for the three solvents will be omitted except to note that the same basic features

Table 15.2 Solvent viscosities and energies of activation for viscous flow[30]

Solvent	Viscosity/cP[a]			Activation energy/kJ mol^{-1}
	20 °C	40 °C	55 °C	
Toluene	0·587	0·465	0·399	9·00
Cyclohexane	0·980	0·704	0·565	12·55
α-Chloronaphthalene	3·33	2·09	1·53	17·78

^{a} 1 cP = 10^{-3} kg m^{-1} s^{-1}.

shown in Figure 15.14 are found for all three solvents. In particular, τ_c became independent of molecular weight at long chain lengths. Figure 15.17 shows Arrhenius plots of the segmental correlation times for PS7 in all three solvents. The most notable feature of these is the non-linearity of the plot for cyclohexane. This may be interpreted in terms of the increasing solvent power of this solvent at high temperatures. However, as the temperature is lowered, polymer-solvent interactions decrease and the chain assumes a more tightly-coiled conformation leading to an increase in the intra-chain steric interactions opposing segmental rotation. Finally, at and below the theta-temperature aggregation occurs and the spectra obtained are more characteristic of solid labelled polymers than of polymers in solution.[30] The limiting activation energy at high temperatures for PS7 in cyclohexane is 27·2 ± 2·5 kJ mol^{-1}. This is to be compared with the

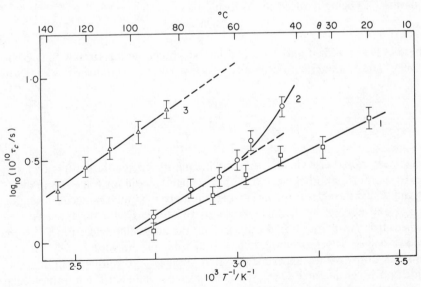

Figure 15.17 Arrhenius plots of log τ_c vs. $1/T$ for labelled PS7: 1, in toluene; 2, in cyclohexane; 3, in α-chloronaphthalene.[30] (Reproduced by permission of the Chemical Society)

values found in toluene and α-chloronaphthalene of 18.0 ± 0.8 and 26.4 ± 1.3 kJ mol^{-1}, respectively.

These results may be discussed in terms of Kramers' theory[32] for the diffusion of a particle over a potential barrier, as modified by Helfand.[33] The latter author discussed the rate of conformational transitions of polymeric chains embedded in viscous media. He found that the rate, k, at which 'particles' (in this case substituent groups on the main chain) diffuse over the barrier is given by:

$$k = [\gamma/2\pi\zeta][\tfrac{1}{2} + (\tfrac{1}{4} + m\gamma/\zeta^2)^{\frac{1}{2}}]^{-1} \exp(-E^*/RT) \qquad (9)$$

In this equation γ is the force constant for the potential barrier and is not to be confused with the magnetogyric ratio for which the same symbol is used in equations (2)–(4) inclusive. The mass of the particle is m, E^* is the height of the potential barrier and ζ is the frictional coefficient. For translational diffusion, ζ is equal to $6\pi a\eta$ where a is the hydrodynamic radius of the particle and η is the viscosity of the solvent. The limit of high viscous damping is defined by

$$m\gamma/\zeta^2 \ll \tfrac{1}{4} \qquad (10)$$

and under these conditions, equation (9) becomes

$$k = [\gamma/2\pi\zeta] \exp(-E^*/RT) \qquad (11)$$

Substituting for ζ gives

$$k = [\gamma/12\pi^2 a]\eta^{-1} \exp(-E^*/RT) \qquad (12)$$

Assuming the usual Arrhenius dependence of η upon temperature leads to equation (12) being reduced to the form

$$k = C \exp[-(E^* + E_\eta)/RT] \qquad (13)$$

where C is a constant and E_η is the activation energy for viscous flow of the solvent. The total activation energy, E_{tot}, for rotational diffusion is therefore

$$E_{tot} = E^* + E_\eta$$

or

$$E^* = E_{tot} - E_\eta \qquad (14)$$

It has been noted elsewhere[30] that the inequality in equation (10) is certainly met by α-chloronaphthalene and is approximately valid for toluene and cyclohexane. E_{tot} is the activation energy pertaining to τ_c at high molecular weights, i.e. τ_{lm}, and equation (14) thus enables us to obtain E^* for the three systems studied. It is gratifying to find that, despite the large differences in E_{tot} for the "good" solvents α-chloronaphthalene and toluene, the values of E^* are the same for both, within experimental error namely, 8.6 and 9.0 kJ mol^{-1}. In contrast, the results for PS7 in cyclohexane give a value for the intramolecular barrier E^* of 14.6 kJ mol^{-1}. This indicates greater steric effects due to a more tightly coiled conformation of the polymer chain in this solvent. In this context some preliminary results on poly(α-methylstyrene) ($\overline{M}_n = 104,000$) are of

Table 15.3 Correlation times τ_{eoe} for end-over-end rotation of PS1 at 55 °C in 1 % solutions [30]

Solvent	$10^{10}\tau_{eoe}$/s	
	Theoretical (eqn. (8b))	Experimental
Toluene	2·0	2·4
Cyclohexane	2·8	2·6
α-Chloronaphthalene	7·5	6·0

interest. For this labelled polymer in toluene solution, $E_{tot} = 21\cdot3 \pm 1\cdot3\,\text{kJ}\,\text{mol}^{-1}$ and hence $E^* = 12\cdot3 \pm 1\cdot3\,\text{kJ}\,\text{mol}^{-1}$. The stiffening effect of the methyl group is clearly reflected by the increased value of E^*.

It was shown in the previous section that a study of the dependence of τ_c upon molecular weight could, using equation (7), lead to an estimate of τ_{eoe}, the correlation time for molecular rotation of the whole polymer molecule. Table 15.3[30] compares the experimental values of τ_{eoe} for PS1 in the three solvents at 55 °C with theoretical values calculated for the non-free-draining coil model using equation (8b). The agreement is excellent but unfortunately is less good for higher fractions. This is probably due to the neglect of normal modes other than the shortest (τ_{lm}) and the longest (τ_{eoe}). Nonetheless, our model satisfactorily explains the main features shown in Figure 15.14.

15.4.1.3 Concentration effects[30]

In order to study the effect that interchain interactions have on segmental reorientation, 1 % solutions of the labelled fractions PS2 and PS7 were made up in toluene. The polymer concentration was increased progressively in each solution by addition of the appropriate unlabelled polymer fraction and τ_c was measured at each concentration and several temperatures. The results are summarized in Figures 15.18 and 15.19 where reduced correlation times τ_c/τ_0 are plotted against log (concentration). The concentrations are expressed as weight per cent of polymer and τ_0 is the limiting value of τ_c as the concentration approaches zero. The usefulness of reduced correlation times lies in the fact that thermally activated processes common to dilute and concentrated solutions are removed from consideration and any remaining temperature dependence will arise from the polymer-polymer interactions.

There are two features common to both Figures 15.18 and 15.19. Firstly, temperature has little effect on the behaviour of the two fractions. Secondly, in both the initial dependence of τ_c/τ_0 upon the concentration, C, is very slight up to a value of C of ca. 10 % by weight. Thereafter a steep rise in τ_c/τ_0 is found for both fractions. The steeply rising portions are almost linear and extrapolation to $\tau_c/\tau_0 = 1$ gives a value of C which may be regarded as being characteristic of the onset of strong concentration dependence. The values thus obtained are approximately 16 and 20 % for PS7 and PS2 respectively. The concentration

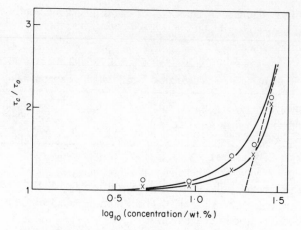

Figure 15.18 Reduced correlation times (τ_c/τ_0) for labelled PS2 in toluene as a function of polymer concentration (in wt. % polymer): ×, 74·5 °C; ○, 34 °C.[30] (Reproduced by permission of the Chemical Society)

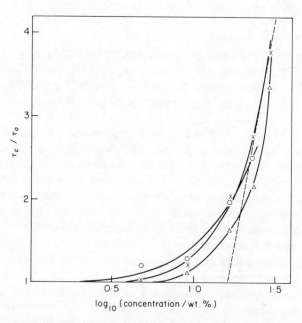

Figure 15.19 As Figure 15.18 but for PS7: △, 74 °C; ×, 54·5 °C; ○, 36 °C.[30] (Reproduced by permission of the Chemical Society)

dependence of the reduced correlation times is markedly similar to the behaviour of the viscosity of concentrated polymer solutions.[34] For example, for polystyrene in toluene there is a critical concentration C_c at which the power dependence of viscosity upon concentration changes from 2·3 to 5·0.[35] It is significant that for PS7 C_c is about 14%[34] which is very close to the value of C found to be characteristic of the onset of strong dependence of τ_c/τ_0 upon concentration. Unfortunately, there is no relevant literature for low polymers such as PS2, although it is clear that for this sample C_c will be higher than for PS7. It seems probable that both the changes in concentration dependence of viscosity and of the reduced segmental correlation time have the same origin.

A common interpretation of the critical concentration in polymer solution viscosities is in terms of the concept of chain entanglement.[34] It is supposed that this leads to a network structure with a consequent sharp increase in viscosity. However, it seems unlikely that this view will account for the spin-labelling results. In PS7 the relaxation is dominated by a local segmental mode which involves the cooperative motion of an unknown but certainly very small number of monomer units on either side of the labelled segment. Since the number density of the labels is very low (ca. 1 label per 600 monomer units for PS7) it is highly improbable that an entanglement will occur sufficiently close to the label to affect τ_c at moderate degrees of entanglement.

An alternative model of concentrated polymer solutions has been proposed by Onogi and coworkers.[35] This is based on the equivalent sphere concept and proposes that the spheres are tightly packed at C_c with the chains having their unperturbed dimensions. The model is in accord with our observations since the intermolecular interactions and conformational changes at C_c are likely to hinder segmental motion. Further, the critical concentration is predicted to be independent of temperature which again is in accord with the present data.

15.4.2 End-labelled poly(methyl methacrylate)

Apart from their intrinsic interest, studies of the dynamics of macromolecules in solution have relevance to certain features of the kinetics of radical polymerizations. In particular we refer to the diffusion-controlled model for the termination step.[36] The rate-determining step is envisaged not as the translational diffusion of two macroradicals together but rather the diffusion, via a segmental reorientation process, of the radical ends to within reaction distance once the propagating macroradicals have come sufficiently close together. Attempts to correlate dynamic measurements with termination rate coefficients have largely been concerned with rotation of segments within the chain as determined by dielectric relaxation techniques, rather than chain ends.[37] The sole exception to this is to be found in the work of North and Soutar[38] who used fluorescence depolarization measurements to measure correlation times for chain-end rotation which were then compared with results obtained by the same technique for segmental rotation. As shown in section 15.2 it is relatively simple to spin-label the chain ends of many synthetic polymers and hence obtain information about the rotational mobility by line-width measurements.

Three samples of end-labelled poly(methyl methacrylate) (6) were prepared and had molecular weights of 81,000, 25,000, and 2300. These will be referred to as PMMA1, PMMA2 and PMMA3 respectively. The spectrum common to these polymers in solution is shown in Figure 15.7. As in the case of randomly labelled polystyrene it was necessary to allow for inhomogeneous broadening of the lines due to interaction with the protons marked with an asterisk in:

$$\begin{array}{c} \overset{*}{C}H_3 \quad C(\overset{*}{C}H_3)_3 \\ | \quad \diagup \\ \text{---}CH_2\text{---}C\text{---}N \\ | \quad \diagdown \\ C{=}O \quad O^{\cdot} \\ | \\ OCH_3 \end{array}$$

For reasons which will be discussed elsewhere,[21] the coupling constant for all these protons was taken to be 0·25 MHz. The various parameters required for

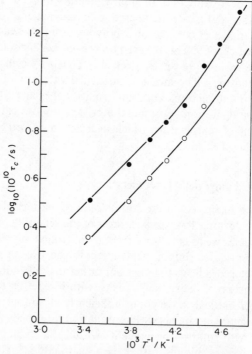

Figure 15.20 Arrhenius plots of the rotational correlation time τ_c for end-labelled poly(methyl methacrylate) (M.W. 81,000) in solution (5% ethyl acetate); ●, τ_c calculated from C coefficient; ○, τ_c calculated from B coefficient (see text)[21]

Table 15.4 E.s.r. spectral parameters for end-labelled PMMA[21]

$a_N = 41.84 \pm 0.18 \, \text{MHz}$	$g_{iso} = 2.00599 \pm 0.00004$
$T_{zz} = 96.21 \pm 0.27 \, \text{MHz}$	$g_{zz} = 2.00191 \pm 0.00014$
$b = 341.3 \, \text{MHz}$	$\Delta\gamma = (5.4 \pm 0.3) \times 10^4 \, \text{rad s}^{-1} \, \text{gauss}^{-1}$

the determination of τ_c were obtained from spectra of the polymers in 5% solution in ethyl acetate both in the liquid state (at room temperature) and as a glass (at 77 K). These are given in Table 15.4. Equation (1) reduces to:

$$T_2^{-1}(m_I) = A + Bm_I + Cm_I^2 \tag{15}$$

Inspection of equation (2) gives $B = -\frac{4}{15}b\,\Delta\gamma\,B_0\tau_c$ and $C = (b^2/8)\tau_c$. It is thus possible to obtain two independent values of τ_c and these are shown in Figures 15.20, 15.21 and 15.22 as Arrhenius plots. For PMMA1 (Figure 15.20) the curves, which are slightly non-linear, follow each other closely. Indeed, if due allowance is made for uncertainties in the values of $\Delta\gamma$ and b, they are coincident to within the limits of combined experimental error. This is in sharp contrast to the behaviour found for the other polymers of lower molecular weight shown in Figures 15.21 and 15.22. This can only be interpreted in terms of anisotropic rotation. For PMMA1 it is reasonable to suppose that the rotational motion of the end-label is solely determined by the mobility of a few segments near the

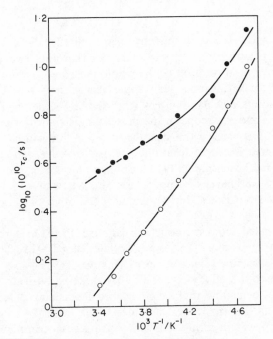

Figure 15.21 As Figure 15.20 but for M.W. 25,000[21]

302

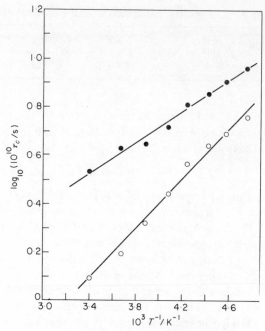

Figure 15.22 As Figures 15.20 and 15.21 but for M.W. 2300[21]

end of the chain and it is easy to show by means of molecular models that such motion can be virtually isotropic. However, for the other two polymers it is likely that end-over-end rotation of the whole macromolecule will contribute to the observed relaxation and our results clearly show that this motion is anisotropic for low molecular samples of PMMA. The detailed treatment of such effects is beyond the scope of the present article. However, it should be noted that equations used in this chapter to relate line widths to correlation times have assumed isotropic rotational diffusion of the spin-label. If this is the case, then the experimentally determined parameter $|B/C|$ should be independent of τ_c[39] and hence of temperature. This is true to within experimental error for PMMA1 but is certaintly untrue for the other two samples (Figure 15.23).

The final point of interest about end-labelled PMMA is the non-linear nature of the Arrhenius plots. This is most obvious for PMMA2 (Figure 15.21) where a fairly well-defined 'knee' is seen. This occurs at -45 ± 1 °C. This is remarkably close to the value suggested by Hughes and North[40,41] who found precisely the same behaviour in the temperature dependence of k_t, the termination rate coefficient for the radical polymerization of MMA in ethyl acetate. Their explanation was in terms of a 'phase transition' from a rigid random coil to a flexible random coil. Confirmation was obtained by studying the dependence of the limiting viscosity number upon temperature.[41]

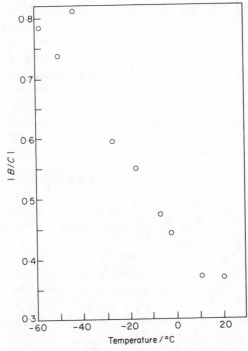

Figure 15.23 The dependence of $|B/C|$ upon temperature for end-labelled poly(methyl methacrylate) (M.W. 25,000) showing the effect of anisotropic rotation[21]

15.4.3 Polyacrylates labelled on ester groups

Three polyacrylates were labelled as described in section 15.2. The molecular weights and densities of labelling are given in Table 15.5. The structures of these labelled polymers (**8**) are to be found in section 15.2 and a typical solution spectrum is shown in Figure 15.8. From the work of Poggi and Johnson,[22] together with the results described in section 15.3, it is clear that due allowance must be made for the unresolved and partially resolved couplings to protons in

Table 15.5 Viscosity-average molecular weights and labelling densities in side-chain-labelled polyacrylates[16,21]

Polymer	$\overline{M}_v$	Monomer units/spin
Poly(methyl methacrylate) (1)	550,000	22,400
Poly(methyl methacrylate) (2)	33,000	13,200
Poly(methyl acrylate)	111,400	1100
Poly(n-butyl methacrylate)	450,000	5200

the piperidine ring. This point has been ignored in previous studies using similar labels.[42,43] There have been two reports of n.m.r. contact shift measurements of the various proton coupling constants in nitroxide radicals similar in structure to those used here as labels. They were by Kreilick[44] and Brière et al.[45] Briefly, those of Kreilick did not result in satisfactory spectral simulations whilst those of Brière et al. did. These findings have been confirmed recently by Whishnant, Ferguson and Chestnut.[46] In the present case, the high-field line was insufficiently resolved to enable two values of τ_c to be determined. However, analysis of the other two lines over a range of temperatures for all the polymers listed in Table 15.5 (5% solutions in toluene) gave the following results.

(i) Both samples of PMMA gave values of τ_c which fitted closely a common Arrhenius plot with an activation energy of 20 ± 1 kJ mol^{-1}. Clearly, contributions to the relaxation from end-over-end rotation of the chains are negligible. This result is somewhat lower than the value obtained by dielectric relaxation measurements,[37] namely, ca. 27 kJ mol^{-1}. Ironically, the use of Kreilick's data[44] in the spectral simulations results in an activation energy of ca. 28 kJ mol^{-1} which is in excellent agreement with the dielectric results.

(ii) Poly(methyl acrylate) and poly(n-butyl methacrylate) gave activation energies of 19 ± 3 kJ mol^{-1} and 14.6 ± 0.6 kJ mol^{-1} respectively. The value for poly(methyl acrylate) is again lower than that obtained from dielectric measurements,[37] namely 23 kJ mol^{-1}.

(iii) At 298 K, the spin-labelling data for PMMA give a value of τ_c of 3.4×10^{-10} s. On the other hand fluorescence depolarization[38] and dielectric relaxation measurements[37] give values of 1.3 and 1.4×10^{-9} s respectively. It seems that we may have encountered a problem which must be borne in mind in all labelling studies. This is that the motion of the label may not precisely reflect that of the polymer. In this case it seems that rapid rotation may occur about the bond between the ester carbon and the oxygen atom or between the γ-carbon of the piperidine ring and the oxygen atom (see 8). Hower, Henkens and Chestnut[42] have used the treatment of Hubbell and McConnell[47] for such a situation with some success to account for the results of their studies of the spin-labelled bovine carbonic anhydrase enzyme. Briefly, if rapid rotation occurs about one particular axis then a new effective spin Hamiltonian may be defined showing axial symmetry about the axis of rotation. The elements of the hyperfine and g tensors are replaced by time-averages, e.g. the parallel and perpendicular components of the 'new' tensor, $T_{\parallel}$ and $T_{\perp}$, are given by

$$T'_{\parallel} = \overline{\alpha^2} T_{xx} + \overline{\beta^2} T_{yy} + \overline{\gamma^2} T_{zz} \tag{16}$$

and

$$T'_{\perp} = \tfrac{1}{2}(1 - \overline{\alpha^2})T_{xx} + \tfrac{1}{2}(1 - \overline{\beta^2})T_{yy} + \tfrac{1}{2}(1 - \overline{\gamma^2})T_{zz} \tag{17}$$

where $\overline{\alpha^2}$, $\overline{\beta^2}$ and $\overline{\gamma^2}$ are the time-averaged squares of the direction cosines of the angles between the rotation axis and the x, y and z axes of the original tensor. Similar equations hold for the g tensor.

Preliminary calculations suggest that replacing the values of b and $\Delta\gamma$ in equation (3) of section 15.3 by b' and $\Delta\gamma'$ calculated from T'_{ii} and g'_{jj} values will do much to remove the discrepancy between the values of τ_c noted above. It should be noted that whilst the modification suggested above may well bring the spin-labelling τ_c into line with that obtained by other techniques *at one temperature*, the activation energies will be unaltered. However, the discrepancies are not too serious and the correlation time is certainly much longer (~ 10–100 times) than that expected for rotational diffusion of the unbound label.

15.4.4 Relaxation in solid polymers

Törmälä and Lindberg[5] have given a brief review of the methods of obtaining data from labelled species which have relaxations with τ_c in the range $10^{-9} < \tau_c < 10^{-6}$ s. The spectra obtained in this relaxation region closely resemble the rigid limit spectrum shown in Figure 15.10. However, as τ_c becomes shorter the extrema separation decreases from its rigid-limit value of $2T_{zz}$ and this behaviour is shown as a function of $1/T$ in Figure 15.11. Freed, Bruno and Polnaszek[48] have used the stochastic Liouville method, with the retention of pseudo-secular terms, to simulate spectra over a range of correlation times for three rotational models. These are the Brownian, 'moderate jump' and 'large jump' diffusional cases. Later, Goldman, Bruno and Freed[49] showed that the results of such calculations are well-described by the analytical expression

$$\tau_c = a(1 - S)^b \tag{18}$$

where $S = T'_{zz}/T_{zz}$ and $2T'_{zz}$ is the extrema separation for the slowly diffusing radical having the correlation time τ_c. The parameters a and b are characterized by the diffusion model adopted and the intrinsic line width of the spectrum.

We have examined[50] two of the narrow fractions of labelled polystyrene, described in section 15.4.1, in the solid state between 77–340 K. The fractions were PS1 and PS7. No significant systematic difference was found in the behaviour of the fractions and mean values of T'_{zz} and T_{zz} were used in subsequent calculations. Clearly, the relaxation does not depend upon molecular weight. Using an intrinsic line width of 8·4 MHz, the following values of a and b were used in the calculations.

Brownian diffusion:

$$a = 5{\cdot}4 \times 10^{-10}\,\text{s}$$

$$b = -1{\cdot}36$$

'Moderate jump' diffusion:

$$a = 1{\cdot}10 \times 10^{-9}\,\text{s}$$

$$b = -1{\cdot}01$$

'Large jump' diffusion:

$$a = 2{\cdot}55 \times 10^{-9}\,\text{s}$$

$$b = -0{\cdot}615.$$

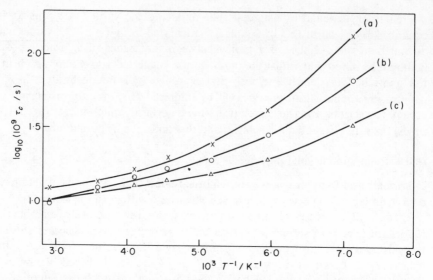

Figure 15.24 Arrhenius plots for rotational correlation times in solid spin-labelled polystyrene calculated by the method of Goldman *et al.*[49] : (a) Brownian, (b) moderate-jump and (c) large-jump diffusion.[50] (Reproduced by permission of John Wiley and Sons, Inc.)

Arrhenius plots for the three models are shown in Figure 15.24. All three are non-linear but limiting values of the activation energy E_a at the low-temperature end of the scale were found: Brownian diffusion 8·4, moderate jumps 6·3 and large jumps 4·6 kJ mol^{-1}. Whilst there are no *a priori* grounds for choosing any one of these three models it seems that a jump model should be more appropriate to the present case. In this connection it is interesting to note that the activation energy for the moderate jump case is very close to that quoted by Yano and Wada[51] for the δ-relaxation in atactic and isotactic polystyrene, namely, 6·7 kJ mol^{-1}. For relaxations in solid polymers it is common practice to quote values of f_m rather than correlation times where f_m is the frequency at which maximum dielectric or mechanical absorption takes place. Accordingly, the spin-labelling values of τ_c were converted into fictitious values of f_m for the moderate jump diffusion model using the relationship $f_m = (2\pi\tau_c)^{-1}$. Figure 15.25 shows a composite Arrhenius plot for the δ-relaxation in polystyrene.[51-57] It seems certain that, at low temperatures, we have observed this relaxation which is discussed by Yano and Wada[51] in terms of a model involving tacticity defects. The reasons for the curvature in the Arrhenius plots of Figure 15.24 are discussed elsewhere.[50]

Computer programs incorporating anisotropic rotations are now available for spectral simulations and should enable more information to be obtained from subtle changes in spectral shape.[39,49] This aspect of the spin-labelling technique seems to us to be of considerable potential in the study of relaxation in solid polymers and slowly tumbling rigid polymers in solution.

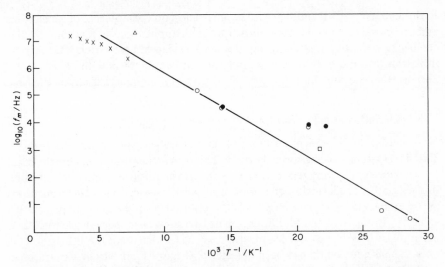

Figure 15.25 Composite plot for the δ relaxation in polystyrene: $\triangle$, n.m.r. T_1 minimum; $\bigcirc$, $\bullet$, mechanical loss peaks; $\square$, dielectric loss peak; $\times$, e.s.r. spin-labelling results. All points are for atactic samples except for the solid circles ($\bullet$) which are for isotactic material. Sources of data are given in reference 50. (Reproduced by permission of John Wiley and Sons, Inc.)

15.5 SPIN-LABELS IN STRUCTURAL STUDIES

In section 15.2, which dealt with the synthesis of spin-labelled polymers, one of the main labelling processes involved reaction of 2-methyl-2-nitroso-propane (MNP) with polymeric carbanions and/or lithiated compounds. MNP has found wide application as a 'spin trap' for identifying transient radical species[58,59] where it forms the more stable nitroxide radical in a single stage reaction:

$$R\cdot + \text{t-BuNO} \rightarrow \text{t-Bu}-\underset{\underset{O\cdot}{|}}{N}-R$$

The synthetic applications of this compound discussed in section 2 underline an earlier suggestion[17] that MNP may also function as a carbanion trap forming the nitroxide after oxidation of the intermediate hydroxylamine:

$$R^- + \text{t-Bu}-\text{NO} \rightarrow \text{t-Bu}-\underset{\underset{O^-}{|}}{N}-R$$

$$\downarrow \text{hydrolysis/oxidation}$$

$$\text{t-Bu}-\underset{\underset{O\cdot}{|}}{N}-R$$

In this capacity MNP could provide information on the structure of carbanions and organo-alkali metal compounds. Two such applications in macromolecular chemistry were investigated during the work reviewed here. These were: (i) the lithiation reaction of styrene homopolymer and copolymers with halostyrenes, and (ii) the decomposition products in the anionic polymerization of PMMA.

15.5.1 Lithiation of polystyrene and styrene-halostyrene copolymers

In section 15.2 the synthesis of *m*-lithiated polystyrene, using the *n*-BuLi/ TMEDA system, was mentioned briefly. This is in fact a somewhat controversial subject since conflicting reports exist regarding the position of metalation using this reagent. Metalation in the *ortho* and *para*[60] as well as *meta*[61] positions and even backbone metalation, has been reported.[62] In the latest paper on this subject[63] it was reported that lithiation at two sites, *meta*:*para* in the ratio 2:1, occurs. Using MNP as the diagnostic reagent for detecting the lithiation site, only the *meta*-nitroxide has been detected[15] (Figure 15.4). Minor variations in reaction conditions for lithiation (temperature—20° to 60 °C, solvents— benzene, cyclohexane and heptane) have not produced any detectable variation in the final nitroxide-labelled polymer which has always been the *meta*-

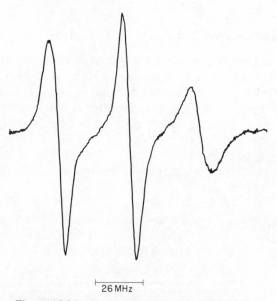

26 MHz

Figure 15.26 E.s.r. spectrum of labelled polymer produced by lithiation of styrene-*m*-chlorostyrene copolymer with *n*-BuLi/TMEDA followed by addition of MNP.[16]

compound.[16] It is quite certain that 10% or less of the *para*-derivative would be detectable in the e.s.r. spectrum of the *meta*-compound. It is perhaps worth emphasizing that the results obtained using MNP as a carbanion trap relate to very small degrees of lithiation, that is to say, the initial stages in the reaction between *n*-BuLi/TMEDA and polystyrene. It is therefore not too surprising that *meta*-substitution appears to be favoured since this reaction is nucleophilic. However, under more forcing conditions *ortho*- or *para*-substitution could occur and it should be remembered that migration of metal atoms on the aromatic ring is quite probable on prolonged exposure of the polymer to the reaction medium.

As an alternative to direct *meta*-lithiation of polystyrene the metal-halogen exchange reaction may also be employed as it was in the *para*-lithiation reaction at *para*-iodinated styrene units. It has been reported[64] that *ortho* and *meta* chlorostyrene units in copolymers can be lithiated using the BuLi/TMEDA system. However, when the reaction was carried out on a styrene/*m*-chlorostyrene copolymer the spectrum of the resulting labelled polymer[16] (Figure 15.26) was not that of *meta*-labelled polystyrene. A similar spectrum was obtained from a styrene/*p*-chlorostyrene copolymer. On the other hand, a styrene/*m*-bromo-styrene copolymer did yield the spectrum of polystyrene carrying t-butyl nitroxide units in the *meta* position. Similarly with a styrene/*p*-bromostyrene copolymer a *para*-labelled product was formed.[16] These results indicate that aromatic bromine and iodine compounds undergo metal-halogen exchange with the *n*-BuLi/TMEDA system, but aromatic chlorine derivatives do not. The spectrum in Figure 15.26 could arise via the reaction:

and

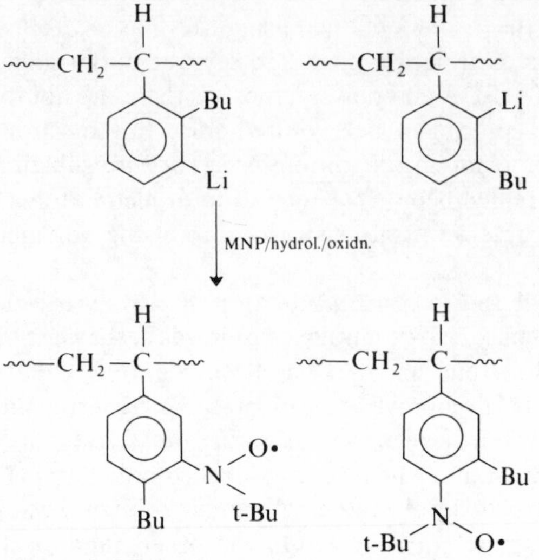

and

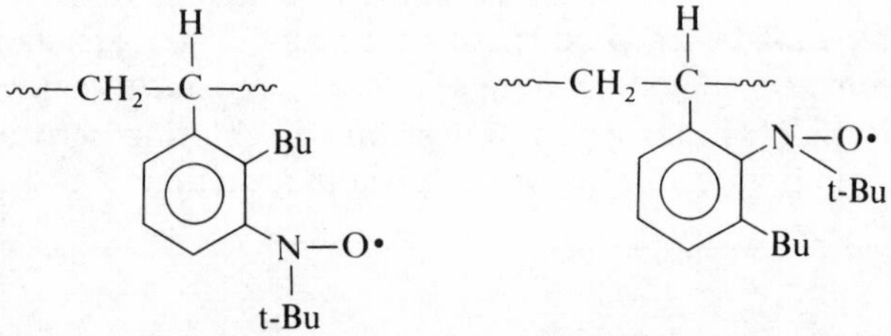

Aryne intermediates analogous to (9) and (10) have been proposed to explain the products of the reaction between organolithium reagents and small molecule chloro-aromatic compounds.[65] The e.s.r. spectra of dialkylphenyl nitroxides show little aromatic proton splitting because rotation of the nitroxide group into the conformation coplanar with the aromatic ring is sterically restricted.[66] The reaction pathway above, although consistent with observation, is not unambiguously proved. The important point in the present context, however, is that the spectrum of the nitroxide product derived from the reaction of BuLi/TMEDA with chlorostyrene copolymers, cannot arise from a simple halogen–metal exchange as seems to happen with the corresponding bromo and iodo analogues. The lithiation of chlorostyrene copolymers is obviously a more complex reaction.

15.5.2 Decomposition products of PMMA anions

During the synthesis of end-labelled PMMA reported in section 15.2 it was observed that if the reaction temperature of the system was allowed to rise above 0 °C before 'killing' the polymeric anions with MNP the e.s.r. spectrum

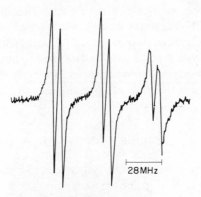

Figure 15.27 E.s.r. spectrum of labelled PMMA from anionic polymerization with n-BuLi in THF at $-70\,°C$. Addition of MNP after 10 min. at room temperature.[19] (Reproduced by permission of I.P.C. Science and Technology Press Ltd.)

of the resulting polymer was different from that of the end-labelled polymer shown in Figure 15.7.[19] For example, if the reaction mixture (THF solution) following n-BuLi initiation was allowed to attain room temperature and held there for about 10 min. before adding MNP the resulting spectrum was the triplet of doublets shown in Figure 15.27.[19] The g value of this species is $2·00587 \pm 0·00006$, identical with the three-line spectrum, while $a_N = 38·83\,MHz$. The doublet splitting is due to a single proton, $a_H = 6·04 \pm 0·11\,MHz$. At $0\,°C$ or after shorter times at room temperature, the product polymer gave a seven-line spectrum—a superposition of Figures 15.7 and 15.27 in which the outermost lines of the two spectra overlap exactly. The triplet of doublets in Figure 15.27 must arise from a radical having the structure:

$$
\begin{array}{c}
H \\
| \\
-C- \\
| \\
N \\
\diagup \quad \diagdown \\
\text{t-Bu} \quad \quad O\cdot
\end{array}
$$

(11)

The early suggestion[20] that this structure arose from rearrangement of the PMMA anion to yield

$$
\begin{array}{c}
\qquad CH_3 \\
\qquad | \\
\sim\!\!-CH\!-\!C\!-\!H \\
\quad | \qquad | \\
\quad N \quad\; CO_2CH_3 \\
\diagup \;\; \diagdown \\
\text{t-Bu} \quad\; O\cdot
\end{array}
$$

(12)

can be discounted on energetic grounds and because deuterated MMA ($CD_2=C(CH_3)CO_2CH_3$) behaves exactly like normal MMA. It is significant that the six-line spectrum was only observed when BuLi was employed as initiator. It is well known that when n-BuLi is used to initiate anionic polymerization of MMA a considerable proportion is lost by reacting with the carbonyl rather than the ethylenic function of the monomer to yield the ketone[67]:

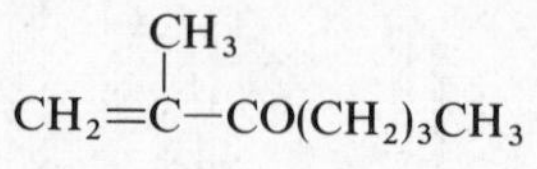

$$\underset{\displaystyle CH_2=\overset{\displaystyle CH_3}{\overset{|}{C}}-CO(CH_2)_3CH_3}{}$$

Any such units incorporated into the polymer could react with polymeric or other carbanions R^- at 0 °C or above:

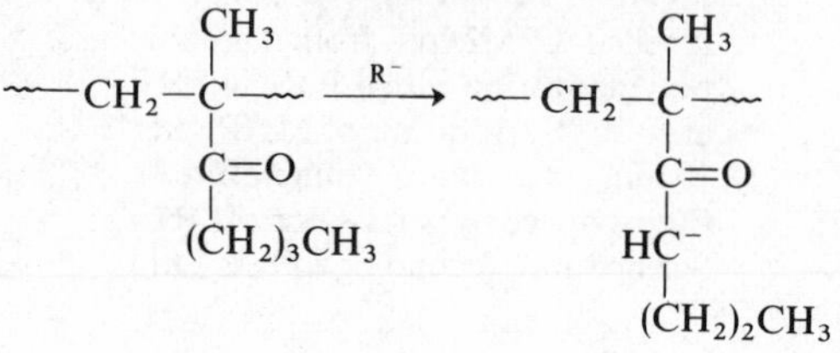

$$\sim\!\!\!-CH_2-\overset{\displaystyle CH_3}{\underset{\displaystyle \underset{(CH_2)_3CH_3}{\overset{|}{C}=O}}{\overset{|}{\underset{|}{C}}}}\!\!\!-\!\!\!\sim \quad\xrightarrow{\;R^-\;}\quad \sim\!\!\!-CH_2-\overset{\displaystyle CH_3}{\underset{\displaystyle \underset{(CH_2)_2CH_3}{\overset{|}{\underset{|}{HC^-}}=O}}{\overset{|}{\underset{|}{C}}}}\!\!\!-\!\!\!\sim$$

$$(13)$$

The hydrogen atoms on the methylene group in the α-position to the carbonyl group are fairly acidic and this reaction is quite feasible. Subsequent reaction of (13) with MNP would yield, after hydrolysis and oxidation, the nitroxide

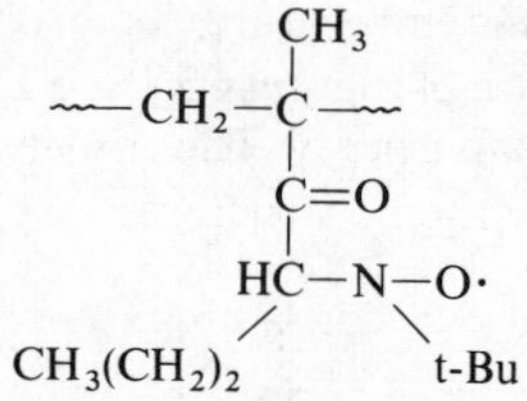

$$\sim\!\!\!-CH_2-\overset{\displaystyle CH_3}{\underset{\displaystyle \underset{\underset{CH_3(CH_2)_2 \qquad\qquad t\text{-Bu}}{\diagup\quad\diagdown}}{HC-N-O\cdot}}{\underset{|}{\overset{|}{C}=O}}}\!\!\!-\!\!\!\sim$$

$$(14)$$

whose spectrum should consist of a triplet of doublets as in Figure 15.27. Further support for this suggestion comes from the facts that fluorenyl lithium initiator, which does not attack the ester group, yields the three-line spectrum at all temperatures of termination with MNP and that preformed PMMA reacts with BuLi and MNP to produce a product with a six-line spectrum identical to Figure 15.27.[19] Also, when $CH_3(CH_2)_2CD_2Li$ is used in the above reactions a three-line spectrum is observed instead of the six-line spectrum.[16] Final proof for this reaction comes from studies of the reaction of the model compound methyl pivalate

$$
\begin{array}{c}
CH_3 \\
| \\
CH_3-C-CH_3 \\
| \\
C=O \\
| \\
OCH_3
\end{array}
$$

with BuLi. In this reaction BuLi can only attack the ester group and the product after addition with MNP shows a six-line e.s.r. spectrum very similar to that in Figure 15.27.[16] In this case we have

$$
\begin{array}{c}
CH_3 \\
| \\
CH_3-C-CH_3 \\
| \\
C=O \quad O\cdot \\
| \quad\;\; / \\
HC-N \\
| \quad\;\;\; \backslash \\
CH_3(CH_2)_2 \quad t\text{-Bu}
\end{array}
$$

the model analogue of (14). Again this six-line spectrum reverts to a three-line spectrum when deuterated n-BuLi is employed as reagent. There can therefore be little doubt that when BuLi is used to initiate the anionic polymerization of MMA the anion (13) is formed if the temperature approaches 0 °C. Although it is well known that the anionic polymerization of MMA is subject to side reactions the one discussed above has not been previously reported and it is doubtful if it could have been observed without the use of MNP or some similar carbanion trap. The only points that remain to be clarified are whether the n-butyl iso-propenyl ketone is copolymerized at −70 °C or whether these units are incorporated at the chain end when the temperature is allowed to rise to 0 °C or thereabouts, and what proportion of the PMMA carbanions react to yield (13) rather than some other products, e.g. methoxy anions, which are unreactive towards MNP. In this connection the extreme sensitivity of the e.s.r. method should be borne in mind.

15.6 REFERENCES

1. T. J. Stone, T. Buckman, P. L. Nordio and H. M. McConnell, *Proc. Natl. Acad. Sci. U.S.*, **54**, 1010, (1965).
2. See for example: C. L. Hamilton and H. M. McConnell, in *Structural Chemistry and Molecular Biology*, ed. A. Rich and N. Davidson, Freeman, San Francisco, 1968, p. 115; L. J. Berliner, *Spin Labelling. Theory and Applications*, Academic Press, New York, 1974.
3. D. Braun, *Pure Appl. Chem.*, **30**, 41, (1972).

314

4. Y. Miura, K. Nakai and M. Kinoshita, *Makromol. Chem.*, **172**, 233, (1973) and previous papers.
5. P. Törmälä and J. J. Lindberg, p. 255.
6. O. H. Griffith, J. F. W. Keana, S. Rottschaeffer and T. A. Warlick, *J. Amer. Chem. Soc.*, **89**, 5072, (1967).
7. A. R. Forrester and S. P. Hepburn, *J. Chem. Soc.* (*C*), 3322, (1971).
8. T. Kurosaki, K. W. Lee and M. Okawara, *J. Polymer Sci.*, *Chem. Edition*, **10**, 3295, (1972).
9. A. T. Bullock, J. H. Butterworth and G. G. Cameron, *European Polymer J.*, **7**, 445, (1971).
10. G. Drefahl, H.-H. Hörhold and K. D. Hofmann, *J. Prakt. Chem.*, **37**, 137, (1968).
11. A. T. Bullock, J. H. Butterworth and G. G. Cameron, unpublished work.
12. D. Braun, *Makromol. Chem.*, **30**, 85, (1959).
13. A. T. Bullock, G. G. Cameron and P. M. Smith, *Polymer*, **13**, 89, (1972).
14. A. T. Bullock, G. G. Cameron and P. M. Smith, *J. Phys. Chem.*, **77**, 1635, (1973).
15. A. T. Bullock, G. G. Cameron and P. M. Smith, *Polymer*, **14**, 525, (1973).
16. A. T. Bullock, G. G. Cameron and J. M. Elsom, to be published.
17. A. R. Forrester and S. P. Hepburn, *J. Chem. Soc.* (*C*), 701, (1971).
18. A. T. Bullock, G. G. Cameron and P. M. Smith, unpublished work.
19. A. T. Bullock, G. G. Cameron and J. M. Elsom, *Polymer*, **15**, 74, (1974).
20. A. T. Bullock, J. H. Butterworth, G. G. Cameron and J. M. Elsom, Prepr. I.U.P.A.C. Conf. Chemical Transformations of Polymers, Bratislava, 1971, P21.
21. A. T. Bullock, G. G. Cameron and V. Krajewski, to be published.
22. G. Poggi and C. S. Johnson Jr., *J. Magn. Resonance*, **3**, 436, (1970).
23. A. Hudson and G. R. Luckhurst, *Chem. Rev.*, **69**, 191, (1969).
24. For general discussions of the nature of homogeneous and inhomogeneous broadening see (a) A. M. Portis, *Phys. Rev.*, **91**, 1071, (1953) and (b) A. T. Bullock and L. H. Sutcliffe, *Trans. Faraday Soc.*, **60**, 2112, (1964).
25. W. L. Hubbell and H. M. McConnell, *J. Amer. Chem. Soc.*, **93**, 314, (1971).
26. B. H. Zimm, *J. Chem. Phys.*, **24**, 269, (1956).
27. A. M. North, *Disc. Faraday Soc.*, **49**, 286, (1970).
28. D. W. McCall and F. A. Bovey, *J. Polymer Sci.*, **45**, 530, (1960).
29. B. Baysal, B. A. Lowry, H. Yu and W. H. Stockmayer, *Dielectric Properties of Polymers*, F. E. Karasz, ed., Plenum Press, New York, N.Y., 1971, p. 343.
30. A. T. Bullock, G. G. Cameron and P. M. Smith, *J. Chem. Soc.*, *Faraday II*, **70**, 1202, (1974).
31. A. Peterlin, *Polymer Letters*, **10**, 101, (1972).
32. H. A. Kramers, *Physica*, **7**, 284, (1940).
33. E. Helfand, *J. Chem. Phys.*, **54**, 4651, (1971).
34. R. S. Porter and J. F. Johnson, *Chem. Rev.*, **66**, 4, (1966).
35. S. Onogi, T. Kobayashi, Y. Kojuna and Y. Taniguchi, *J. Appl. Polymer Sci.*, **7**, 847, (1963).
36. A. M. North, *Quart. Rev.*, **20**, 421, (1966).
37. A. M. North, *Chem. Soc. Rev.*, **1**, 49, (1972).
38. A. M. North and I. Soutar, *J. Chem. Soc.*, *Faraday I*, **68**, 1101, (1972).
39. S. A. Goldman, G. V. Bruno, C. F. Polnaszek and J. H. Freed, *J. Chem. Phys.*, **56**, 716, (1972).
40. J. Hughes and A. M. North, *Trans. Faraday Soc.*, **60**, 960, (1964).
41. J. Hughes and A. M. North, *Proc. Chem. Soc.*, 404, (1964).
42. J. F. Hower, R. W. Henkens and D. B. Chestnut, *J. Amer. Chem. Soc.*, **93**, 6665, (1971).
43. D. B. Chestnut and J. F. Hower, *J. Phys. Chem.*, **75**, 907, (1971).
44. R. W. Kreilick, *J. Chem. Phys.*, **46**, 4260, (1967).
45. R. Brière, H. Lemaire and A. Rassat, *Bull. Soc. Chim. France*, **12**, 4479, (1967).

46. C. C. Whisnant, S. Ferguson and D. B. Chestnut, *J. Phys. Chem.*, **78**, 1410, (1974).
47. Reviewed by H. M. McConnell and B. G. McFarland, *Quart. Rev. Biophys.*, **3**, 91, (1970).
48. J. H. Freed, G. V. Bruno and C. F. Polnaszek, *J. Phys. Chem.*, **75**, 3385, (1971).
49. S. A. Goldman, G. V. Bruno and J. H. Freed, *J. Phys. Chem.*, **76**, 1858, (1972).
50. A. T. Bullock, G. G. Cameron and P. M. Smith, *J. Polymer Sci., Polymer Physics Ed.*, **11**, 1263, (1973).
51. O. Yano and Y. Wada, *J. Polymer Sci.*, **A-2, 9**, 669, (1971).
52. B. I. Hunt, J. G. Powles and A. E. Woodward, *Polymer*, **5**, 323, (1964).
53. J. M. Crissman and R. D. McCammon, *J. Acoust. Soc. Amer.*, **34**, 1703, (1962).
54. J. M. Crissman, A. E. Woodward and J. A. Sauer, *J. Polymer Sci.*, **A, 3**, 2693, (1965).
55. R. D. McCammon, R. G. Saba and R. N. Work, *J. Polymer Sci.*, **A-2, 7**, 1721, (1969).
56. K. M. Sinnott, *S.P.E. Trans.*, **2**, 65, (1962).
57. V. Frosini and A. E. Woodward, *J. Polymer Sci.*, **A-2, 7**, 525, (1969).
58. C. Lagercrantz and S. Forshult, *Acta Chem. Scand.*, **23**, 708, (1969).
59. M. J. Perkins, P. Ward and A. Horsfield, *J. Chem. Soc.* **(B)**, 395, (1970).
60. A. J. Chalk, *J. Polymer Sci.*, **B, 6**, 649, (1968).
61. N. A. Plate, M. A. Yampolskaya, S. L. Davydova and V. A. Kargin, *J. Polymer Sci.*, **C, 22**, 547, (1969).
62. U.K. Patent 1,121,195, (1968).
63. D. C. Evans, L. Phillips, J. A. Barrie and M. A. George, *J. Polymer Sci., Polymer Letters Ed.*, **12**, 199, (1974).
64. A. F. Halasa, Advances in Chemistry Series, No. 130, *Polyamine-chelated Alkali Metal Compounds*, A.C.S. (1974), p. 177.
65. L. Friedman and J. F. Cherowski, *J. Amer. Chem. Soc.*, **91**, 4864, (1969).
66. A. R. Forrester and S. P. Hepburn, *J. Chem. Soc.* **(C)**, 1277, (1970).
67. N. Kawabata and T. Tsuruta, *Makromol. Chem.*, **86**, 231, (1965).

16

E.s.r. Studies of Dynamic Flexibility of Molecular Chains

K. Shimada and M. Szwarc

SUNY College of Environmental Science and Forestry, Syracuse, New York

16.1 INTRODUCTION

Conformations of any molecular chain in a liquid medium change continuously. For example, in a chain such as $X \cdot (CH_2)_n \cdot X$ the two *unreactive* end-groups, X, perform random movements caused by Brownian motion but restricted by the constrictions imposed by the chain binding them. In the course of this motion they approach each other, separate, approach again, etc., and the frequency of what may be described as encounters is determined by factors such as the length of the chain, temperature, solvent viscosity, etc. Moreover, this frequency depends also on the intrinsic *dynamic* flexibility of the chain, a property determined by the nature of the chain and not by its length. We wish to describe here a method that allows us to investigate the intrinsic dynamic flexibility of molecular chains and to discuss the results.

16.2 EXPERIMENTAL APPROACH AND THE METHOD OF CALCULATION

Our approach to this problem was outlined elsewhere[1,2]; hence, only the salient features of the method will be described. Hydrocarbons $N-(CH_2)_n-N$ with n varying from 3 to 20 were prepared,[3] N denoting an α-naphthyl moiety. These were reduced with potassium to afford radical anions, $N-(CH_2)_n-N^{\overline{\cdot}}$, possessing not more than *one* electron per chain. Their ESR spectra were recorded at several temperatures using hexamethylphosphorictriamide, HMPA, as the solvent, the dilution being sufficiently high ($[N-(CH_2)_n-N^{\overline{\cdot}}] < 10^{-4}$ M) to eliminate any effects arising from *inter*-molecular reactions. For the sake of illustration, the e.s.r. spectra of such radicals with $n = 20, 16, 12, 8, 6, 5, 4$, and 3, recorded at 0° and at 45 °C, are shown in Figures 16.1 and 16.2. Their appearance changes appreciably with the length of the chain and is strongly affected by the temperature of solution.

The variations in the shape of the e.s.r. spectra are caused by *intra*-molecular electron transfers which occur whenever the end-groups having the appropriate orientation approach each other sufficiently close to allow the transfer.

318

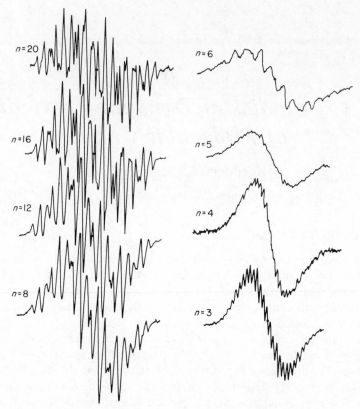

Figure 16.1 E.s.r. spectra of $N-(CH_2)_n-N^{\cdot}$, K^+ in HMPA for $n = 20$, 16, 12, 8, 6, 5, 4, and 3 at $-15\,°C$. Width of spectra: 27 G

Therefore, the frequency of *intra*-molecular 'collisions' of the end-groups could be deduced from the analysis of the spectra. To accomplish this task, all the coupling constants of the odd electron with the 7 aromatic protons and the 2 protons of the adjacent CH_2 group have to be known and this information is obtained from the e.s.r. spectrum of *n*-butyl-α-naphthalenide.[1,2] It has been shown that the coupling constants with the protons of the second, third etc., CH_2 groups are too low to be of significance. See Table 16.4 in the Appendix for the pertinent data. The e.s.r. spectra of $N-(CH_2)_n-N^{\cdot}$ were then computer-simulated by a method outlined by Harriman and Maki.[4] Thus, a desired spectrum is obtained by summation of about 250,000 curves, each representing the e.s.r. spectrum of a species having an odd electron jumping P times per second between two sites characterized by the resonance frequencies ω_A and ω_B. The resonance frequencies ω_A and ω_B are any two of those describing the 384 lines of the e.s.r. spectrum of the $-CH_2N^{\cdot}$ radical, and all the pairs A and B are included in the summation. One more parameter, viz., $1/T_2$, is involved in the calculation. Each of the 384 lines results from the overlap of the multiplet

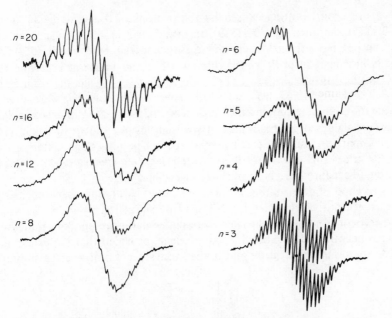

Figure 16.2 E.s.r. spectra of N-$(CH_2)_n$-N$^{\overline{\cdot}}$, K$^+$ in HMPA for n = 20, 16, 12, 8, 6, 5, 4, and 3 at 45 °C. Width of spectra: 27 G

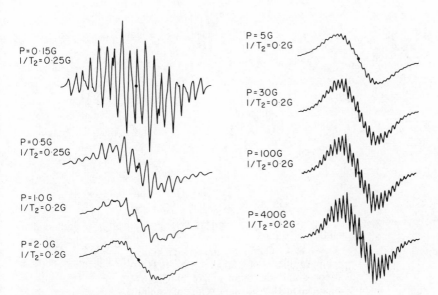

Figure 16.3 Computer-simulated e.s.r. spectra of N-$(CH_2)_n$-N$^{\overline{\cdot}}$ for P values varying from 0.15 G to 400 G. Width of spectra: 27 G

lines arising from splitting caused by the protons of the second CH_2 group ($a = 0.13$ G), the third ($a = 0.045$ G), and the fourth ($a = 0.035$ G). Such lines were computer-simulated. Their shape is Lorentzian and width (0.20 G) is virtually independent of P, provided the width of the input line is greater than 0.02 G while P is greater than 1 G ($\equiv 1.76 \times 10^7$ s^{-1}) but smaller than 500 G. For $P < 1$ G, one finds $1/T_2 \approx 0.25$ G. These values of $1/T_2$ were used to generate the computer-simulated spectra of $N—(CH_2)_n—N^{\cdot}$ radicals calculated for various values of P. Examples of such curves are shown in Figure 16.3. Their comparison with the experimental spectra permits their matching, and thus a definite value of P is ascribed to each observed spectrum. Details of the matching procedure have been outlined elsewhere.[5]

This method of determining P was augmented by an alternative technique. The e.s.r. spectra of $N—(CH_2)_n—N^{\cdot}$ may be simulated by the *experimental* e.s.r. spectra of n-butyl-α-naphthalenide obtained in the presence of a judiciously chosen concentration of n-butyl-α-naphthalene, provided that the exchange is slow. At the same temperature and in the same solvent both spectra should be

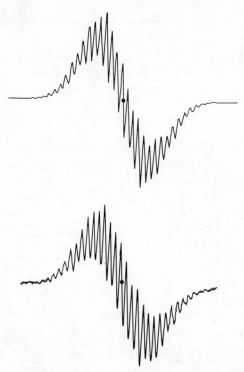

Figure 16.4 Computer-simulated e.s.r. spectrum of the hypothetical dimer $(CH_2N)_2^{\cdot}$ (upper), and the spectrum of $N\text{-}(CH_2)_3N^{\cdot}$ recorded at 45 °C (lower). Width of spectra = 27 G

identical if the *rates* of the *intra-* and *inter*-molecular exchanges are equal. The bimolecular rate constants of the *inter*-molecular exchange were determined for the investigated temperature range[1,2] by the conventional line-broadening technique[6]; see Appendix, Table 16.5. Therefore, the P value pertaining to the rate of *intra*-molecular exchange in a $N—(CH_2)_n—N^{\frac{\cdot}{\cdot}}$ system could be calculated, i.e., they are given by the product of c and the appropriate bimolecular rate constant, k_b, of the *inter*-molecular exchange, c being the concentration of *n*-butyl-α-naphthalene needed for matching the spectrum observed in the *inter*-molecular exchange with the investigated one. Thus derived P values for the slow exchange region agree with those calculated by the above described computer simulation.

In the intermediate and fast exchange regions, the latter method fails. The e.s.r. spectrum of the *inter*-molecularly exchanging system collapses then into one featureless line which sharpens as the exchange becomes faster. In contrast, the e.s.r. spectrum of the *intra*-molecularly exchanging system approaches in the fast exchange limit that of a hypothetic dimer $(—CH_2 . N)_{\frac{\cdot}{2}}$. The spectrum of the latter species was computer-simulated; its coupling constants are $\frac{1}{2}$ of those of $\text{\textasciitilde}CH_2N^{\frac{\cdot}{\cdot}}$ but the number of protons is doubled. For the sake of illustration, the computed spectrum of the hypothetical dimer is shown in Figure 16.4 together with the experimental spectrum of $N—(CH_2)_3—N^{\frac{\cdot}{\cdot}}$ recorded at 45 °C, the latter referring to the fastest observed rate of *intra*-molecular exchange. The agreement between the spectra is excellent.

16.3 RESULTS AND THEIR SIGNIFICANCE

The results of the above investigations give the dependence of P on n (the number of carbon atoms in the chain) and temperature. The dependence of log P on log n at constant temperature is shown in Figure 16.5, while Arrhenius plots of log P *vs.* $1/T$ for constant n are shown in Figure 16.6. Inspection of Figure 16.5 shows that P is a monotonic function of n, that sharply decreases as n increases from 3 to 4 and then to 5. The striking differences in the 'activation energies' of the systems corresponding to different n's are clearly revealed by Figure 16.6. The activation energy is exceptionally high when $n = 3$, i.e., about 44 kJ mol^{-1}, slightly lower when $n = 4$ (37 kJ mol^{-1}), but it remains virtually constant for $n > 5$, i.e. 21 $\pm$ 2 kJ mol^{-1}.

The *intra*-molecular exchange exhibits a feature resembling a bimolecular reaction. The acceptor $—N$ attached to one end of the chain can never be further from the donor, $—N^{\frac{\cdot}{\cdot}}$ attached to the other end, than the length of the extended chain. Hence, the *intra*-molecular exchange may be treated as if it were a 'bimolecular' electron transfer in a system where one molecule of a donor and another of an acceptor are enclosed in a sphere having radius determined by the length of the chain. Such a system becomes 'diluted' as the length of the chain increases, and this dilution contributes in a 'trivial' fashion to the decrease of P caused by an increase of n.

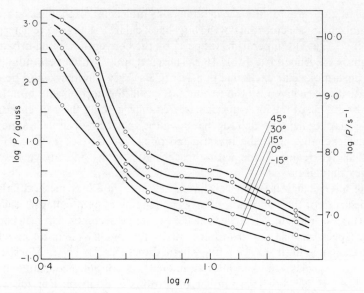

Figure 16.5 $\log P$ as a function of $\log n$ at $-15°$, $0°$, $15°$, $30°$, and 45 °C. Solvent HMPA

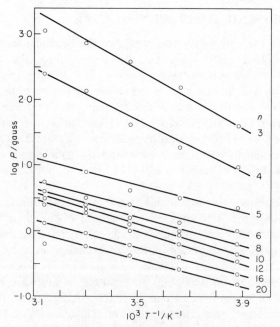

Figure 16.6 Arrhènius plots of $\log P$ *vs.* $1/T$ for constant n. Solvent HMPA

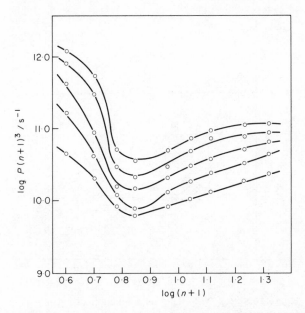

Figure 16.7 $\log P(n + 1)^3$ as a function of $\log (n + 1)$ for constant temperature. From down-up, $-15°, 0°, 15°, 30°$ and 45 °C. Solvent HMPA

The 'trivial' dilution factor should be excluded from our consideration when a deeper significance of P is contemplated. In the most naive treatment, the radius of the hypothetical sphere is assumed to be proportional to the length of the extended chain, i.e., proportional to the number, $n + 1$, of bonds separating the naphthyl moieties. The P values could be then normalized by multiplying them by $(n + 1)$.[3] The plots of $\log P(n + 1)^3$ vs. $\log (n + 1)$, graphed for a series of increasing temperatures, are shown in Figure 16.7. The resulting curves reveal an intriguing minimum for $n = 6$, but for $n \geq 6$, $\log P(n + 1)^3$ rises continuously and no approach to a plateau is seen for large values of n. However, intuitively one feels that the properly normalized P should become constant for large n and hence the above normalization procedure seems unlikely to be correct. We found empirically that if normalization is carried out by multiplying P by $(n + 1)^{\frac{3}{2}}$, the plots of $\log P(n + 1)^{\frac{3}{2}}$ vs. $\log (n + 1)$, shown in Figure 16.8 do indeed approach asymptotic values and the minima seen in Figure 16.7 disappear, at least at the lower temperature range, although the curves drawn for $T = 30°C$ and 45 °C show shallow minima and maxima. It is uncertain whether these are real or whether they reflect some experimental errors. We are inclined to accept the latter alternative and consider the average values of $P(n + 1)^{\frac{3}{2}}$ obtained for all values of n greater than 6 as characteristic constants providing a measure of the intrinsic dynamic chain flexibility at each particular temperature. These constants are therefore independent of n, being

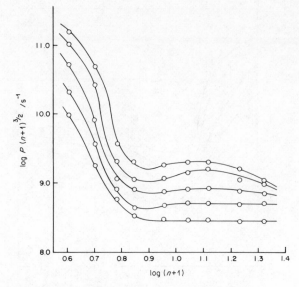

Figure 16.8 $\log P(n + 1)^{3/2}$ as a function of $\log(n + 1)$ for constant temperature. From down-up, $-15°$, $0°$, $15°$, $30°$, and $45°$C. Solvent HMPA

determined by the nature of a chain, temperature of solution, viscosity of the solvent, and the character of the end groups.

16.4 DYNAMIC OR STATIC MODEL OF INTRAMOLECULAR EXCHANGE?

It has been remarked previously that the frequency P of intra-molecular exchange may be treated as a rate of *inter*-molecular electron transfer occurring at appropriate concentration of the acceptor. Assuming that the frequency of transfer between two interacting groups having a particular mutual orientation is the same whether the reaction is *inter*- or *intra*-molecular, we find this effective concentration of the acceptor to be given by the ratio P/k_{ex}. Thus calculated values of $C_{eff} = P/k_{ex}$ are listed in Table 16.3. However, since the acceptor $-N$ can never be further away from the donor $-N^-$ than the length of the *fully extended* chain, its *minimum* 'concentration', C_{min}, is given by $3000/4\pi l^3 N$, expressed in mol litre^{-1}, where l denotes the length of the fully extended chain and N is the Avogadro number. These minimum concentrations, given for each chain in the last column of Table 16.3, are substantially larger, by a factor of 2–5, than C_{eff}.

A static model of *intra*-molecular reaction assumes that the *equilibrium* distribution of mutual configurations of the interacting groups determines its rate. Denoting by K_i the configurational equilibrium constant of each orientation of the interacting groups and by f_i the frequency of electron exchange between them when locked in this orientation, one finds the observed frequency of the *intra*-molecular electron transfer to be $P = \sum K_i . f_i$. Obviously, K_i depends on the length and nature of the chain linking the interacting groups.

It is highly unlikely that such a model would yield values of P which are substantially smaller than k_{ex}/C_{min} for values of the n in the range 8–20. This would demand that the fraction of favourable configurations permitting the exchange be smaller for $N—(CH_2)_n—N^{\bar{}}$ than for a freely moving $—N$ moiety enclosed in a sphere having the radius of a fully extended chain while $—N^{\bar{}}$ is located in its centre. It seems therefore that the chain retards the motion of the interacting groups and this effect is reflected in the values of P. The factor $(n + 1)^{\frac{3}{2}}$ accounts for the changes in C_{eff}, at least within the range of n studied by us, and hence we may repeat again that $P(n + 1)^{\frac{3}{2}}$ provides us with a measure of dynamic flexibility of a chain.

16.5 THE EFFECT OF THE VISCOSITY OF THE SOLVENT

To obtain information about the effect of solvent viscosity we investigated the kinetics of the previously studied *inter-* and *intra*-molecular electron transfers proceeding in dimethoxymethane (DME) containing 2% HMPA. This study covered a wider temperature range, viz., from $-75\,°C$ to $+15\,°C$.

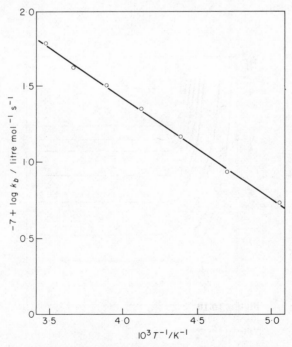

Figure 16.9 Arrhenius plot of $\log k_b$ *vs.* $1/T$. k_b is the bimolecular rate constant of electron transfer between *n*-Bu-$N^{\bar{}}$, K^+ and *n*-Bu-N in DME ($+2\%$ HMPA)

326

Table 16.1 The bimolecular rate constants, k_b, for electron transfer between Bu-N$\bar{\cdot}$, K$^+$ and Bu-N in DME (+2% HMPA)

T/°C	$10^{-7}\,k_b$/litre mol^{-1} s^{-1}	Viscosity/cP
−75	5·6	
−60	8·8	1·69
−45	15	
−30	23	0·93
−15	32	
0	42	0·61
+15	60	

$E_b = 12\cdot6$ kJ mol^{-1}, $A_b = 1\cdot0 \times 10^{11}$ litre mol^{-1} s^{-1}, $-E_\eta = 8\cdot2$ kJ mol^{-1}.

The bimolecular rate constants are given in Table 16.1 and the respective Arrhenius line is shown in Figure 16.9. The activation energy is higher than that of the viscosity of the medium (see Table 16.1), but the k values determined at 0° or 15 °C are only slightly greater than those in HMPA. The rates of the intramolecular process, P, are listed in Table 16.2 and graphically displayed in Figure 16.10 as plots of $\log P(n + 1)^{\frac{3}{2}}$ versus $\log (n + 1)$ at a series of temperatures. Such plots reach limiting values for $n > 6$, this being their most significant feature.

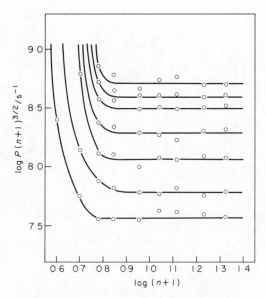

Figure 16.10 $\log P(n + 1)^{3/2}$ as a function of $\log (n + 1)$ at constant temperature. From down-up, $-75°$, $-60°$, $-45°$, $-30°$, $-15°$, $0°$, and 15 °C. Solvent DME (+2% HMPA)

Table 16.2 *Intra*-molecular electron exchange rate constant/s^{-1} for N—(CH$_2$)$_n$—N$^{\cdot-}$ in DME (+2% HMPA)

	Temperature/°C						
n	-75	-60	-45	-30	-15	0	15
3	6.3×10^7	2.4×10^8	3.1×10^8	4.0×10^8	4.7×10^8	4.6×10^8	4.4×10^8
4	5.1×10^6	1.3×10^7	5.6×10^7	1.1×10^8	2.8×10^8	3.1×10^8	3.9×10^8
5	2.5×10^6	5.2×10^6	8.8×10^7	1.6×10^7	2.5×10^7	3.5×10^7	5.0×10^7
6	1.9×10^6	3.5×10^6	6.9×10^6	1.2×10^7	2.0×10^7	2.5×10^7	3.3×10^7
8	1.3×10^6	2.2×10^6	3.6×10^6	7.2×10^6	1.1×10^6	1.3×10^7	1.7×10^7
10	1.2×10^6	1.6×10^6	3.3×10^6	5.0×10^6	9.8×10^6	1.1×10^7	1.5×10^7
12	8.8×10^5	1.4×10^6	2.4×10^6	3.6×10^6	6.7×10^6	8.8×10^6	1.3×10^7
16	5.7×10^5	8.1×10^5	1.8×10^6	2.9×10^6	4.5×10^6	5.6×10^6	7.0×10^6
20	3.9×10^5	6.4×10^5	1.3×10^6	2.2×10^6	3.4×10^6	4.2×10^6	5.2×10^6

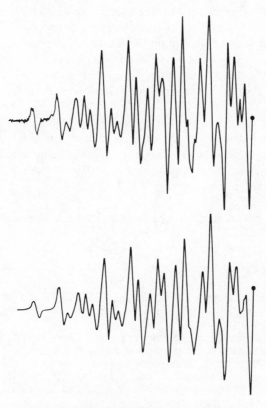

Figure 16.11 Experimental (upper) and computer-simulated (lower) spectrum of n-Bu-α-N$^{\cdot-}$ K$^+$ in **HMPA**

Comparison of P or $P(n + 1)^{\frac{3}{2}}$ determined at $0°$ and $15\,°C$ in both solvent systems shows that they are virtually identical, although the viscosities of these media are greatly different. This could be taken as a further evidence favouring the dynamic model of this *intra*-molecular reaction in which the resistance of the chain, and not the viscosity of the solvent, determines the motion of the interacting groups. Unfortunately, bimolecular rate constants are also similar in the two solvents and this cast doubt on the above conclusion. We hope to clarify these problems in the course of further work.

16.6 CONCLUDING REMARKS

The dynamic flexibility of a chain should be distinguished from its static flexibility. The latter determines those properties of a chain that depend on the multitude of its conformational states *but not on time*, e.g. its average end-to-end distance or the probability of ring closure between two designated points on its chain etc. In contrast, the dynamic flexibility is a measure of the *rate* of conformational changes, which could be reflected in the frequency of the *intra*-molecular collisions between the end groups of a linear polymer provided the studied reaction is sufficiently fast. Of course, the probability of ring closure is an important factor determining the frequency of 'tight' collisions but not the only one. It is possible to visualize two polymeric molecules of the same length, one being more flexible than the other in the static meaning of that term, but less flexible if the dynamic criterion of flexibility is used in the comparison.

In this work the electron transfer process was chosen as a tool to study chain flexibility but we have not been concerned with the intimate details of this process. It suffices to say that the investigated reaction occurs during encounters of the two end-groups and is not mediated by the chain—a possibility discussed by other investigators.[8,9,10]

The support of this study by the National Science Foundation is gratefully acknowledged.

16.7 REFERENCES

1. H. D. Connor, K. Shimada and M. Szwarc, *Macromolecules*, **5**, 801, (1972).
2. M. Szwarc, 22nd Nobel Symposium, p. 291; Almqvist and Wiksell, Stockholm, 1973.
3. P. Caluwe, K. Shimada, and M. Szwarc, *J. Amer. Chem. Soc.*, **95**, 6171, (1973).
4. J. E. Harriman and A. H. Maki, *J. Chem. Phys.*, **39**, 778, (1963).
5. K. Shimada, Ph.D. Thesis, Syracuse, N.Y., 1974.
6. R. L. Ward and S. I. Weissman, *J. Amer. Chem. Soc.*, **79**, 2086, (1959).
7. K. Shimada and M. Szwarc, *Chem. Phys. Lett.*, **28**, 540, (1974).
8. H. M. McConnell, *J. Chem. Phys.*, **35**, 508, (1961).
9. F. Gerson and J. M. Martin, *J. Amer. Chem. Soc.*, **91**, 1883, (1969).
10. P. J. Flory, *Principles of Polymer Chemistry*, Cornell University Press, 1953.

16.8   APPENDIX

The attached tables and graphs provide additional data pertinent to the aforegoing results. Tables 16.5 and 16.6 as well as Figure 16.12 demonstrate that the exchange in HMPA is virtually diffusion-controlled.

Table 16.3   The effective concentrations, C_{eff}, of the acceptor $-N$ in the *intra*-molecular electron transfer, $N-(CH_2)_n-N^{\bar{}} \rightleftarrows {}^{\bar{}}N-(CH_2)_n-N$ proceeding in HMPA. $C_{eff} = P/k_{ex}$. Units mol litre^{-1}

$T/°C$ n	-15	0	15	30	45	C_{min}^a
8	0·055	0·060	0·057	0·061	0·071	0·284
10	0·040	0·053	0·047	0·055	0·057	0·155
12	0·033	0·037	0·037	0·050	0·045	0·094
16	0·020	0·023	0·022	0·022	0·024	0·042
20	0·013	0·018	0·015	0·014	0·011	0·022

^{a} C_{min} is the minimum concentration of the acceptor calculated as 1 molecule in a sphere of radius equal to the length of the *extended* chain.

Table 16.4   Coupling constants of *n*-butyl-α-naphthalenide (cf. Figure 16.11)

Proton	a/G in THF at 10 °C	a/G in HMPA at 25 °C	a/G in DME^a
2	1·60	1·53	
3	1·77	1·68	
4	4·10	4·46	
5	5·20	5·09	
6	1·65	1·62	
7	2·03	2·02	
8	5·10	4·89	
1'	2·58	2·78	2·80
2'	0·155	0·130	$-0·155$
3'	0·045^a	0·045^a	0·045
4'	0·035^a	0·035^a	$-0·035$

The assignments are based on the analogy with the a constants reported for α-methyl-naphthalene$^{\bar{}}$ (R. E. Moss, N. A. Ashford, R. G. Lawler and G. K. Fraenkel, *J. Chem. Phys.*, **51**, 1765, (1969); and α-ethyl-naphthalenide$^{\bar{}}$ (G. Moshuk, H. D. Connor and M. Szwarc, *J. Phys. Chem.*, **76**, 1734, (1972).
^{a} E. de Boer and C. MacLean, *Molec. Physics*, **7**, 191, (1965). The coupling constants for 3' and 4' protons were assumed to be the same in THF and HMPA as those reported by de Boer for DME.

Table 16.5 Rate constants for electron transfer k_b in HMPA

$$n\text{-Bu-}\alpha\text{-N}^{\overline{\cdot}} + n\text{-Bu-}\alpha\text{-N} \rightarrow n\text{-Bu-}\alpha\text{-N} + n\text{-Bu-}\alpha\text{-N}^{\overline{\cdot}},$$

and for spin–spin exchange k_2. η is the HMPA viscosity

$T/°C$	$10^{-8} k_b/\text{litre mol}^{-1}\,\text{s}^{-1\,\text{a,c}}$	$10^{-8} k_2/\text{litre mol}^{-1}\,\text{s}^{-1\,\text{a,b}}$	η/cP
-15	2·0	2·3	—
0	3·0	3·4	6·23
15	4·9	5·0	4·24
30	7·1	7·3	3·03
45	9·8	10·7	2·22
60			1·71

$$E_b = 18\cdot0 \text{ kJ mol}^{-1} \qquad E_2 = 16\cdot7 \text{ kJ mol}^{-1} \qquad -E_\eta = 16\cdot3 \text{ kJ mol}^{-1}$$

[a] The rate constants were determined from the concentration dependence of the width of the end line of the e.s.r. spectra. The values for the linewidth were corrected for the effects due to overlap with the next line.

[b] The spin exchange studies were performed by varying the radical concentration from $2\cdot6 \times 10^{-3}$ M to $16\cdot1 \times 10^{-3}$ M.

[c] $[n\text{-BuN}^{\overline{\cdot}}] = 8 \times 10^{-4}$ M. These data have been recalculated. They differ slightly from those published earlier.[8]

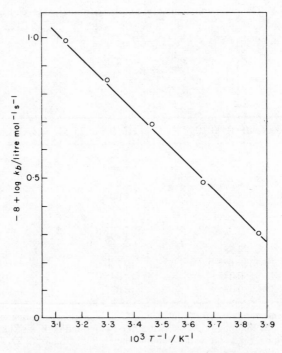

Figure 16.12 Arrhenius plot for the bimolecular rate constant of electron transfer between $n\text{-Bu-}\alpha\text{-N}^{\overline{\cdot}}$ and $n\text{-Bu-}\alpha\text{-N}$ in HMPA

Table 16.6 Comparison of *inter*-molecular electron exchange in HMPA

System	n-Bu-α-N	α-EtN	β-EtN	N[a]
$E/\text{kJ mol}^{-1}$	18·0	17·6	17·2	17·2
$10^{-11}A/\text{litre mol}^{-1}\,\text{s}^{-1}$	7·8	8·3	5·9	6·9
$10^{-8}k_{30}/\text{litre mol}^{-1}\,\text{s}^{-1}$	7·1	7·0	6·9	7·3

n-Bu-α-N = n-butyl-α-naphthalene $\qquad$ β-EtN = β-ethylnaphthalene
$\qquad$ α-EtN = α-ethylnaphthalene $\qquad\qquad$ N = naphthalene
[a] These values were redetermined recently. They are more reliable than those previously published.[7]

Subject Index

334